On Amphibians and Reptiles
in Linnaeus's Students' Dissertations

Wahlgreniana 4

On Amphibians and Reptiles in Linnaeus's Students' Dissertations

Esteban O. Lavilla

International Society for the History
and Bibliography of Herpetology
www.ISHBH.com

2026

Wahlgreniana is named in honor of Richard Wahlgren (1946–2019), a founding member and first Chairman of the *International Society for the History and Bibliography of Herpetology*. Without Richard's tireless dedication to ISHBH, the society could not have made it through its early years. *Wahlgreniana* is a series of book length works complementing the ISHBH journal, *Bibliotheca Herpetologica*. Books in this series are published on an irregular basis and are sold separately from ISHBH subscriptions.

International Society for the History and Bibliography of Herpetology

Printed on demand by IngramSpark global distribution network
Layout & design by Breck Bartholomew

ISBN: 979-8-218-88776-6 (hardcover)

www.ISHBH.com

Contents

Introduction

Carl Linnaeus (1707–1778), the well-known Swedish botanist, physician, and zoologist, stands as a towering figure in the history of science, celebrated for his groundbreaking contributions to taxonomy. Despite his immense contributions, Linnaeus harbored a deep-seated aversion to amphibians and reptiles, a sentiment that permeates his work. This aversion may stem from his profound religiosity (see Lavilla, 2012), as both amphibians and reptiles have been vilified since the origins of Judeo-Christian culture. According to these traditions, the devil took the form of a serpent to tempt Eve, leading to humanity's expulsion from paradise (Genesis 3:1-24); a plague of frogs or toads was one of the seven that struck Egypt (Exodus 8:1-15); lizards are deemed unclean according to Leviticus (11:29-31), along with numerous other examples where these animals are depicted unfavorably[1]. Linnaeus does not discuss these concepts directly, although he may have felt some conflict when disproving the existence of the Hydra of Hamburg, previously presented as a tangible representation of the monster mentioned in St. John's Apocalypse (12:3; 13:1).

This distaste is explicit in the *Systema Naturae* in two ways. In the first five editions, Linnaeus (1735, 1740a, b, 1744, 1747a) included a comment expressing his gratitude for the benevolence of the Creator for not allowing the proliferation of Amphibia, suggesting that an abundance of Hydras, Dragons, Basilisks, and other such beings would have made life impossible for the human race[2]. Starting from the tenth edition, Linnaeus (1758, 1766, 1767) expressed his displeasure for this group, defining Amphibia as frightening animals with cold bodies, sallow (or pale, cadaverous) color, cartilaginous skeletons, nasty skin, grim faces, brooding eyes, offensive smells, hoarse voices, miserable postures, and terrible venom...[3] He concluded that, consistent with these characteristics, the Creator did not create them in large numbers. He noted that of all animal groups, Amphibians are the least abundant, despite Seba's (1734) attempt to increase their numbers.[4]

Despite his personal distaste for amphibians and reptiles, Linnaeus played a pivotal role in fostering scientific inquiry into these organisms through his supervision of doctoral dissertations. As a prolific thesis advisor, he guided and inspired numerous students, overseeing 185[5] doctoral dissertations throughout his career at the University of Uppsala between 1743 and 1776. These theses, often overlooked in the vast Linnaean legacy, offer valuable insights into the scientific understanding of amphibians and reptiles during the 18th century.

Two of these theses, *Siren lacertina* and *De Morsura*

[1]On the other hand, passages such as Numbers 21:8-9 ("The Lord said to Moses, 'Make a snake and put it up on a pole; anyone who is bitten can look at it and live.' So Moses made a bronze snake and put it up on a pole. Then when anyone was bitten by a snake and looked at the bronze snake, they lived.") and John 3:14-15 ("Just as Moses lifted up the snake in the wilderness, so the Son of Man must be lifted up, that everyone who believes may have eternal life in him.") can be viewed as exceptions.

[2]While there are subtle differences between the various editions, the commentary in the first (Linnæus, 1735) indicates: "...*Amphibiorum classem ulterius continuare noluit benignitas Creatoris; Ea enim si tot Generibus, quot reliquæ Animalium Classes comprehendunt, gauderet, vel si vera essent quæ de Draconibus, Basiliscis, ac ejusmodi monstris οἱ τεγραλόγοι fabulantur, certe humanum genus terram inhabitare vix posset...*". ["...The benevolence of the Creator did not wish to continue the class of amphibians any further; for if it enjoyed as many genera as the other classes of animals, or if what the storytellers fabricate about dragons, basilisks, and such monsters were true, surely the human race could scarcely inhabit the earth..."]

[3]Linnæus (1758, p. 194). "...*AMPHIBIA pleraque horrent Corpore frigido, Colore lurido, Sceleto cartilagineo, Cute fœda, Facie torva, Obtutu meditabundo, Odore tetro, Sono rauco, Loco squalido, Veneno horrendo; non itaque in horum numerum sese jactavit eorum Auctor...*". ["... Most amphibians are dreadful with a cold body, a lurid color, a cartilaginous skeleton, a filthy skin, a grim face, a brooding gaze, a foul odor, a hoarse sound, a squalid habitat, and terrible venom; therefore, their Creator did not boast of their number..."].

[4]Linnæus (1758, p. 195). "...*AMPHIBIOLOGI omnium paucissimi sunt, nullique veri. Seba numero stupendo sibi ignota pulchre collegit & delineavit, sed multiplicavit, minimeque descripsit. Catesbæus nec paucos Serpentes pulchrius delineavit, quam notavit...*". [The translation here encounters a challenge with *Amphibiologi*. This term could refer to either amphibians or specialists who study amphibians (amphibiologists). Depending on the interpretation, the phrase suggests "...Amphibians are very few, and no one knows their true number. Seba, with an astonishing number, collected and beautifully illustrated species unknown to him, but he exaggerated their number and described them minimally. Catesby, on the other hand, depicted several serpents more beautifully than he described them..."].

[5]It is common to speak of 186 theses, but the last, on the genus *Hypericum*, defended by Carl Niclas Hellens (Hellenius, 1776), was directed by Carl Linnæus the Younger.

serpentum, deal with amphibians and reptiles as central topics. Another three, *Amphibia Gyllenborgiana*, *Museum Adolpho-Fridericianum*, and *Surinamensia Grilliana*, describe collections that were duly donated to the University, all of which have a significant component of Amphibia in the Linnaean sense. Two more, *Lignum colubrinum* and *Radix senega*, deal with the identification and uses of supposed antivenins of vegetal origin. In addition to these seven, amphibians and reptiles were occasionally mentioned in another 25 dissertations.

Consensus indicates that Linnaeus was the author of those dissertations (i.e., Jackson, 1913; Ramsbottom, 1959; Smit, 1979; Stafleu, 1971; Stearn, 1966), but the issue still warrants discussion. This attribution is justified by the presumed fact that Linnaeus often proposed the subject, dictated the content, and provided the material, an idea probably originating from Acrell (1796: 156)[6] and later translated and transcribed by Jackson (1913: 102), which asserts that Linnaeus dictated the theses in part in Swedish and in part in Latin, and that composing an essay took no more than three hours... According to this view, the student was relegated to a largely unscientific role—merely demonstrating fluency in Latin through translation and defense, covering the printing expenses, and paying for the celebration banquet. Another justification is found in the Swedish educational system of the time, which granted the director intellectual property of the thesis. Jackson (1913) also addressed dissertations published under Thunberg's direction between 1787 and 1801[7], plus the *Aphorismi Botanici* of Agardh (1817-1826). The system changed in 1852 when, according to Ramsbottom (1959), new provisions established that the defender of the thesis was its author.

Further evidence is found in the discovery of manuscripts among Linnaeus' papers preserved at the Linnean Society of London, coinciding with the dissertations *Anthropomorpha* and *Pan Svecicus*, justifying Stafleu's (1971: 144) categorical assertion that Linnaeus was the author of these theses. This is easy to say but difficult to sustain in some cases.

Some authors (i.e., Jackson, 1913; Heller, 1983; Krok, 1925) recognize that the role of some students went beyond mere translation or financing, indicating that their presentations were the result of their own studies and observations. This category includes the dissertations defended by Bjerkén (1757), Hasselquist (1747), Löfling (1749), Söderberg (1771), Sparrman (1768), Tillaeus (1771), and Wänman (1768). Curiously, this selection does not appear to be impartial, favoring those disciples who subsequently achieved individual renown. One might ask why the theses defended by Aphonin (1766) and Karamyschew (1766), explicitly based on Russian examples, do not fall into this category. Curiously, these two overlooked students, along with Acharius, Avellan, Boecler, Dahlgren, Gahn, Hedin, Heidenstam, Holmberger, Murray, Salomon, Westring and Wickman, among others, placed *'auctor,' 'auctor et respondens,' 'auctor defendente,'* or similar before their names on the title pages of their respective dissertations. Examples can be multiplied *ad nauseam*.

Other authors, such as Stearn (1957), Ramsbottom (1959), Heller (1983), Lindell (2012), and Krecsák & Bauer (2025), among others, sought a compromise solution by assigning joint co-authorship to Linnaeus and the respondent for the original dissertation, and solely to Linnaeus for the publication in *Amoenitates Academicae*, a very curious dual stance that assigns different authorships to two intrinsically similar texts. But this, too, must be taken *cum grano salis*.

Statements in the preceding paragraphs are speculative, but there are some overlooked facts in these discussions. In the list of authors of the first edition of *Fauna Svecica*, Linnaeus (1746) explicitly mentioned Hast as the author of *Amphibia Gyllenborgiana*, reserving for himself the role of advisor (*praese*). Similarly, in the *Elenchus auctorum et abbreviatorum nominum delucidatio* of the ninth edition of Systema Naturae (Linnaeus, 1756), he did the same with Hast (*Amphibia Gyllenborgiana*), Dubois (*Taenia*), Kähler (*Crystallorum generatione*), Balk (*Museum Adolpho-Fridericianum*), and Sundius (*Surinamensia Grilliana*), and his own name does not appear related to any of these theses. The situation is different when Linnaeus refers to his own works. In this case, the word *Auctoris* is placed between the abbreviation and the full title[8].

Contemporary authors also acknowledge the authorship of the dissertations to the students. Gronovius (1754, 1756) attributed the key dissertations for their herpetological content to Hast, Balk, and Sundius, who, in turn, did the same with Hast and Balk.

However, the *Elenchus auctorum* of the ninth edition (Linnaeus, 1756) was replaced in subsequent editions (Linnaeus, 1758, 1766) by the *Ratio Editiones*, the official list of print runs (*Systema Naturae iteratis vicibus prodiit*), which included the analyzed collections—some with references to their published catalogs (*Collectanea*) and travels (*Itinera*) by both Linnaeus and his students (*discipulorum*)—but there are no explicit references to the dissertations. Instead, they are encompassed under the general reference of *Amoenitates Academicae*, in which all these theses were subsequently reprinted. Krecsák and Wahlgren (2008: 3) mentioned that Linnaeus published the dissertations collectively as a second edition under his own name in *Amoenitates Academicae*, sometimes updated with changes in nomenclature and other data, but this is partially incorrect. In the index of all volumes, either edited by Linnaeus (1 to 7) or Schreber

[6]"...Alla Disputationer skref han dictando, dels på Svenska, dels Latin, hvilkas i ordning fattande och städande, blef Respondenternas sak: och ehuru han ej sjelf bekymrade sig om Latiniteten uti dem, gaf han nog sit nöje tilkänna, när de voro väl skrifne, och tvärtom. At skrifva en Disputation behöfdes således knapt 3 timmar; ty i och för sig sjelf var den ej annat än en Lection öfver ämnet, don Respondenten antecknade..." ["...All Disputations he composed by dictation, partly in Swedish and partly in Latin; and it fell to the Respondents to gather them into due order and to set them in fair form. And though he did not himself concern his mind with the purity of the Latinity therein, yet he made his contentment sufficiently manifest when they were well written—and likewise the contrary. To write a Disputation, therefore, scarce required three hours; for in itself it was nothing more than a Lecture upon the subject, which the Respondent set down in notes..."].

[7]Twenty two in a series entitled *Museum naturalium Academiæ Upsaliensis*, and 16 as the *Nova Genera Plantarum*.

[8]Fn. *Auctoris fauna Suecica. Lugd. Bat. 1746. 8º*. It. Œl. *Auctoris Iter Oleandicum & Gothicum, 8º.Suetice, Stockh. 1742*. It. W. Goth. *Auctoris Iter Westrogothicum. 8º. Suetice, Stockholm. 1747*. Syst. Nat. *Auctoris Systema Naturæ*.

(8–10), the title of the dissertation is followed by the name of the student, and in each chapter, the title of the dissertation is followed by the phrase *sub praesidio D:n. Doct. Caroli Linnaei* (or equivalent), followed by the name of the student.

Richard Pulteney, the first biographer of the Swedish naturalist, makes some noteworthy comments (Pulteney, 1781). On page iv of the Introduction, he clearly states that the *Amoenitates Academicae* are not strictly Linnaeus's own work, although his influence on them is remarkable[9]. Pulteney also points out that Linnaeus was not the original compiler of the first volume[10], but upon recognizing its value, he reissued this volume and became the editor of the subsequent tomes. William George Maton, in Pulteney (1805), referring to the ninth volume of the *Amoenitates*, edited by Schreber in 1785, notes that since there was no evidence Linnaeus interacted with the students whose dissertations were included, Schreber simply indicated the author, title, and year of defense in his analysis[11].

This misconception arises because post-Linnaean authors considered each volume of the *Amoenitates Academicae* as a single-authored monograph, when in fact, the ten volumes have a structure analogous to today's scientific journals, with varied and multi-author content, in which Petrus Camper (vol. 1, ed. Lugduni Batavorum), Linnaeus (vols. 1–7), and Johann Christian von Schreber (vols. 8–10) acted as successive editors, as already noted in late 18th-century bibliographic references (i.a., Desallier d'Argenville 1780: 365, footnotes). Accordingly, in the following analysis, the student's name will be used when referring to the authors of the dissertations, as originally recognized by Linnaeus himself.

The presentation of the analyzed dissertations offers several options for organization (thematic, chronological, etc.). It was decided to begin with those specifically addressing amphibians in the Linnaean sense, followed by the analysis of the collections donated to the University of Uppsala's museum, then the examination of plant-based antivenins, and concluding with a chronological list of theses that mention amphibians and reptiles, even if only briefly.

The translations from Latin to English have aimed to be as precise as possible (in fact, the specific descriptions are verbatim). However, in some cases, such as certain poems and dedications originally published in both Latin and French or Swedish, some liberties were taken to preserve the meter. It should be noted that the analyzed theses were those electronically available at the Hunt Institute for Botanical Documentation, a research division of Carnegie Mellon University, and no comparisons were made with the reprints in *Amoenitates Academicae*.

Nota bene: Following the provisions of Art. 8.2. of the ICZN (1999), the names (*nomina*) cited in this work have value in their historical context but are not published for purposes of zoological nomenclature.

Acknowledgments: I would like to express my sincere gratitude to Breck Bartholomew, editor of the journal, for his meticulous editorial work and invaluable guidance throughout the publication process. I also extend my heartfelt thanks to reviewers Aaron Bauer and Laszlo Krecsák for their thorough and insightful evaluations, which significantly enhanced the quality of the manuscript. Their dedication and expertise are deeply appreciated.

[9]Pulteney, 1781: iv: "...The *Amœnitates Academicæ*, although strictely speaking they are not Linnæus's own works, have yet so large a share of his authority stamped upon them, are so intimately connected with his writings, and, it is presumed, are so much less known than they deserve to be, that it was judged proper to give a brief account of the whole collection..."

[10]Pulteney refers to the edition edited by Petrus Camper and published by Cornelius Haak in Leiden (Linnæus, 1749b).

[11]Pulteney, 1805: 487: "...As Linnæus was not known to have communicated to the respective respondents any part of the dissertations contained in this volume, it scarcely comes within our province to notice them in the present work. As many of the subjects are interesting, however, it may be desirable to the reader at least to be made acquainted with their titles, and the authors names...". Included dissertations are *Hæmorrhagia uteri sub statu graviditatis* (Elff, 1749); *Methodo investigandi vires medicamentorum chemica* (Hiortzberg, 1754); *Consectaria Electro-Medica* (Zetzell, 1754); *Pulsus intermittens* (Wählin, 1756); *Cortex Peruvianus* (Petersen, 1758); *Ambrosiaca* (Hideen, 1759); *Hemoptysis* (Gråberg, 1767); *Vena Resorbentes* (Thunberg, 1767); *Febrium Intermittentium curatio varia* (Tillæus, 1771); *Hæmorrhagia ex Plethora* (Heidenstam, 1772); *Suturæ Vulnerum* (Bœcler, 1772); *Medicamenta purgantia* (Rotheram, 1775); *Perspiratio insensibilis* (Avellan, 1775); *Canones Medici* (Hedin, 1775) and *Scorbutus* (Salomon, 1775).

S. A. N.

SIREN LACERTINA,

DISSERTATIONE ACADEMICA
ORBI ERUDITO DATA
QUAM VENIA NOB. ET EXPERIENT. FAC. MED.
AD REG. ACAD. UPSAL.

PRÆSIDE

VIRO NOBILISSIMO et EXPERIENTISSIMO

D:NO DOCT. CAROLO
A LINNÉ,

EQUITE AURATO DE STELLA POLARI;
S:Æ R:Æ M:TIS SUEC. ARCHIATRO,
MED. ET BOTAN. PROFESSORE REG. ET ORD. ACAD. PARIS.
PETROPOL. IMPER. N. C. HOLM. UPSAL.
LOND. ANGL. FLOR. BEROL. MONSP. TOLOS. BERNENS. EDINB.
NIDROS. MEMBRO,

PUBLICE VENTILANDAM SISTIT
STIPENDIARIUS REGIUS

ABRAHAMUS ÖSTERDAM,

HOLMIENSIS

IN AUDITORIO CAROLINO MAJORI
DIE XXI JUNIJ ANNI MDCCLXVI.
H. A. M. C.

UPSALIÆ.

Österdam, A. 1766. *Siren lacertina.* Upsaliae. S. n. [3] + 1-15 + [1] + 1 plate.

Österdam's dissertation contains, as noted by Pulteney (1805), the "...complete history of the *Lizard Siren*, or *Mud-Inguana* [sic], of Carolina, a new amphibious, biped, eel-shaped animal, furnished both with gills and lungs; the former placed entirely without the body. This animal is so singular in its structure, as to have occasioned Linnaeus to form a new order, under the term Meantes, which is placed between the Amphibia and Nantes. It is sometimes seen two feet long, and sends forth a cry somewhat like that of the young of the Duck kind, but more acute and clear. It is described and figured by Mr. Ellis, in the *Phil. Trans.* vol. LVI. p. 189."[12] The dissertation is dedicated to Carl F. Piper and to Olave Acrell and is divided into nine sections, including the conclusion. The description of the animal is restricted to section VI and is of remarkable precision.

H. D. W.[13] Siren lacertina, Academic Dissertation given to the learned world with the permission of nobles and experts of the Medical Faculty at the Royal Academy of Upsala, under the Presidency of the most Noble and Experienced Dr. Carol Linné, Knight of the Polar Star; S:ae R:ae M:tis Suec, Archiatro, Physician-in-chief of the Royal Swedish and French academies, Professor of Medicine and Botany at the Royal Academy of Paris, Member of the Imperial Academy of St. Petersburg, the Royal Society of London, the Academies of Florence, Berlin, Montpellier, Bern, Edinburgh, and Nidaros, presented publicly for defense by the Royal scholarship holder Abraham Österdam, from Stockholm, in the Grand Carolinian Auditorium on the twenty-first of June, in the year 1766. At the customary hours before noon. Upsala.

[s.n.]

To My Lord, Count Charles Frederic Piper, President of the Royal College of Finance and Knight and Commander of the King's Orders.

My Lord,

Your outstanding qualities and your love for the Fatherland, to which you have rendered the greatest services, are respected by everyone. If I were [s. n.] to paint a picture of them, I would not be bold enough to apply my weak brush to it. It is only, my Lord, the benefits that you have deigned to bestow upon me that I would like to celebrate worthily. However, the condescension, kindness, and generosity with which you encourage people of my class, never separating yourself from the source of these virtues, the greatness of your soul, make this task even more daunting for me. Therefore, my Lord, allow me to simply present these beginnings of my academic studies to you, as a small token of my humble gratitude before the eyes of the public. This gratitude is too lively to ever be extinguished, and I judge myself unworthy of life if I fail to be with submission and the deepest respect,

My Lord,
Your very humble and obedient servant.
A. Österdam

[s. n.]

Most esteemed and renowned Sir Olaf Acrel, Professor of Surgery, Chief Surgeon of the Hospital and Equestrian Flag of the Legion, Member of the Swedish Academy of Sciences and Parisian Surgical Society, most Illustrious Patron

Many people who know and understand a great deal often speak highly of you, most renowned Sir, and they have so often extolled your virtues that I feel compelled to do the same, for they are truly well deserved. Your generosity, which you have shown to me day by day, is so great that I find myself wishing to find the words to express my deepest gratitude. I owe you, Sir, not only for all that I have learned in the art of surgery, but for my entire fortune as well. I humbly offer you these pages as a token of my gratitude and deep appreciation, and I earnestly hope that as you continue to enjoy great success and good fortune for many years to come, you will also remember me kindly. I remain, and will always remain,

Of Your Most Illustrious Name
Your most devoted admirer,
A. ÖSTERDAM

[s. n.]

---

[12]See Ellis, 1766.

[13]S.A.N. = *Summo Annuente Numine* in the original; translated as "By the Highest Divine Will"; thus, H. D. W.

Time Destroys Comments and Opinions.

§. I.

It is a well-known fact that much of our current scientific knowledge is a legacy left to us by the Greeks. The variety of fields in which they excelled is easily demonstrable, from Hippocrates in medicine to Euclid in geometry, Epictetus in ethics, Homer in poetry, Demosthenes in eloquence, and several others in architecture and nautical subjects. Although they recognized Zeus as the supreme ruler of the universe, they attributed to him all the bad properties they feared; his unpleasant fornications resulted in fictional gods and goddesses intermingled in a degrading progeny. The Romans, who received the empire from them, adopted the knowledge, customs, and religious practices of those whom they conquered. [p. 2] But they went even further than this, so that there was scarcely a mountain, river, or spring which did not have its own Nymph, its own protecting deity assigned to it. However, it would have been preferable if they had made greater progress in other disciplines; but as a warlike people, so concerned with extending the boundaries of their empire, they came to esteem no one except a man of military prowess; hence after the death of the Emperor Marcus Aurelius, the highest affairs were entrusted only to the most warlike men, and soldiers who could hardly write their own names were chosen to be Emperors; learned men were considered worthless compared to the ignorant, and literature perished in barbarous ages; while only monks, enclosed within monasteries, had leisure for study, those who were secluded from the world, disdained the contemplation of nature, and deluded by idle fancies to please the ignorant masses, held knowledge of nature in hatred, and regarded its students as magicians. They even called this book of nature, though written by the finger of God, directly opposed to another divine revelation. From this came the many miracles performed before the ignorant masses and the number of deified men, which exceeded the number of gods among the ancient Romans. Even in these times, stories and legends of Mountain Spectra, domestic deities, wild fauns, dark histriones, and sea serpents and sirens[14] inhabiting the waters, along with other companions of Lucifer, who were thought to have been cast down from the sky, have continued to be retained. Among the common people in less cultivated lands, this belief has become so strong that it seems almost impossible to uproot. Moreover, they devised countless monstrosities and prodigies of nature. After the darkness of barbarism, the dawn of the Enlightenment first arose, and from there a Theater of Nature presented itself to horrified spectators, entirely removed from the divine. By way of example during this weighty time, let Cuba be presented as an author who immediately after that miserable night of barbaric superstition published the *Hortus sanitatis*[15], where the appearance of the natural history of this age can be gathered from the following: [p. 3]

CERASTUS. A serpent with eight horns, long as an ox.
JACULUS: A serpent with feathered wings.
PEGASUS: A horned bull with feathered wings.
GRIPHES: A winged lion.
OPIMACHUS: A winged hare with a bird's beak.
LEPHUS: A man with a dog's head.
ONO: A centaur with a donkey's head.
HARPYA: A bird with a human head.
DRACONCOPEDES: A serpent with a human head.
BASILISCUS: A rooster with a snake's tail.
ZITIRON: A helmeted soldier that ends in a fish.
NEREIS: A scaly human with a fish tail.
PLATANISTA: A virgin with a fish tail.
SIRENA: A long-haired, mammal-breasted woman with a fish tail.
MONACHUS marinus: A monk-headed man that ends in a fish.
LEPUS marinus: Marine hare entirely covered in scales.
KOKY: A scaly goat with human fingers.
EQVUS MARINUS, the horse ending fish, and a hun-

[14] *Spectris montanis, Laribus domesticis, Faunis sylvestribus, Scurris tenebrarum*, & *Neptunis* atque *Sirenibus* in the original.

[15] Published in German as *Gart der Gesundheit* in 1485 and translated into Latin, as *Hortus sanitatis* in 1491, in edition of Jacob von Meydenbach.

dred other things that shame and disgust me to even mention. In the following centuries, investigators of nature, believing only what their own eyes observed, eliminated the myths of the ancients. However, it was difficult for old ideas to be eliminated. Monstrous dragons were still painted by Kircher as living in the subterranean world[16], and in Scheuchzer's Helvetian journey[17] it was believed that Scottish geese grew on trees and were transformed from barnacles into ducks. Seba dared to propose a seven-headed *Hydra* or deformed dragon, and Belon and Rondelet, careful authorities, judged that the *Bonarum marinarum*, or Bishop fish not entirely impossible. [p. 4] Such myths adhered for a long time to people living a solitary life, such as hunters, fishermen, and sailors, who easily convinced themselves and others of absurd things. Exploring the depths of the sea was very difficult, so sea monsters remained hidden in darkness for a long time. However, they saw wondrous creatures such as whales, rays, and other animals of unusual shape. The chimera ejected onto the shore was considered a miracle, with a forehead marked by Phrygian and Daedalian twists[18], which was meant to discourage such embellishments in humans. In their early voyages to the Indies, people used to tell us many wondrous things in their travel accounts, and here they mentioned sirens, as described by the Greeks, sitting on the water, resembling virgins with large breasts and human faces, two arms and hands, long hair, and with the lower part resembling a fish. See Barchewitz in *Hodaeporico India Orientalis* 1730 8vo, as well as Hasaeus' *dissert. Philolog.* p. 520[19], and *Historical-Political Geography* London 1729.[20] Also, Barthol. *hist. anat.* I. p. 188. tab. 188[21], depicts a Siren captured in the Brazilian sea and dissected by Pavius of Leiden, with a bald head, otherwise human-like, with two arms and hands, but webbed like a goose, with breasts on the chest, protruding ears with a neck, and with the lower part contracted into an oblong shape without a tail (see our figure[22]). Indeed, these convinced our immortal Artedi[23] to classify the Siren among the species of fish, as shown on p. 81. A few years ago, in the public news[24], it was told with many convincing details that a siren was captured in Denmark, near Nycopia (Nykøbing?) and kept alive in a large bucket; for this reason, our Nob. D.[25] Director was ready to travel to see this natural miracle, had it not been for the long hair attributed to this prodigy, which he realized could not exist in a marine animal. Therefore, having sent a letter to the Royal Academy of Sciences [p. 5] and to our representative in the Danish Court, he asked that this be investigated more carefully; after receiving the response, he found that it was a mere myth, and we have not heard anything about new Sirens since then.

§. II.

It is not feasible to recount all the wonders that ancient scholars, and others, have documented about sirens. To summarize, in ancient tales, Sirens were depicted as marine creatures with two arms, each having hands and fingers, that ended in a fish-like tail without legs. They could swim like humans and possessed a powerful and melodious voice. Many ancient accounts suggest that their singing was a soothing and enchanting sound, similar to that of Apollo's lyre-playing for the dolphins. Some even believe that the sirens may have lured dolphins to them. However, it is highly likely that the dolphins were never involved with sirens, and instead, the sea creatures that were mistaken for them were Manatees, as they are the only known marine animals that can produce a musical sound that captivates and delights the human ear.

---

[16]*in mundo subterraneo* in the original; see Kircher (1664; 1665).

[17]*itinere helvetico* in the original; see Scheuchzer (1723).

[18]The phrase describes a forehead that is deeply furrowed with lines that resemble the intricate coils or twists associated with Phrygian art and the craftsmanship of Dædalus.

[19]Barchewitz (1730: 381) characterizes Der Manaten oder See-Kühe, comparing with See-Männer. Österdam did not see this book personally, and the reference is based on Hasæi's (1731) authority. The latter analyzes the Manathi in point 6 (p. 519), and in section 7 (p. 529) deals with the Dœjong, citing Barchewitz.

[20]Paschoud (1729: 227) noted: "...In the Sea surrounding these Isles is frequently seen a sort of Fish, or Sea-Mounster, about a bigness of a Calf, which in Shape dœs much resemble the ancient Sirens, or Mermaids, whence Sailors call it a Woman-Fish, because its Face, Head, Neck, and Breast, are somewhat like those of the fair Sex...".

[21]Bartholini (1654). Historia XI, *Sirenis seu marini hominis anatome* starts at p. 162; the plate follows p. 164. One figure of this plate was included by Österdam in the illustration of his dissertation.

[22]See Plate I.

[23]Artedi (1738a).

[24]The reference to "public news" about the capture of a mermaid off the coast of Nycopia appeared in several journals in 1749. Among others, on p. 428 of vol. XIX of The Gentleman's Magazine of September, 1749, is advertised that "...At Nykoping, in Jutland, was lately caught a mermaid, which from the waist upward had a human form, but the rest was like a fish, with a tail turning up behind, the fingers were joined together by a membrane; it struggled, and beat itself to death in the net..." (Urban, 1749). Regarding Danish sirens, Bartholini (1670) wrote an article (Observatio XXIII) on *De Sirene Danica*.

[25]*Nob. D.*, or *N. D.* Latin *Nobile Domine*, Noble Lord.

§. III.

In our time, our understanding of natural history has reached new heights and has revealed that God's work is a uniform chain, proceeding steadily and deliberately from one step to the next, without sudden leaps or jumps. Nature has covered Mammals with hair and provided them with four legs; if they live in water, their hind legs are joined not into a fish's tail, but into a tail-like shape, and all of them have mammary glands, unique in the entire animal kingdom. Birds are clothed only in feathers, they walk on two feet, their front limbs are in the form of wings and their beaks are naked and elongated. Fishes are covered with overlapping scales and have gills instead of lungs for breathing; indeed, manatees and some snakes seem to have scales, but they are formed [p. 6] and decorated in a completely different way. Amphibians in some respects closely resemble fish, such as having cold blood and often gills, but they are nevertheless distinct from them in having lungs. Worms are naked and have not a single bone inside. Insects, also devoid of internal bones, are armored and can be distinguished by the presence of lateral pores or spiracles and their unique antennae. Among all of these [animals], a truly wondrous quality occurs which has greatly exercised the minds of curious naturalists, namely metamorphosis, which is no less paradoxical than that of Ovid[26]. This quality is shared by most insects, except for the Aptera; it is rarely observed in worms, but can sometimes be seen in certain Cyprian species, as well as in amphibian lizards, Ostracion, Diodon, and a few others. Through this type of metamorphosis, we observe that the same animal is presented in this theater through various forms of larvae. Who, unless they have experienced it, would believe that a larva, resembling a smooth, heavy, and voracious stick armed with numerous spines and six legs, could transform into a light butterfly, with four wings boasting beautiful colors and adapted only for gathering nectar? Who would claim that the delicate Ephemera, flying through the air, was transformed from a larva resembling a gluttonous dragon-worm? Who would affirm that the Myrmeleon, resembling a spider with extremely voracious mandibles and antler-like projections, can transform into a delicate neuropteran? That frogs once had a tail like a lizard, and that aquatic lizards had a tail like a spear? The changes observed in nature should be accurately followed in a sequence that proceeds forward, not backward, as was believed by Merian and Seba[27], who thought there was a retrograde transmutation of *Rana paradoxa* into a fish. Certainly, such animals in their larval stage are less recognizable than humans wearing masks in theatrical performances. However, it is regrettable that curious minds have not yet discovered any [p. 7] theory that can determine the perfect species that the unknown larva will give. We still do not know whether *Gryllus perspicillatus* will have wings or not, and no one would have taken *Gryllum pedestrem*, very similar in appearance and structure to the pupa, for a perfect animal unless it had been observed during copulation. Moreover, bed bugs[28] are indeed only larvae or pupae, which are never fully developed and do not develop wings, but reproduce as wingless adults. Whether they are found with wings in warmer climates, as stated by Mr. Scopoli in his Carniolica[29], we leave to the experts. Perhaps the same is true for them as for *Ruellia clandestina, Amannia latifolia*, and some other flowers with mutilated petals that do not fully expand in our climate except during the height of summer.

§. IV.

Therefore, considering the theater of the universe, I cannot help but marvel at the wonderful works of the Supreme Artisan in the Plant Kingdom, which have been discovered by the most diligent nature researchers, previously unknown in our century; in East India by Rumphius and Burmann; in the West by Plumier, Browne, and Jaquin, in North America by Clayton and Kalm, in North Asia by Gmelin, Steller, and Messerschmidt, and at the Cape of Good Hope by others, which make an incredible number. Truly amazing properties are presented here. *Ceratocarpus* is rolled by the wind from one end to the other in the open fields of Tartary, just like *Anastatica*, which is tossed from shore to shore in the Red Sea. *Sarracenia* as *Nymphaea*, has its leaves transformed into funnels, so that it can store water like an aquatic plant on the mainland. *Bromelia Karatas* hides its fruits on the ground with menacing thorns until they mature, then offers them spontaneously. *Laurisolia* Sloan. jam. tab. 168[30], brings forth its bark in the form of snowy, latticed coverings, as if they had been carefully crafted by the hands [p. 8] of skilled women artisans. *Vallisneria* male flowers, separated from their plant, swim around their female counterpart like living insects. But to dwell on each one individually would take too long. Animals, so greatly neglected by the ancients, have been discovered in a number ten times greater in our time, to the admiration of Divine wisdom. First came the whales, the largest animals in the group of Mammals, and then *Sorex minutus*, the smallest of all mammals, weighing no more than 3 scruples[31]. We are familiar with the *Gryphus*, the largest of all birds, and with a whole flock of *Trochilus* birds, the smallest of which weighs as much as one scruple. D. Jaqvin discovered that sparrows[32] serve as warning for hens against

[26]Reference to *Metamorphoseon*, Latin pœm written by Publius Ovidius Naso around the year 8. Unidentified edition; see Posthius (1563).

[27]Merian (1726); Seba (1734).

[28]*Cimices lectularii* in the original

[29]Scopoli (1763).

[30]Refers to *Laureola minos sylvatica foliis ex adverso nascentibus, baccis rubris* of Sloane (1696).

[31]*Scrupulum*, unit of weight in the apothecaries' system, equal to 20 grains, or one-third dram, and equivalent to 1.296 grams.

birds of prey, just as dogs guard sheep. Meanwhile, *Psittacus pendulinus* sleeps suspended from branches with one foot. *Buphaga* remove gadfly larvae from the backs of cattle. The mites torment the *Crotophaga*, while the *Psophia* sings with its mouth and resonates with its anus. Snakes are attracted by *Aristolochia*, and smaller amphibians fascinate animals, so that they fly into their mouths. *Gymnotus electicus* produces a remarkably powerful electric impulse, surpassing that of the *Torpedo*. Among fish, *Cyprinus auratus* nearly surpasses the shine of gold. *Fulgora* shines like a torch. *Cancer ruricola* uses its claws to violently crush its prey. *Gordius* worms passes through clay like a fish does through water. *Hydra* and many other animals are composed of many parts. *Corallia* are constructed by animals. *Gorgonias* are a type of plant, but their flowers are animals. *Volvox* carries offspring, grandchildren, great-grandchildren, and even those not yet born within itself. *Ustilago* on barley is alive. *Chaos redivivus* can be dried out for the entire year, but it can still revive through water like moss. These are some examples of the diverse organisms that exist in nature. For brevity, I shall not say anything more about the remainder.

§. V.

The class of fishes has received few new additions, but it has been greatly enriched by the contributions of the illustrious Master Garden[33], who lives in South Carolina in America. By generously sending [p. 9] almost all the fishes from the northern part of America to our N.D. Director, he has earned an immortal name. At the request of our N.D. Director, he anatomically dissected the branchiostegous fishes to investigate whether they possess not only gills but also lungs. He found them to be present and, after dissecting and properly preserving them, sent them here. Recently, this eminence[34] sent an aquatic animal whose form is so unique that it defies comparison with any other known species. Most notably, it possesses only two front feet and none on the back, a feature that has not been observed in any other animal. For whales, their hind legs are fused together in the shape of a tail, as previously stated; this tail is actually a fin, resembling that of an eel, and bears some resemblance to the mythical Sirens of ancient times; furthermore, they have the talent to vocalize, which, except for men, is unique to birds. Therefore, I deemed this animal worthy to be shown to the curious, because it is new and differs from all others.

§. VI.

To convey a clear idea of this, I prefer to first present its figure[35] and description to the reader.

The body is eel-like, scaleless, about a foot long or slightly less, as thick as a thumb, somewhat rounded, obscure gray, sprinkled with numerous paler spots. There are wrinkles along the sides, between the head and the anus, forty on each side. Two lateral lines on each side. The head is like a lizard's, not thicker than the body, ovate, covered with naked skin. There are two small eyes, covered with common skin. Two nostrils on the sides of the snout. The upper lip line is small. The jaws are covered with lips, the lower one shorter. The upper jaw has a single marginal tooth, somewhat notched. Additionally, there is a row of very small teeth [p. 10] on each side of the palate. The lower jaw has a single tooth extending along the entire jaw with a sharp edge, and towards the sides, there is a series of very small teeth towards the throat. The tongue is somewhat round, smooth. In place of ears, there are gill covers on each side, solitary, pendulous, fleshy, tripartite, of which the upper lobe is trifid, and the others only crenate. The gill opening is very small. The anterior feet are very short, near the head or behind the gill covers, or shorter than the head, with arms extended outside the body. The ulna is about the size of the arms, somewhat rounded. The hands are fleshy, with four very short fingers, the outer one shorter than the others. The claws are awl-shaped, sharp, very short. There are no hind feet at all. The anus is located two-thirds from the head. The tail is almost as high as the body, but compressed; above and below it is double-edged, with a keel line parallel, ending behind rounded or barely pointed.

I add to this description, which Mr. Garden himself made in its native place, according to the living animal, from which also the internal delineation and Anatomy can be clearly seen.[36]

Aquatic animal with a fierce and insolent livid expression. Head depressed, wide, spotted, oblong, and serpent-like. Mouth wide. Snout barely extended, with the upper lip receiving and covering the lower one at the corners. Nostrils solitary on each side, situated at the junction of the upper lip with the snout, large and distant from each other. Eyes small, blue-gray, located on the upper plane of the head, on the sides. No branchiostegal membrane. Branchial opercula present on both sides, namely three appendages that are longer, thicker, wider, very smooth, hanging on the

[32] *Pattan* in the original.

[33] Alexander Garden (1730–1791), Scottish physician, botanist and zoologist active in South Carolina (USA) from 1752 to 1782.

[34] *Phœnix* in the original.

[35] See Plate I.

[36] This description is, apparently, in a letter that could not be located, sent before May 18, 1765. At that time, Garden wrote to Linnæus saying, among other things, that '...Besides these, I have sent you the characters of a very extraordinary animal, which I never saw till lately, though they are very common here. I have likewise sent you two specimens of it, though they are much less than the one from which I made out the characters, which was between two and three feet in length. I have sent it to Mr. Ellis for the museum. It appears to me to be a middle link between the *Lacerta* and the *Muræna*, having some things in common with both, and yet differing from both...' L3591 in linnæus.c18.net.

sides, and have ears that somewhat resemble those of a dog. Four gills on each side, with two free gills in the middle and two other gills attached to the side of the upper or lower gill opening. Externally smooth, without feather-like fringes. Internally provided with some rough, sharp, [p. 11] long, and tooth-like tubercles arranged in two rows. Only one tooth, undivided, that extends throughout the entire lower jaw, curved, black, horny with a sharp, smooth edge. On the sides of the same lower jaw, a row of very small, sharp teeth extends longitudinally towards the throat on both sides. The middle and anterior part of the upper jaw is also occupied by a single tooth of the same consistency and color as the one in the lower jaw, but much smaller and with a slightly serrated edge. The palate is covered on both sides with numerous small, sharp teeth arranged in rows. The tongue is simple, soft, smooth on all sides, and attached to the bottom of the lower jaw at the roots. It is round at the front and free and loose at the apex and edges. The body is thick and somewhat cylindrical, with a slightly flattened back in adults; the belly is flat and the anus is somewhat round, and from here it tapers to a compressed tail. Tail with fins on both margins. Skin mucous and very slippery, with many spots scattered here and there. Scales so thin and imperceptible that they cannot be seen unless the fish is dried, at which point they appear attached and sparse, covered by the skin. Two lateral lines on each side, made up of long, longitudinal dark spots; one originates from the axilla of the pectoral fin and the other from the axilla of the branchiostegal operculum. The anus, wide and long, is located in the posterior part of the abdomen, towards the middle of the fish. Two feet close to the gills but sloping towards the belly, distinctly composed of thighs, shanks, and feet; four-toed, with each digit equipped with a conspicuous claw. Anterior region of thorax fortified with hard cartilage, which protects the heart and facilitates the passage of the rough arteries; houses the gold-colored heart, which lacks an atrium and is proportionally small for the size of the fish. Lungs very large, separated, and distinct; extend from the thorax to the anus throughout the entire abdomen, closely approximating the back; each has its own trachea, running through the thorax and maintaining [p. 12] communication with the gills. The liver is the largest of all, although much shorter than the lungs, yet thicker and heavier. The gall bladder is quite long, suspended around the middle of the liver's neck. The intestine, not much longer than the fish itself, has few convolutions and extends towards the anus, gradually widening in its course to the cecum, but not having a notable stomach or any appendages.

NOTE. Two other organs are found near the anus, each divided into two parts and thick, whose use is unknown to me. This animal lives in the watery swamps of southern Carolina, where it lives both in and out of the water. It climbs onto logs and branches that have fallen into the water, and when the swamps dry up during certain times of the year, it sings with a plaintive voice, similar to that of a young duck but sharper and clearer, as D. Garden relates in his letters.

The figure shows natural magnitude. See the plate[37].

§. VII.

Having presented the idea of this animal, it remains for us to consider it systematically, and also in relation to other animals. From its possession of lungs and their associated respiration, it is clear that it is an Amphibian; for no other animals are equipped with lungs except Mammals, Birds, and Amphibians. This cannot be a Mammal, lacking hair and mammary glands, nor a Bird without feathers and a toothless protruding jaw; therefore, it must be an Amphibian, naked, with two front legs and no hind legs. Amphibians are divided into Reptiles, which have four legs without gills, into Snakes, which have no legs or wings or gills, and into Nantes, which have fins and gills. *Anguis bipes* indeed has two hind legs, but is without anterior legs, if one can call the rudiments of legs, [p. 13] which are like filaments with two digits, legs. In this classification, our *Siren* is closer to Reptiles, and among them to Lizards, as it could be considered a *Lacerta bipes*, if it were not for the gills, which are typical of Nantes; therefore, we must exclude it from the first order of Amphibians. Nantes indeed have lungs and gills, just like our *Siren*, but because they have true legs with claws, they cannot be included in this natural order. Its sound or song is very singular. *Rana arborea* and all other frogs do indeed have a sound, but one that cannot be compared to the voice of our animal, which excels among Amphibians. Therefore, if we add this to the System, a new order must be established after Reptiles, which contains animals that have two legs, breathe through lungs and gills, and are capable of living in both water and on land, even more so than Nantes.

§. VIII.

I have previously touched upon the astonishing metamorphosis of certain animals and amphibians. This is clearly observed in lizards[38]; while they are larvae, they live in the water, resembling fish, with two pectoral fins near the head. When they transform, they shed their outer skin and are delighted with their legs[39]. When this animal from Carolina was sent to the Honorable President by Mr. Ellis in London, this very acute and observant student of nature wrote that this larva is likely that of the Iguana Lizard[40], also pointing

[37]See Plate I. Not in "natural magnitude"

[38]Note that all known salamanders in the XVIII century were included in the genus *Lacerta*.

[39]The allusion to the appearance of the limbs due to the loss of an outer layer of skin reflects the ideas that Swammerdam (1693) had on the subject.

[40]*Lacertæ Igvanæ* in the original.

out that is the one which the Americans name Mud-Iguana[41]. But the matter is not yet clear, as the larva has digits with claws, which no known lizard larva has. *Lacerta Aquatica, Palustris, Punctata*, and *Salamandra* have mute digits; also, the voice of this animal does not seem to match that of the larva. The four digits on its hands show that this animal is not a known lizard larva, let alone an Iguana, [p. 14] which has five digits on its front feet. However, nature is multifarious, and the laws written about it are often in vain.

Conclusion

If, in fact, this creature is not a larva, as has been demonstrated, it cannot belong to Nantes, since they have fins rather than arms with clawed digits, and therefore it cannot be classified as belonging to that group; furthermore, it cannot be classified as a new bipedal genus of reptiles, since it possesses both lungs and gills, which are entirely absent in reptiles, including snakes. Therefore, it remains to establish a new order in the class of amphibians[42], which includes animals with legs, equipped with both lungs and gills [p. 15]. This order will contain a new and distinct genus, characterized as an Amphibious animal with two legs, which could aptly be called *Siren*, name used by the ancients to describe an aquatic animal that had two hands and produced vocal sounds, and whose posterior body ended in a fish-like tail. The descriptions of sirens having a human face, a mammalian chest, a hairy head, and so on, seem completely fabulous. Finally, we add a trivial name: *Lacertina*, due to its resemblance to lizards, especially with regards to its tail, similar to aquatic lizards: specific difference is unnecessary, since no such animal has been detected in that genus. Nevertheless, we do not believe that this creature can be placed in a systematic classification of animals until further observations have been made in its native habitat. For us, it suffices to have described such a rare animal and to have spurred the curious to a more detailed consideration of it.

That's all.

To our Respondent, a most excellent Friend.

While you are climbing the steep heights of Pindus with swift steps,
Take the just rewards that are owed to you.
May glory not be lacking for illustrious virtues, I wish,
And may you gather many benefits of wealth and structure.
Always rejoice in the help of the kindly divinity,
And may Love's unwavering protection guard you.

J. S.

[s. n.]

To our RESPONDENT and most respected Friend.

Given the present occasion, my dear friend, as you prepare to publish your elegant dissertation, as a commendable proof not only of your genius but also of your solid progress in the shrine of Natural History, I cannot help but express my congratulations for the friendship that has remained unbroken between us for so long. I would wish to give you praise proportionate to your merits, as reason dictates, but my limited eloquence is inadequate. In the meantime, if any sincerity or candor exists in me, I congratulate you, my dearest friend, on this important specimen of your knowledge, coupled with the many embellishments that will win over everyone who appreciates virtue and the liberal arts, and that will secure your unshakable happiness. And if, as I confidently expect and hope, you reap plentiful fruits from your labors in the future, then I pray that you will accept my heartfelt. gratitude and that of others for them, and I shall continue to pray for this for as long as I live

Christ. Ern. Böecler.

[41]In English in the original. In an undated letter (probably written in August, 1765, fide Smith, 1821), John Ellis wrote to Linnæus: "... I received the skin of the Mud Iguana of Dr. Garden. This animal is certainly a larva. Those that were small, of about seven or eight inches long, have three appendages on each side, in the room of fins, but they are not opened sufficiently, as in the large skin, which is 30 1/2 inches long. Here they look like the fins of *Lacerta vulgaris* in its larva state, and are pinned in both sides; but the reason why he takes it for a new genus is, that there is no appearance of the hind legs in the least, but the fore ones are very distinct and plain, with bones and nails. It has a remarkable palate, full of rows of teeth. I believe you will find, when I send you a specimen, that there is nothing new in it...". See also Ellis (1766).

In another letter, dated at Grays Inn, London, on January 31, 1766, Ellis commented: "... Dr. Solander was yesterday at our famous anatomist's Dr. Hunter's, to see four specimens of this new animal Mud Iguana; but none of them are so large as the dried specimen that I have, which is 31 inches long. He has dissected one of them, and says it has the same kind of teeth, and is, in every respect, like our *Lacerta* (called in english Eft or Newt) in its aquatic or larva state, and differs in nothing from it, but in having only two feet. Dr. Hunter says that the larvæ, both of our *Lacerta*, and the Iguana, have gills and lungs...".

[42]The order suggested by Österdam was established by Linnæus in the Appendix to volume II of SN 12 (Linnæus, 1767), with the name of Meantes.

Plate I

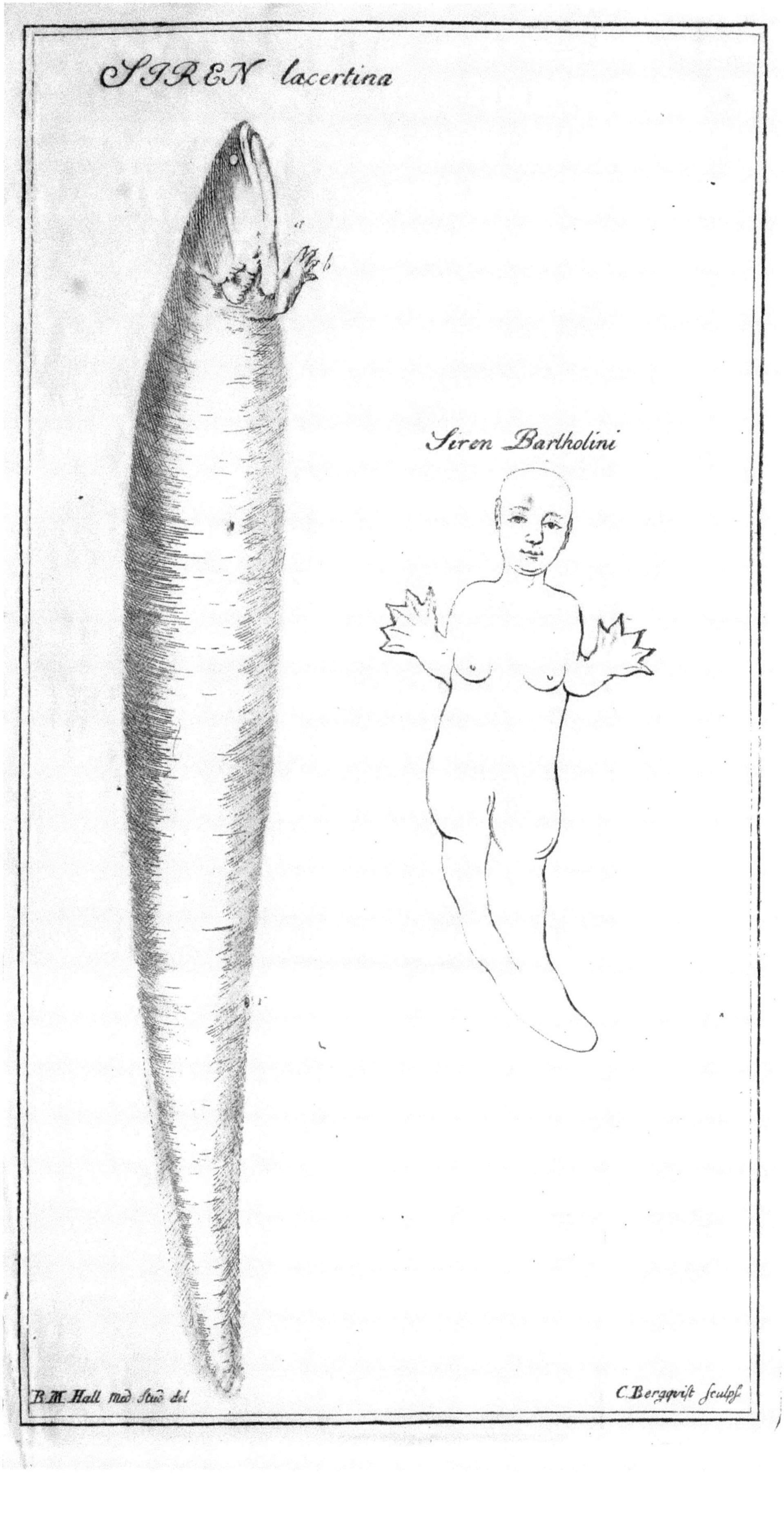
SIREN lacertina
a
b
Siren Bartholini
B.M. Hall Med Stud del
C. Bergqvist Sculps.

Q. F. F. S.

DISPUTATIO MEDICA,

DE MORSURA SERPENTUM,

QUAM,

INDULTU NOB. ATQUE EXPER. ORD. MED.
IN SUPREMO AD SALAM LYCÆO,

SUB PRÆSIDIO

VIRI NOBILISSIMI NEC NON EXPERIENTISSIMI

D:NI DOCT. CAROLI LINNÆI,

EQUIT. AURAT. DE STELLA POLARI, S:Æ R:Æ M:TIS ARCHIATRI, MEDIC. ET BOTAN. PROF. REG. ET ORD. ACAD. SCIENT. UPSALIENS. HOLMIENS. PETROP. BEROL. IMP. NAT. CUR. LOND. MONSPEL. TOLOS. ET FLORENT. SOCII,

PUBLICE VENTILANDAM EXHIBET,

STIPENDIARIUS REGIUS,

JOH. GUSTAVUS ACRELL,

Regiæ Academiæ Scientiarum Adſcriptus,

HOLMIENSIS.

IN AUD. CAR. MAJ. AD D. XVI. JUNII, MDCCLXII.

Horis Ante Meridiem Conſvetis.

UPSALIÆ.

Acrell, J.G. 1762. De Morsura serpentum. Upsaliae. S.n. [4] + 2-19 + 1 fig. + [1].

Johann Gustav Acrell defended his thesis on 16 June 1762, and it was re-edited by Linnaeus in the sixth volume of *Amoenitates Academicae* (Acrell, 1763). The thesis a comprehensive compilation of contemporary knowledge on snakebites, their effects, and potential treatments. Acrell provides detailed descriptions of snake anatomy, venom apparatus, and envenomation symptoms, including comparative analyses of various snake species. He critically evaluates ancient and contemporary antidotes, offering empirical observations on their efficacy. Additionally, the text includes discussions on the nature of snake venom and its mode of action, as well as an exploration of folk beliefs surrounding snake charming.

M.I.B.F.P.[43] A Medical Disputation on the Bite of Serpents, *which*, by the permission of the Noble and Experienced Medical Order, was exhibited in the highest Salamis Lyceum, under the guidance *of a most Noble and Experienced man* D:ni Doct. Caroli Linnaei, Knight of the Order of the North Star, Royal Chief Physician, Professor of Medicine and Botany, member of the Royal and learned Academy of Sciences in Uppsala, Stockholm, St. Petersburg, Berlin, Imperial Naturalist Society of Moscow, Royal Society of London, Montpellier, Toulouse, and Florence. *Presented publicly for discussion* by the Royal Stipendiary Joh. Gustavus Acrell, *enrolled in the Royal Academy of Sciences, Stockholmense. In the lecture hall of the Royal Palace on June 16, 1762, at the usual morning hours.* Uppsala.

[s.n.]

H. R. M.[44] Man of highest faith Senator of the King and Kingdom of Sweden, President of the Royal Chancellery, Chancellor of Uppsala University, Knight and Commander of the Orders of the Realm, *most distinguished, most noble and excellent* Baron and Lord Dom. Anders Johan von Höpken,[45] Greatest Patron.

[s.n.]

If it was neither unworthy nor ungrateful during the most prosperous era of Rome to frequently read about its birds, about which snakes kill and which herbs help[46], as attested to by Ovid, the most famous author, about the poet Macro[47], and even in the presence of the great Caesar Augustus[48] and his intimate friend Maecenas[49], who distinguished himself by his glory in the name of literature and its study, as we know from the monuments of his time, then perhaps you will not blame me, or at least I hope you will not, if humbly, Most Excellent Maecenas, I dare to dedicate this brief work on snakebites to your sublime name. Indeed, by a singular blessing of fate, we live in an extremely favorable time for our country, in which not only the natural sciences flourish and develop, along with the study of letters,

[43]Q.F.F.S., *Quod Felix Faustumque Sit* in the original; the sentence can be translated to mean "May it be fortunate and prosperous", thus, M.I.B.F.P.

[44]The original S:æ R:æ M:tis stands for *Suæ Regiæ Majestatis*, translated here as His/Her Royal Majesty; thus, H. R. M.

[45]Anders Johan von Höpken (1712-1789), Swedish Count, University Chancellor, Court Marshal, and one of the founders of the Royal Swedish Academy of Sciences in 1739. He was known for his political influence and his support of the arts and sciences.

[46]"*...sæpe suas volucres legere, quæque necet serpens, quæ juvet herba...*" in the original. This is a verbatim transcription from Ovid. *Tristia*, Book IV, X, verses 43-44: "*...sæpe suas volucres legit mihi grandior ævo, | quæque nocet x serpens, quæ iuvat herba, Macer...*"; it was translated as "...Ofttimes Macer, already advanced in years, read to me of the birds he loved, of noxious snakes and healing plants..." by Arthur Leslie Wheeler (Ovid, 1939).

[47]Æmilius Macro or Macer (?-16), Roman pœt, author of *Ornithogonia* (from Balio) and *Theriaca* (from Nicander of Colophon. See Bardon (1956: 44).

[48]Cæsar Augustus (born Gaius Octavius; 63 BC-14 AD), first Roman emperor.

[49]Gaius Cilnius Mæcenas (c. 70 – 8 BC), political advisor of Cæsar Augustus and patron of arts.

almost without exception, but also, what we cannot but feel with the deepest sense of devotion and joy, is cultivated, fostered, and exalted by the very Princes of the Kingdom and their successors, as well as by eminent patriotic figures and patrons. And in this you, Most Excellent Maecenas, stand out for your supreme and divine zeal in all the sciences, and for the grace you have earned in the field of letters, being one of the most celebrated in our country, without the need for any introduction, although it would be more appropriate to admire and venerate you than to describe it with such imperfect words. However, I cannot help feeling inspired by this vision, especially when you have been kind enough to dedicate your time and attention to these academic matters. And there is another reason, which far exceeds all my expectations, but which wonderfully attracts me: I have undoubtedly perceived, through the indications of your great affection for my uncle, that you, Most Excellent Maecenas, have thought of me on more than one occasion. Although all this should be praised with worthy words, I dare not and cannot bear it. [s. n.] However, I beg you with all my heart that, in your great kindness, you allow me the honor of being under your patronage. And I beg this of you with the same devotion that I feel when I think of your kindness. And with this patronage, I ask that you accept my sincerest wish for health, prosperity, fortune, and a distinguished family, and I assure you that I will never cease to pray for you as long as I live.

Of Your Most Excellent Name
Your most devoted vassal,
Johannes Gustavus Acrell.
[s. n]

Introduction.

If we analyze carefully the various elements that conform the three kingdoms of nature, we see that each of them can be considered both, as harmful to health and life of man, or as a support and provider of sustenance; on the basis of its effects on us, we say that those in the first category are poisonous, and those of the second, healthy. But as the saying goes, all things in excess are dangerous, and thus a healthy meal can become the worst poisons. So the wine, the nectar of the gods, drunk in moderation refreshes the body and mind, and their vapors applied to the nose revive those who have fainted, rescuing them from the world of the dead; but when drinking in excess, nature, by its own strength, is determined to break free of him with a nauseous and acid vomit; its corrosive qualities can produce fever, and often we see examples of melancholy and premature death among gluttons. The opposite is also true, as the worst poisons, [p. 2] when used properly and skillfully applied, become highly effective drugs: So, we use various preparations derived from Opium[50], Elaterium[51], Scammony[52], Euphorbium[53], and Mercury[54].

We define poison as something that, taken or applied externally to the human body in very small amounts, has the ability to produce a violent effect that destroys the health and life; but it is not possible to assign essential differences between the healthy and the poisonous, as used in excess, any wholesome substance becomes poisonous.

Poisons derived from different natural kingdoms are

---

[50]Dried latex obtained from *Papaver somniferum* (Papaveraceæ).

[51]Purgative, extracted from the juice of *Ecballium elaterium* (Cucurbitaceæ).

[52]Purgative, from the root of *Convolvulus scammonia* (Convolvulaceæ).

[53]Resinous juice of *Euphorbia resinifera* (Euphorbiaceæ).

[54]Represented with the symbol ☿ii in the original.

different. Thus, among the Minerals, arsenic, orpiment[55], and numerous metal preparations stand out. In the Vegetal kingdom, there are the following natural orders: the CONTORTARUM, such as *Asclepias* and *Cynanche*; the LURIDARUM, such as *Hyoscyamus*, *Tabacum*, *Atropa* and *Datura*; the TRICOCCARUM, such as *Euphorbium*; the MULTIFILIQUARUM, such as the *Hellebore* and *Nigella*; the UMBELLATARUM ▽*m.*, such as the *Cicuta*, *Sium*, and *Ethusa*. In the Animal kingdom, there are Serpents, Frogs, and Toads, just to name a few.

Having briefly discussed the topic in this introduction, we now turn to consider the venoms in those animals that, armed with different instruments, wound by puncture or bite.

It is with disbelief that we admired the weapons with which the Supreme Creator has provided animals for defense, especially those that are poisonous. For some, there are teeth and bites, such as bats and soricids; *Trachinus draco* injures with the anterior spine of its dorsal fin, and various jellyfish burn like nettles. But our wonder is enormous when faced to these weapons that are infected with brutal poisons, such as those attached to the tails of some insects, such as scorpions, wasps[56], and bees, just to name a few, [p. 3] or those located in the mouth of the snakes, which are the subject of this dissertation.

AMPHIBIA.

Snakes terrify almost all humans and other animals with their venom and other singular properties that elude our understanding, and their unexpected proximity can scare even the bravest. First, we shall see some peculiarities of these animals. In the class *Amphibia* we find animals endowed with some unique features that differentiate them from others: most of them can live on both land and in water; many hibernate half the year, appearing as though they are dead; their breathing differs markedly from that of birds and mammals, as they inhale air without a perceptible exhalation. Their bodies are rigid in nature, with abundant cold blood. They have a double *penis*, which is not found in any other animal, except, perhaps, in some insects. They have a hoarse voice, with a hideous appearance and a repugnant odor that our senses detect. Some are oviparous while others viviparous. Similar to crabs, they possess a cartilaginous skeleton, and they molt every spring so that their growth is unlimited[57]. It is also worth mentioning their particular way of capturing prey. Lacking limbs, they cannot pursue their prey, so they stay hidden in the shade and among the roots of trees until an animal passes by. Then they suddenly jump out and capture and eat it.

The *Boa constrictor*, the largest among snakes, ambushes hares, deer and other animals in the thickets, suffocating their prey by coiling around and later swallowing it after covering it with saliva, as attested by the travelers into India[58]. However, since prey capture does not occur frequently, and lacking [p. 4] true molars, the Author of Nature has given them jaws that are not articulated by gum, but only connected at their base by ligaments, so that they can expand their jaws to devour the body of an animal twice their thickness or more. As noted by Cleijer, *Boas* in Java can swallow an entire buffalo, while other travelers mention that Boas have devoured a whole deer, and the antlers coming out of their mouths are quite a spectacle[59].

Moreover, our *Colubri Beri* are frequently encountered with distended bellies, and upon dissection, they are found to have swallowed frogs, mice, and on one occasion, I discovered one that had devoured seven young birds that had not yet been digested.

There is another phenomenon, observed by residents of Philadelphia and confirmed by reliable sources, that has yet to be fully explained and requires further study. According to reports, the American *Crotalum* is a lethargic snake that lurks in the shade of trees with its jaws agape and bright eyes fixed on any passing bird or squirrel. It is said to possess a remarkable power of fascination. The squirrel complains with a sharp, plaintive sound, as if aware of its fate, and jump from branch to branch as if trying to escape. But, for some inexplicable spell, jumps down from the tree directly and voluntarily to the jaws of the enemy. Observations by some English[60] appear to support this claim. They locked up a

[55]Arsenic sulfide (As_2S_3). From L. *auripigmentum*, in turn, from L. *aurum*, gold + L. *pigmentum*, coloring, dye, pigment, tint or paint, reflexting the deep orange-yellow color of the mineral.

[56]*Sphegum* in the original.

[57]Amphibian characteristics are almost a verbatim transcription of Linnæus (1758: 194). The mention of a cartilaginous skeleton is curious, at least in this context. This assertion can be traced back to Aristotle who, both in *De Historia Animalium* Liber X, 5:5 and in *De Generatione Animalium* Libri V-3:3 (Aristoteles apud Gaza, 1533) included "*Rana*" among the cartilaginous, oviparous animals that lay perfect eggs. This was a misunderstanding of Linnæus. In fact, Aristotle referred to *Rana marina*, a fish today in the genus *Lophius* (Lophiiformes: Lophiidæ). Since Aristotle is not among the primary sources of the tenth edition of Systema Naturæ, and his works were not in Linnæus's personal library (Anonymous, 1896), it is assumed that the information is a transcription from Gesner (1604: 817) "... *Rana pisces est cartilagineus*, Aristoteles..."

[58]According to Brand (1781), this comment refers to Baldæus (1672), who, in Chapter LII, states that "... The serpents of Ceylon are not altogether so large as those of Java and Banda; at Batavia there was once taken a serpent, which had swallowed an entire stag of large size; and one taken at Banda had done the same with a Negro Woman..."

[59]Refers to Cleyer (1684). Andreas Cleijer (1634–1697/98), surgeon at the Batavia Castle, reported in a letter to Christian Mentzel, surgeon and librarian of the Great Elector Friedrich Wilheim, that in the body of a snake of 25 *pedes operarios (vulgo Rinlandicos)* of length he found a deer of middle age; in other there were a wild he-goat, and a third one had a porcupine with its quills. Refers also that in the Moluccan Island of Amboin a gravid woman was degluted by a snake, while in the kingdom of Arcan, in the border with Bengal, the inhabitants keep these snakes as a protection against wild buffalos. According to Bails (1790), the *pedes Rinlandicus* is equal to 0,9667 times the foot of the King of Paris, equivalent, in turn, to 1/6 of a French Toise. Corresponds to about 31.36 cm.

[60]See, among others, Jones (1700) and Beverly (1705).

mouse with fascinating *Vipera caudisona* in an iron box; the mouse was sitting in a corner and the snake in the opposite, and fixing the eyes on it continually, with the sheer force of his gaze forced the mouse to jump into his mouth. The same experiment [p. 5] was replicated in Italy using a pregnant female *Colubri Beri* and a *Mure Musculo*, with identical results. What's more, similar behavior can be observed among our slow toads, which are related to snakes. During the summer, they rest in the shade of trees and wait for flies and butterflies to fly into their open mouths[61].

Snakes' Weapons

Few topics have sparked as much philosophical controversy as the composition of snake venom, how it operates, and the effects it produces. In ancient times, several nonsensical conjectures were made about the poisonous apparatus of snakes, which still persist, firm as a rock, among the common people. These beliefs include a fictitious sting attached to the tail of some snakes, which they supposedly use to attack. Some painters even represented this falsehood. Others say that the black and split tongue, which the snake stirs at either side, has the power to harm. Finally, some, who want to appear wiser, attribute injuries and deaths to the teeth in general. These are errors of great magnitude that disappear upon closer examination[62]. Within the lips of poisonous snakes, there is a small bone attached to the upper jaw that has the ability to move forward and back, to which are stuck two or three fangs[63] larger than the others. In much the same way as the claws of a cat, with the help of that little bone, an angry snake projects the fangs to bite, or retract them to hide. These fangs, called by the vulgar "larger teeth of snakes", were excellently described by Tyson *Act. Angl. p. 144*[64] in the anatomy of the rattlesnake: "*they have a large foramen at the base, and near it a long slit clearly visible, similar to the pen of a writer; and that part of the tooth* [p. 6] *from the slit to the base is clearly hollow; when the gums are gently pressed the poison ascends through the tooth cavity and flows out of the slit. These teeth are very sharp, solid and firm towards the end, and the slit is located in the outer, convex part, not in the internal, so nothing could have been more conveniently designed to cause injury or to infuse the poison*"[65].

Each fang is surrounded by a vesicle provided with glands that secrete certain humor; when the vesicles are pressed the liquid flows out of the end of the fang. When the snake is angry, it raises its head, extends the small bone provided with fangs already discussed, and attacks the enemy with the combined strength of its body and muscles, injuring with the extended fangs and allowing the venom to flow into the wound immediately. Experience has shown that if the teeth are broken with a pair of tweezers, it is possible to handle the snakes safely, as some American charlatans do. After carefully extracting the venomous fangs, they allow the rattlesnakes to bite with the remaining teeth until the blood flows freely, attributing fatal effects to these now harmless weapons.

Venoms

So far, we have discussed the way that snakes produce wounds. Now we shall analyze the origins and nature of the venom, which is no small matter. In ancient times it was believed that anger was aroused by black bile, and this myth, devoid of truth, was unhesitatingly applied to all species. This provided the basis for a hypothesis that the venom only acts when the snake that bites was irritated, inventing the existence of a channel that conducts the bile from the gallbladder to the mouth, [p. 7] from which it flowed into the wound, producing the lethal symptoms.

Leaving aside these ancient beliefs, we will review what has been said on the subject by modern, more cautious and attentive researchers. Towards the end of the last century, medicine and other sciences received new insights, dispelling the dark guesses that covered them until then. Additionally, our knowledge about snakes received a powerful impulse under the auspices of the Grand Duke of Tuscany, Ferdinand the Second[66]. This prince, to whom literature owes so much, was eager to answer questions about the nature of serpents and invited Steno, Redi, and other eminent philosophers to his court. After gathering a large number of poisonous snakes, Redi conducted numerous experiments with them, in which he discovered many previously unknown features, some of which are of our interest. When he made a living snake bite a dog, or when he struck it with the tooth of a freshly dead snake (where the venom remained active), he observed that the results were similar. If the bites were repeated, the effects were weaker and weaker until the venom contained in the gland was depleted. The explanation lies in the gland, which is located at the base of the tooth, and when compressed, it releases a fluid that leaks at its end.

[61]This belief was probably taken from SN 10. Linnæus (1758: 210; 213), when referring to *Rana bufo* stated: "...Attract insects into its mouth through enchantments...", while in *Rana arborea* wrote: "...Attracts flies to its mouth...". It has not been possible to locate the origin of the comment on anurans that attract insects into their mouths through incantations. Following Pennant (1769: 324), it is provisionally assigned to Linnæus himself, based on local people's stories.

[62]These erroneous beliefs were already corrected by Redi (1664; 1666; 1670a, b).

[63]Acrel used the noun *cuspide*, with the meaning of dart or lance; here is translate as fang.

[64]Tyson (1683). 144 is the work's number; the description that follows is on page 46.

[65]As part of the content of Tyson's description was lost in the translation from English to Latin by Acrel, and back to English here, it is advisable to present a verbatim transcription of Tyson (1683: 46): "...In all these Teeth, especially the larger, we took notice of a pretty large *Foramen* or Hole towards the root of it, and towards the point there was a plain visible and large Slitt, like the cut of a Pen sloping; and that part from the Slit to the Root was perfectly hollow; which first of all was discovered to us, by pressing gently with our finger the side of the *Gumme*; for then we did perceive that the Poysou (sic) did readily arise through the hollow of the *Tooth*, and issued out of the *Slit*. This we tryed several times; which tryals, as likewise our searching for all the *Teeth* wee could here find, did spoyl our Enquiry into the *Baggs* and the *Glands* that furnish them with that liquor...".

[66]Ferdinando II de' Medici (1610-1670), Granduca di Toscana from 1621 to 1670.

When the venom that flows in this way is collected in a bread crumb or in a sponge, an animal bitten by that snake does not suffer more damage than the wound itself, which is similar to the puncture of a needle, until the venom is restored after several days. However, when an animal was wounded with the tip of a needle dipped in venom, their suffering was similar to if it had been bitten by the snake itself. Redi said, with great admiration, that the venom, placed on a glass, [p. 8] sun-dried, and then diluted in water and injected with the tip of a needle, has the same effect as if it were fresh. At that time, Tozzi[67] surprised the experts in natural history with a demonstration worthy of a snake charmer. It was discussed, in the presence of the prince, whether there was a danger to life if a person drank poison instead of alcohol or water. Confident in his art, Tozzi drank a good portion of venom. This astonished everyone, and although the spectators expected his instant death, he emerged unscathed, as if had drunk only water. This fact, which stunned the prince and the naturalists who were with him, was well-known to the ancients. A memorable recording of this is documented in Lucanus *Libr. IX, v. 614*[68].

> *Noxia serpentum est admisto sanguine pestis*
> *Morsu virus habent, et fatum dente minantur,*
> *Pocula morte carent.*[69]

We should not overlook the harmful effects of this external poison, which has been known to barbaric nations both now and in the past. These nations have dipped their arrows in various types of poison, such as the juice of *Euphorbia*, putrid flesh, tobacco oil, and especially the venom of vipers, which is particularly effective and horrifying. Some modern-day Indians still practice this, and in the past, it was done by the Scythians, whose notable actions were recorded by Pliny in his Natural History: *The Scythians*, says, *permeate his arrows in snake venom and human blood, an irreparable crime, because death occurs at the slightest touch*[70]. [p. 9]

But let us return to the words of our Florentine, who sought to confirm Tozzi's findings through new experiments. He collected a substantial amount of poison and administered it to different animals in various ways, always without any subsequent harm. However, as soon as the poison was applied to an external wound, it produced all the dreadful symptoms that arise from an actual bite, such as malignant and fatal inflammations and fevers, unless the body, by some spontaneous action like a hemorrhage, expelled the harmful material. These experiments led to such an advanced level of knowledge that it allowed the learned individual to penetrate deep into the hidden recesses of nature with much greater certainty.

While Redi and his colleagues were diligently studying nature through experiments in Tuscany, others were indulging in mere speculation elsewhere. In France, Moyses Charas, a well-known first Surgeon of the King, worked tirelessly to support his theory that the venom of snakes resides in the spirits of angry animals. He argued that it was impossible for a snake to deliver a lethal bite unless it was provoked beforehand[71]. It cannot be denied that emotional states can produce effects similar to those of poison, as evidenced by the livid face of an angry person. However, in instances where sudden terror strikes, it often leads to pale faces and diarrhea as the body expels humors from the intestinal tract, similar to what happens in cholera. This is also observed in nursing mothers, whose milk turns poisonous when they are irritated and causes fatal diarrhea when ingested by a child. Regardless, all of these factors fail to support Charas' view. Although the vague hypothesis of animal spirits and anger might have found a comfortable refuge in the ignorance of the past, today it is doomed to oblivion. [p. 10]

Let's put aside the Charas hypothesis and share our opinion. It is more challenging to establish a new system with positive evidence than to destroy an old one when it comes to such an essential and obscure matter. However, even if this doesn't settle the question or exclude future research, it's better to proceed with principles that hold at least a kernel of truth and some analogy to what experiments have found so far. With this belief, let's venture a new hypothesis. In the past century, some doctors, primarily in Britain, sought to discover new methods of administering drugs without losing their effectiveness through the chyle[72]. They developed a technique involving the infusion of medicine directly into an open vein[73], with proponents touting it as a revolutionary advancement. Despite some initial suc-

---

[67]Luca Tozzi (1638–1717), Italian physician. Some of his contributions on snakes are in Tozzi (1693; 1747).

[68]Lucanus (1469) *Pharsalia*. Lucan's Pharsalia (also known as De Bello Civili) is a grim, anti-heroic epic recounting the Roman civil war between Cæsar and Pompey. Rejecting traditional epic grandeur, Lucan emphasizes chaos, despair, and the unnatural violence of civil conflict. One of the pœm's most striking episodes appears in Book 9, where Cæsar's troops, marching through Libya, are afflicted by a horrific plague of venomous snakes. Soldiers die grotesque, agonizing deaths from exotic serpents—an image rich in gruesome detail and symbolic of nature itself turning against mankind.

[69]The passage is part of Cato's harangue to his soldiers when, in the desert, they discovered a copious spring infested with snakes. An expanded paragraph is reproduced here to clarify Acrell's citation, particularly the reference to the "cup," a metaphor for the water source. Cato said: "*...Vana specie conterrite leti, / Ne dubita, miles, tutos haurire liquores. / Noxia serpentum est admixto sanguine pestis; / Morsu virus habent, & fatum dente minantur / Pocula morte carent...*" In Duff's translation (Lucan, 1962), "...Soldiers, the appearance of death that terrifies you is delusive; do not hesitate to swallow the harmless water. The poison of snakes is only deadly when mixed with the blood; their venom is in their bite, and they threaten death with their fangs. There is no death in the cup...". The Latin *pocula* can be translated as either "cup" or "drink"; hence, I prefer rendering *Pocula morte carent* as "those who drink will not die."

[70]Plinii (1582, Liber XI: liii). "*...Scythæ sagittas tingunt viperina sanie, & humano sanguine: irremediable id scelus mortem illico adfert leui tacto...*".

[71]Charas (1684). See *Caput IV. De veneni a vipera morsu inflicti.*

[72]χυλοποιητιχας in the original.

[73]See Peters (2009).

cesses, the outcomes were often unfavorable, either because new drugs were untested or older medicines were not used with the necessary caution. It is well-known that the infusion of distilled spirits is lethal, but it is surprising to note that even undiluted milk, a staple of human infancy, can lead to swift death when injected into the veins. The reason for this is likely due to the immediate mixture of milk with blood, bypassing the digestive system. Nature does not tolerate heterogeneous mixtures. If scholars and science support this deduction, they will have no difficulty in finding an analogy between the infusion of milk and of snake venom. [p. 11] To our knowledge, no other explanation has been put forth that appears as plausible or true. Suppose a small amount of poison is applied to a wound, absorbed into the veins, and mixed with the blood. As with a small amount of yeast that can leaven a large amount of dough, the multiplying force of nature can corrupt the total volume of blood to varying degrees. The poison mixed with blood may increase exponentially and corrupt everything, although we cannot determine whether this happens through coagulation, corrosion, or some other process. An illuminating experiment was described by Ray in *De Quadr.* p. 317. A gentleman in India, seeking to demonstrate the different types of snake poisons to his guests, had a native bring over various snakes. First, he used a large snake that he claimed was harmless. To prove it, a ligature was placed on his arm, and the snake was teased to bite him. He collected about half a tablespoon of blood from the bite and spilled it over his bare thigh. Then, he took a smaller snake, known as the Cobra de Capello (*Colubrum Najam*), and spoke of its venom's awesome power. To demonstrate it, he tightly held the snake's neck, squeezing the bladders of the jaw until it left about half a drop of its venom, which he put on the clotted blood on his thigh. Immediately after mixing, a large fermentation occurred, like boiling and effervescence, [p. 12] and the blood changed instantly to yellow fluid[74]. This experiment sheds light on how the poison acts and confirms the famous observation that snake bites can produce jaundice.

When we compare this scenario with the experiments conducted by Redi and Tozzi, we find that their findings align perfectly with this hypothesis. Redi's experiments clearly showed that the liquid extracted from the vesicles that moisten the fangs of snakes is poisonous only when mixed with blood through a puncture or wound. Similarly, Tozzi's experiments revealed that the venom is harmless when ingested or applied externally. Additionally, there was another experiment that showed that animals killed with arrows poisoned with snake venom could be safely consumed. The only plausible explanation for this phenomenon is the immediate action of snake venom mixed with blood, which causes a strong fermentation. Due to the significant diversity of snake species, different symptoms can arise as a result. However, we must confess that we are unable to identify the specific cause of these differences. It is not something that can be expected of us, as we believe that we cannot provide a detailed explanation for these variations.

Symptoms of Snake Bites

Therefore, it remains for us to generally and specifically list the more notable varieties of these snakes.

Our *Beri* is known to be accompanied by the following symptoms: Pain in the bitten body part, as if continuously pierced by a needle, accompanied by swelling, paleness, twitching of the eyelids, tears, chills with shivering, weak and sometimes irregular pulse, thirst, delirium, cold sweat, [p. 13] back pain, cramps, diarrhea with vomiting, asthma, anxiety, drowsiness, and convulsions, with death being the ultimate outcome.

We are not familiar with the *Haemorrhous*, which is described by Lucan as causing blood to flow from all over the skin after biting.

Dipsas, which is said to flee from humans, was also described by *Lucan Libr. IX. de bello Pharsalico.* It is said that when it hisses, it ignites an unquenchable thirst. Here are Lucan's own words, vividly describing the effects of the venom on Tuscum[75], the flag-bearer of Cato:

Non Decus imperii, non moesti jura Catonis
Ardentem tenuere virum, quin spargere signa
Auderet, totisque furens exquireret agris
Quas poscebat aquas sitiens in corde venenum.[76]

And, a few lines later

Scrutatur venas penitus squalentis arenae
Nunc redit ad Syrtes & fluctus accipit ore,
Aquoreasqe placet, sed non sibi sufficit humor,
Nec sentit fatique genus, mortemque veneni,
Sed putat esse sitim; ferroque aperire tumentes
Sustinuit venas, atque os implere cruore.[77]

---

[74]The same story was already told by Darelius (1749) (*vide infra*).

[75]In Book 9 of Lucan's *Pharsalia*, the focus shifts to the deserts of Libya, where Cato the Younger leads the remnants of the Republican army after the defeat at Pharsalus. As his men march across the burning sands, they are struck by a terrifying plague of serpents. Lucan vividly describes a host of exotic, venomous snakes—including the seps, dipsas, prester, and basilisk—each with horrifying effects: flesh melting from bones, insatiable thirst, bodies swelling and bursting. One of the most gruesome deaths is that of Aulus, the standard-bearer, who is bitten by a seps. His body liquefies from within—skin, muscle, and bone dissolving into a formless mass—leaving only a pool of decay beneath the still-standing standard. In this context, *Tuscum* is a gentilic adjective indicating Aulus's place of birth, in the region of Etruria (modern-day Tuscany).

[76]*Pharsalia*, Lib. IX: 747-750. "...neither national pride nor the authority of grief-stricken Cato could stop him: boldly he threw down the standard and searched everywhere in his frenzy for the water which the thirsty poison at his heart demanded..." (Translation by J. D. Duff; Lucan, 1962).

[77]*Pharsalia*, Lib. IX: 755-760. "... [*Aulus*] searches for water deep down in the barren sand, and then returns to the Syrtes, and swallows the brine; the sea-water gives him pleasure, but there is not enough of it. The nature of his suffering and his death by poison were unperceived by him: he thought it was merely thirst, and ventured to open his swollen veins with his sword, and fill his mouth with the blood..." Translation by J. D. Duff; Lucan, 1962).

The *Aspis* snake appears in ancient monuments in three types[78]:

I. *Phtyas*, with a calloused forehead, a greenish-gray color.

II. *Chersea*, very long, gray-green with red eyes.

III. *Chelidonia*, black, with a white belly: its bite is said to produce two or three hiccoughs, changes of color, stupor, cold, sleep and palpitation. Among these adders is one called *Phtyas*, which kills producing drowsiness and lethargy, as is clear from the writings [p. 14] which bring to mind the species that *Belon. Itin. p. 203.* called *Nintipolonga Zeylanica Raj. Anim. p. 332*[79], *&* in *Systema Nat. 216, N:o 173*[80] is called *Amodytes* (sic). Cleopatra, the queen of Egypt, is said to have also used the venom of this asp (*Phytia*). For, as we learn from reliable authors, wishing to end her soft and delicate life with a gentler death whenever fate should decree it, this most resolute woman instructed her physicians to prepare a poison that would bring death without the sensation of dying. After trying various poisons on those condemned to death, they finally hit upon the asp. Its bite, without any previous sickness or convulsion, only a soft perspiration on the face, a dulling of the senses and drowsiness, hastened death, almost without the victim's being aware of it. The queen was informed by those who knew of the poison, and she either offered her own arm to the asp to bite or else pricked it with a needle dipped in poison which she had kept for this purpose, and so died, as some say, overcome by the long sleepiness that followed the puncture. The *Systema Naturae* enumerates many kinds of venomous snakes, including, besides many others, all the rattlesnakes:

Vipera	*Atropos*	*Leberis*	*Ammodytes*
Berus	*Chersea*	*Prester*	*Aspis*
Lebetinus	*Severus*	*Stolatus*	*Lacteus*
Naja	*Atrox*	*Niveus*	*Corallinus*
Dipsas	*Mycterizans*		

The venom of *Naja* strikes humans so vehemently that death can occur within an hour, causing the flesh to liquefy and separate from the bones, leading some to believe it is the same serpent known as Sepe in ancient times. Although there are many species [p. 15] of venomous snakes that produce a wide range of symptoms, all share the commonality that their venoms are only fatal when injected into the bloodstream through a bite or sting. Thus, we must now turn our attention to remedies against these deadly poisons.

Cure

To comprehensively cover all the remedies men have employed against snakebites due to necessity would greatly expand this work. Therefore, our analysis focuses only on some of the most significant antidotes based on their historical use, efficacy, and recommended application for removing poison from the affected area. Various methods have been tried over time, and the oldest and simplest involves suctioning the wound with lips. However, this method poses some risks and is unpleasant, as the person performing the suction must ensure they have healthy gums and no wounds to avoid further harm. As such, it's better to avoid this approach and instead make a scalpel cut in the wound, open the vessels, and let the contaminated blood flow freely. Applying a suction cup can help eliminate any remaining traces of poison. Afterward, apply poultices, warm emollients, and spirituous elements like Ammonium Salt or Theriac, and if unavailable, a mixture of plain soap and vinegar on the injured area. Sudoriferous drugs and emetics for internal use like [p. 16] *Tartarus emeticus*[81], *Ipecacuanha*[82], etc., are effective, as well as strong sudorifics. In recent times, three new highly potent antidotes from Europe, Asia, and America have been introduced to counteract the most severe snakebite effects.

Europeans use a great deal of olive oil, as well as other pressed oils. The patient should drink the oil in copious amounts, sometimes up to half a pound or more, while simultaneously applying the same oil liberally to the affected area. In *Acta Anglicana p. 443.*[83] and *Acta Paris. 1737*[84], there are several documented instances of successful treatment with this method. However, while we do not doubt the efficacy of the oil in treating the bites of *Beri*[85] and *Presteri*[86], we must acknowledge that the properties of this antidote have not been fully studied or understood, and we cannot make definitive claims about its effectiveness in all cases. We offer an unexpected testimony, related by the distinguished Director: while in Scania, he encountered a woman who had been bitten by a *colubri Cherseae*[87]. Despite being administered olive oil according to the prescribed formula by the noble doctor, the woman died in great agony[88]. It is possible

[78]The comment on the three types of *Aspis* is coincident with Matthioli (1570: 955).

[79]Ray (1693: 332), "11. *Serpens Indicus ex fusco nigricans cum maculis albicantibus* Nintipolonga Zeylon. i.e. *Sumnum lethalem morsu inducens*, Serpens hypnoticus."

[80]Linnæus (1758). No snake with number 173 (or with a count of 173 scales) exists on page 216. *Ammodytes*. 142-32 has number 174.

[81]$K_2Sb_2(C_4H_2O_6)$

[82]Extract of the dried root of *Cephælis ipecacuanha*.

[83]Mortimer (1741). See also Williams (1738) and Dufay (1741).

[84]Geoffroy & Hunauld (1740). Presented at the Académie Royale des Sciences on August, 7 1737.

[85]*Coluber berus* in SN 10.

[86]*Coluber prester* in Linnæus (1761); in the synonymy of *C. berus*.

[87]*Coluber chersea* in SN 10.

[88]Linnæus (1745a) wrote, on August 2, that "... A woman had been stung by an adder; I prescribed a large quantity of olive oil for her to be taken often and in large quantities but a few days later I heard that this remedy, much praised by the English, had no effect whatsœver, although the woman was bitten by a grey snake (*Vipera*) only 6 hours prior to treatment...".

that the oil does not work against all European snakes, or perhaps we have overlooked some necessary observations regarding its application, both in terms of medicine and the individual patient. We therefore strongly recommend further experimentation.

Let us now turn our attention to the American remedies, with the powder of the *Senega* root being considered the most excellent. For a long time, the indigenous people of North America have been using this root [p. 17] against the venomous bites of rattlesnakes[89], ascribing almost miraculous properties to it. However, these determined and envious natives kept this secret so closely guarded that not even a king's ransom could pry it from them, until Tennent brought it to light for the benefit of all humanity[90]. Tennent taught that the root of *Polygala senega* is infallible, except when a large vein is affected, as the venom's sudden and violent progression doesn't allow enough time for the antidote to take effect. This can be found in the dissertation *radice Senega* in *Amoen. Acad.* V. 2.[91] We have a remarkable example of the effectiveness of this antidote locally, in the past decade in a suburban area near Uppsala, where a maid who was defecating behind some bushes was bitten by a snake in her most vulnerable female parts. The resulting symptoms were excruciating, and even the most skillful physicians were unable to provide effective relief. Our President[92] was consulted, who quickly dispatched the only two doses of Senega that he possessed. These were administered with such timely precision that within two days, the maid was almost completely cured. It is desirable to set aside all remedies of dubious effectiveness and introduce this exceptional medicine into our apothecaries. Besides this root, the Americans have many other remedies, such as *Eryngium foetidum, Eryngium aquaticum, Veratrum luteum, Aletris farinosa, Uvularia perfoliata, Sanicula canadensis, and Aristolochia serpentaria*, among many others whose virtues remain undiscovered.

We have thus arrived at another equally celebrated Antidote of the Asian people, easily one of the most ancient in eastern India. It is indeed the *Ophiorrhiza mungos*, whose history, and almost divine use against the bites of the most venomous *colubri Najae*, [p. 18] was brilliantly discussed in the *Amoenitates* by Kaempfer[93]. We will not dwell much on its description since we have the leisure to rely on the very elegant Dissertation of the honorable Advisor Darelius[94]. However, it is worth mentioning that since this medicine is so effective, it is much better to use the true *lignum colubrinum* instead of *Strychnos colubrini*, whose use is not entirely free from doubt, at least according to what we have observed.

The snake stone[95], also known as *Pedra de Cobras di capello*[96], is said to be taken from the head of the *Colubri Najae*, famous throughout India. It is proclaimed by Garzia[97], Kirkero[98] and others as a specific remedy against the bite of snakes. It is said that when applied to the venomous wound, it will absorb all the poison until it falls off on its own. However, its use is suspected by others, and Redi[99] was unsuccessful in his many experiments with it. Many claim that this stone is artificial and can only be made from the burnt horn of a deer.

Enchantments

In antiquity, the Ophiogenes, like the Psilos of Libya and the Marsians in Italy, were peoples of snake charmers capable of curing their bites, as testified by Ovid, Pliny[100], and other ancient writers. They were able to handle extremely poisonous snakes without being hurt, and they

[89] *Crotalo* in the original.

[90] Tennent (1738).

[91] This dissertation, by Jonas Kiernander, was originaly published in 1749 and reprinted in pp. 126-153 of the second volume of *Amœnitates Academicæ* (Kiernander, 1751); see below.

[92] *Nob. D. Præsidem* en el original.

[93] Kæmpfer (1712).

[94] See the translation of Darelius, 1749, below.

[95] *Lapis serpentum* in the original.

[96] As in the original.

[97] Horto (1593).

[98] Kircher (1667a, b).

[99] Redi (1671).

[100] Plinii (1582, Liber VII: ii), "...Crates of Pergamus relates, that there formerly existed in the vicinity of Parium, in the Hellespont, a race of men whom he calls Ophiogenes, and that by their touch they were able to cure those who had been stung by serpents, extracting the poison by the mere imposition of the hand. Varro tells us, that there are still a few individuals in that district, whose saliva effectually cures the stings of serpents. The same, too, was the case with the tribe of the Psylli, in Africa, according to the account of Agatharchides; these people received their name from Psyllus, one of their kings, whose tomb is in existence, in the district of the Greater Syrtes. In the bodies of these people there was by nature a certain kind of poison, which was fatal to serpents, and the odour of which overpowered them with torpor: with them it was a custom to expose children immediately after their birth to the fiercest serpents, and in this manner to make proof of the fidelity of their wives, the serpents not being repelled by such children as were the offspring of adultery. This nation, however, was almost entirely extirpated by the slaughter made of them by the Nasamones, who now occupy their territory. This race, however, still survives in a few persons who are descendants of those who either took to flight or else were absent on the occasion of the battle. The Marsi in Italy are still in possession of the same power, for which, it is said, they are indebted to their origin from the son of Circe, from whom they acquired it as a natural quality...". (Translation by John Bostock and H. T. Riley; Pliny, 1855).

extracted the poison from the body by sucking it out[101], as Hasselquist has seen done recently in Egypt[102]. Even today their secret is jealously guarded and they do not reveal it for any price. The vulgar believe, since ancient times, that if a man spits on a snake, it will die before sunset, and Pliny mentions that the saliva of men is deadly for snakes[103]; also, that the *Viverra mungos* when he goes to fight with a snake, chews rue which, doubtless, is the *ophiorrhiza* mentioned in *Amoen. Acad. 2 P. 100.*[104], and then spits it on the snake [p. 19], paralyzing and making it unable to counterattack and inflict wounds[105]. Recently D. D. Jacquin, returning from the West Indies, reported in a letter to our President that he bought the secret of snake-charming with gold; we still don't know if chewing a plant he called *Aristolochiam anguicidam*, or by some other new method, but we hope that soon D.D. Jacquin communicates this secret to the public, thereby satisfying our desires and curiosity[106].

That's all.

The figure[107] shows the skull of the common viper, with the jaws, from the side:

a. venomous fang: this, when moved, works with the bony peg c, which, connected to the bone b, removes this during its movement and retraction below its joint.

f. shows the lower jaw and d, e, its two supports, from which the necessary range of motion for swallowing prey depends.

---

[101]Comment included by Linnæus in SN 12: 348.

[102]*Hasselq. Itin. 70.* in the original; see Hasselquist (1757: 70).

[103]Plinii (1582, Liber VII: ii), "...However, all men contain a poison available as a protection against snakes: people say that snakes flee from contact with saliva as from the touch of boiling water, and that if it gets inside their throats they actually die; and that this is especially the case with the saliva of a person fasting...". (Translation by John Bostock and H. T. Riley; Pliny, 1855).

[104]Darelius (1751).

[105]The origins of this belief may be traced to Aristotle and Ælian. Aristotle, in The History of Animals, Book IX: vi, notes "...The weasel, when it fights with a snake, first eats wild rue, the smell of which is noxious to the snake..." (Translation by D'Arcy Wentworth Thompson; Aristotle, 1910) Ælian, in turn, in On the characteristics of animals, Book IV: 14, notes "...The marten is an evil creature, and an evil creature is the snake. And so when the marten means to fight with a snake, it chews some rue beforehand and then gœs out boldly to battle, as though fortified and armed. The reason is that for a snake the rue is utterly abhorrent..." (Translation by A. F. Scholfield; Ælian, 1958).

[106]The properties of *Aristolochia anguicida* were later published by Jacquin (1763: 232).

[107]The figure, approximately 60 x 70 mm in size, was printed separately and glued only in some copies of the dissertation, although its caption appears in all of them.

B. c. D.

AMPHIBIA GYLLENBORGIANA,

DISSERTATIONE ACADEMICA,
Cum
Consensu Ampliss. Facultatis Medicæ
In Regia Academia Upsaliensi,
PRÆSIDE
VIRO EXPERIENTISSIMO & CELEBERRIMO;
D:O DOCT. CAROLO
LINNÆO,
Med. & Botan. PROFESS. Reg. & Ord.
Descripta
Et publico examini subjecta
A
BARTH. RUDOLPHO HAST,
OSTROBOTHNIENSI,
STIPENDIARIO REGIO.
In Audit. Carolino Majori die XVIII. Junii.
ANNI MDCCXLV.
Horis, ante Meridiem, solitis.

UPSALIÆ.

Hast, B.R. 1745. Amphibia Gyllenborgiana. Uppsaliae. S. n. [6] + 1-34 + [2].

B. R. Hast studied the amphibians and reptiles of Carl Gyllenborg's donation to Uppsala University. Hast identified nine species of snakes, ten species of lizards and salamanders, one frog, and four turtles. His dissertation, re-issued in the first volume of the *Amoenitates Academicae* (Hast, 1749), served as a key reference and was extensively cited in the tenth edition of Linnaeus's Systema Naturae (1758). Hast's dissertation introduced a method of species snake classification/identification based on the number of scuta and squama, marking a significant advancement from the previous reliance on color for species differentiation. This classification method was subsequently published by Linnaeus and was used in the Systema Naturae.

All specimens of the original donation were present in the times of Thunberg (Radloff, 1787), while Lönnberg (1896) recorded all snakes and frogs but only five "*Lacerta*" (including lizards and salamanders) and two turtles. Today, according to Wallin (2001), Uppsala Museum houses one species of frog (*Rana arborea*), eight species of snakes (*Coluber ahaetulla, C. angulatus, C. annulatus, C. berus, C. cobella, C. domicella, C. lemniscatus, C. natrix*, and *C. petola*), ten species of lizards and salamanders (*Draco volans, Lacerta ameiva, L. cordylus, L. crocodilus, L. gecko, L. iguana, L. lemniscata, L. marmorata, L. salamandra*, and *L. teguixin*), and two species of turtles (*Testudo geometrica* and *T. mydas*) from the original donation. Of note is the fact that Linnaeus (1758) did not include a reference to *Lacerta cauda tereti corpore sesqui longiore, pedibus pentadactylis, dorso lineis longitudinalibus striato* in the characterization of *Lacerta lemniscata.*

W. G. h.[108] Amphibia Gyllenborgiana, an Academic Dissertation with *the Approval of the Esteemed Medical Faculty at the Royal Academy of Upsala*, presided over *by the Most Experienced and Celebrated* Master and Doctor Carlolus Linnaeus, Professor of Medicine and Botany, Royal and Ordinary, *Described and submitted to public examination* by Barthold Rudolph Hast, *of Ostrobothnia*, a Royal Stipendiate. In the Great Caroline Auditorium on the 18th of June, 1745, *at the usual morning hours. Uppsala.*

[blank]
[s. n.]

H. R. M.[109] To the man of utmost trustworthiness, Senator of the *Swedish-Gothic King and Kingdom*, President of the *Royal Chancellery*, Chancellor of the *University of Upsala*, *The most noble and illustrious* Count and Hero, Lord Carl Gyllenborg, Supreme Patron.

[s. n.]
Most Illustrious and Prominent Count

Our gratitude dares to consecrate this humble dissertation to you, Most Illustrious Count, because it is owed to you alone, since the animals which provided its subject matter were collected through *your* efforts and donated to the muses of Uppsala through *your* generosity.

Others will glorify Your great Name with hope and expectation of favors; we celebrate the[s. n.] benefits received. For us, who study the art of Machaonia[110] and delve into the depths of nature, it is due to your care and under your auspices, since you began to embrace our muses[111] with singular patronage, that we have achieved things that surpass all our praises and acclamations. We owe to you the physical instruments, to you the anatomical preparations, to you

[108] The original B. c. D., *Bonus cum Deo*, can be translated as "With God's favor" or "With God's help." Thus, W. G. h.

[109] The original S:æ R:æ M:tis stands for *Suæ Regiæ Majestatis*, translated here as His/Her Royal Majesty; thus, H. R. M.

[110] "The art of Machaonia" refers to Macháōn (Μαχάων), son of Asclepius, the god of medicine, and a skilled healer and physician himself. He was one of the leaders of the Greek army during the Trojan War and was said to have been responsible for healing many of the wounded soldiers, including several of the major herœs like Menelaus, Diomedes, and Ajax.

[111] "...*ex quo Camenas nostrasœpisti...*" in the original. In ancient Roman mythology, the *Camenæ* were a group of goddesses associated with springs, wells, and fountains, as well as with poetry and prophecy. In the given context, "*Camenas nostras*" would refer to the muses or pœtic inspiration of the writer.

the greenhouses of plants, brought from other parts of the world in such enormous numbers, with which our Botanical Garden already flourishes. However, these many great proofs of a highly generous spirit are still insufficient to satisfy your affection for us, so you granted us your own Museum, collected at your cost and care, especially for our use. Things never seen before call attention here: here there are dreadful Snakes; here are the Lizards, Frogs, Turtles, here are very beautiful embalmed Birds that inhabit the Far East; here the most elegant Corals, Lithophytes and their relatives, brought from the depths of the ocean; here the most select stones, fossils, minerals and petrifications and, finally, here we have admired the various parts of Quadrupeds and other animals. All of these [items] your generosity has given to promote our studies and to enhance our knowledge.

We, who have attempted to describe only a portion of this museum, the one that concerns amphibians, [s. n.] dedicate this catalog to you with the utmost reverence. If our resources had not been so inadequate for our desires, each of these animals, which we could only sketch in writing, would have been represented in a more perfect form and presented to the public. Nonetheless, we hope that you, the benefactor of this gift, will deem it worthy of your grace. The times to come will celebrate your generosity with immortal praise, and those who, in the future, share our love of science and our desire to promote it, will follow your example. I will remain, faithful

to Your Most Illustrious and
Sublime Name.
Your most devoted worshipper
Barth. Rudolph. Hast.

[s. n.].

Prologue

Here is presented before your eyes, dear Reader[112], a description of certain Amphibians, which, brought from foreign lands to our north, are particularly remarkable for their unique nature and rarity. They form a part of that precious gift, which not long ago, the generosity of the illustrious Hero and Count, Senator of the King and Kingdom of Sweden, President of the Royal Chancellery and Chancellor of the University of Uppsala, Mr. Carolus Gyllenborg, bestowed upon this seat of the Muses[113].

It is not with one, but with many proofs, that this noble patron of science and the liberal arts, and in particular of our own Muses, has shown his most generous and benevolent will. Since the moment they were admitted under his patronage, he has taken them under his closest care. As a result, many things have been happily seen: some renewed and enlarged, and some newly added to the ornament that adorns the Helicon of Uppsala[114]. I speak of the Hippodrome built and restored under his auspices and benefactions, from which the most noble discipline of horsemanship has been reinstated. I speak of the Astronomical Observatory built and enriched with the most accurate instruments[115],[s. n.] seen to outclass any others in Europe. I speak of various and excellent supports of the natural sciences: I speak of the Academic Hospital, restored and equipped with the most outstanding anatomical preparations[116]. I speak of the Botanical Garden, now blossoming with its charming species, and soon to be greatly enlarged with an immense number of rare plants brought from regions with a different sun, conspicuous with its new director and recently constructed and perfected hypocausts, so that it can confidently compete with many others in Europe[117]. Finally, I speak of the Gustavian House of our Muses, which craftsmen have already begun

[112]"*Subjicitur oculis tuis, B.L. descriptio...*" in the original. The abbreviation "*B.L.*" in a dedication and commendation note in Latin from the 18th century may refer to "*Bene Lectori*," which also translates to "To the kind reader" or "To the reader who is well-disposed." It was a common phrase used to address readers in the introduction or preface of many books during that time period.

[113]Gyllenborg made the first donation of his collection and library in two installments, in 1744 and 1745. See Wahlgren, 2011.

[114]The *Upsaliensis Helicon* is a metaphorical reference to the University of Uppsala's literary and artistic traditions. In Greek mythology, Mount Helicon was considered a source of inspiration for the Muses, who were associated with pœtry, music, and other creative arts.

[115]Astronomiska observatoriet i Uppsala, founded in 1741.

[116]Akademiska sjukhuset, originally established in 1708.

to work on, in order to be refurnished and adorned with a more elegant structure, and to expand the space of the growing public library[118]. Into this illustrious list of evidence for the Most Favorable Count's support of the University of Uppsala, there is at last added his own and most elegant collection of rare amphibians, of insects, embalmed birds, and of the most beautiful variety of corals and minerals[119], which he himself was accustomed to delight his literary leisure with, which he generously granted, to be preserved in the house of the Muses, specially assigned to the Prefect of the Botanical Garden[120], and to be celebrated with immortal praise for his incredible efforts in promoting natural history.

Therefore, in order for some public benefit to come from this great gift, which we privately marvel at and admire, and at the same time for the memory of the benefit to be enduring, I, captivated by the magnificent name of the donor, have undertaken to describe and present to the public eye some part of it, namely these amphibians, with singular pleasure in natural history, especially zoology, while hoping that the other parts of this treasure will not lack the diligent attention of others to be communicated in the same manner.

[s. n.]

Amphibia.

Among animals, the class Amphibia[121] is entirely unique. Indeed, we see some of them standing on four legs and equipped with claws no different from some mammals[122]; others crawl without legs like eels; others are completely naked; and others are covered with scales like fish. All of them have cold blood; almost all of them can live both on land and in water; many of them hide in underground crevices during the winter to hibernate. However, the sight of almost all of them is frightening, and they seem to threaten humans with great danger, as these animals are most hostile to those who encounter them. Therefore, one of the main blessings of our northern region is the fact that very few amphibian species can be found here, considering the great number of these creatures that inhabit the southern lands, among many other more pleasant and delightful creatures of benign nature, but which can cause serious harm and even pose a threat to life itself. Their condition will certainly appear less enviable to us, as we ourselves observe or see descriptions of various journeys of travelers, not without horror, so many different harmful animals of this kind, such as snakes, lizards, and toads, which grow in those places and are usually unknown to us.

Therefore, in order to depict them properly, I have decided, based on the diversity of amphibians, to divide this treatise into four chapters according to the types: the first one about snakes, the second about lizards, the third about frogs, and finally the fourth about turtles.

Chap. I.

Snakes.

Snakes are easily distinguishable from other amphibians, as they are devoid of legs and crawl like eels, although some of them can climb trees in a spiral motion. However, the All-Wise Creator has not left them defenseless against the attacks of other animals, even though they lack limbs. Indeed, they are equipped with a terrifying weapon in their mouths, where hollow fangs transmit the poisonous liquid collected in a vesicle located above the base of the teeth. As a result, the most terrifying and deadly diseases and pains often arise from their bites. Nevertheless, snakes of different kinds produce harmful effects in dissimilar ways:

Haemorrhois, for instance, expels blood from all parts of the body with its bite, as Lucretius (*sic*)[123], sings in Book 9:

Impressit dentes haemorrhois aspera Tullo
Magnanimo Juveni miratorique Catonis,
Utque solet pariter totis se effundere venis
Corycii presura Croci[124]*; sic omnia membra*
Emisere simul rutilum pro sanguine virus,
Sanguis erant lachrymae, quaecunque foramina novit,
Haemor ab his largus, manat cruor, ora redundant
Et patule nares, sudor rubet; omnia plenis
Membra fluunt venis, totum est pro vulnere corpus.[125]

---

[117]The Botaniska trädgården, Uppsala Universitet, was founded by Olaus Rudbeck in 1655 and destroyed in the city fire in 1702. It remained abandoned until 1741, when its reconstruction began under the direction of Linnæus, who explained his ideas in Nauclér (1745) dissertation, *Hortus Upsaliensis*.

[118]Today, Museum Gustavianum.

[119]See note 113.

[120]Refers to Carolus Linnæus.

[121]Taxon that includes nowadays amphibians and reptiles.

[122]*Quadrupedia* in the original.

[123]Refers to verses 806-814 of *Pharsalia, seu belli civilis libri*, by Marcus Annæus Lucanus, not by Lucretius as noted. This epic pœm, unfinished by Lucano's death in 65, was frequently cited by Linnæus and his disciples and had countless editions since its *editio princeps* in Rome, 1469, by Sweynheym and Pannartz.

[124]Sprinkling saffron was used to perfume theaters and it seems that statues also sprayed it through holes made for this purpose. «Corycian», for Mount Corico, in Cilicia, where a saffron of excellent quality was grown (Holgado Redondo *in* Lucano, 1984).

[125]In a free translation, "...The rough hemorrhoids impressed their teeth in Tullus, / The magnanimous youth and admirer of Cato, / And as Corycian saffron rushes forth, / Pouring equally from all its veins; so too, every limb together Sent forth a red fluid in place of blooming blood, / Tears became blood; from every opening, / A generous hemorrhage flowed, and blood oozed, / Mouth and nostrils, sweat reddened; / All limbs flowed with veins full, the entire body was a wound..."

Aspis, also known as *phtyas*, with its bite, causes numbness, pleasant fatigue, sleep, and even death without any pain. [p. 3] Therefore, Cleopatra, the Queen of Egypt, after the death of Mark Antony, did not want to be taken in triumph to Rome by the victorious Emperor Augustus, and chose to die by this sweet type of death, just as she had indulged in pleasures throughout her life.

Chersea causes hiccups, a change in color, coldness, sleep, palpitation of the heart, and severe pain.

Heritinandel, a snake of Malabar, with its bite corrupts the flesh of the entire human body, so that it putrefies, falls away, and the wounded person dies after a thousand torments. However, a remedy is prepared for this evil in a watery decoction of *Antidesma*[126], taken in large quantities.

Nintipolonga, a certain snake from Ceylon, induces lethargy with its bite.

Seps from Lybia[127] causes putrefaction and gangrene with its bite. It is said that Agathocles[128] received a toothpick infected with its secretion, causing gangrene and death throughout his body. Boerhaav.[129]

Vipera Caudisonae[130], found in America, can kill a person within six to twelve hours of being bitten by its venom. However, its antidote has been found there in our time, the root of *Polygala*[131], *caule simplici, erectos; foliis ovato-lanceolatis, alternis, integerrimis; racemo terminatrice erecto*, Gronov Flor Virgin 80. [Polygala with a simple, erect stem, plus ovate-lanceolate, alternate, entire leaves, and a terminal, erect raceme, *Gronov. Flor. Virgin.* 80].

The venom of both *native and exotic vipers* is well known to cause pain, swelling, fever, jaundice, insomnia, and occasionally gangrene, but rarely death.

Cobra de Capello[132] is said to kill humans with its venom within an hour, as reported by the travelers' journals. [p. 4] The constraints of the present treatise prevent us from discussing more about snakes. Therefore, it is worth noting that it is especially necessary for physicians to have expertise in natural history of snakes, in order to know which ones are harmful and which ones are harmless, and what effects each of them produces, so that they can provide appropriate remedies to the injured.

Snakes, especially among amphibians, have until now been obscurely treated by authors, causing great difficulty for everyone due to their lack of limbs and uniform appearance. As a result, SEBA, deceived by colors in his *Thesaurus*, drew and described the same species, the *Cobra de Capello, seu Anguem conspicillo insignitam,* or snake marked with a badge, thirteen times. See his Thesaurus, vol. II, tab. 85, fig. 1, tab. 89, fig. 1-4, tab. 90, fig. 1-2, tab. 94, fig. 1, and tab. 97, fig. 1-4.[133]

Therefore, with much concern for the true and consistent specific markings of snakes, and attributing inconsistent and false markings to their colors, we have finally discovered the true distinguishing features. Namely, the transverse abdominal scutes, counted from the throat to the anus, constitute the true specific differences, which careful serpent classifiers should strive to attend to in the future. However, since many species have nearly the same number of abdominal scutes, I have added the number of paired scales covering the underside of the tail, counted from the anus to the tip of the tail, which shows the proportion between the abdominal scutes and the tail, and provides a true marking along with the preceding information[134]. [p. 5]

1. *ANGVIS scutis abdominalibus CXLIV, caudalibus XXXIX.* [Snake with 144 abdominal and 39 tail scutes].[135]

α. *Vipera anglica nigricans*. Petiv. mus. pag 17. n. 104.

Vipera. Gesner. serpent. pag. 124. Aldrovand. serp. pag. 108. Jonst. serpent. pag 7.

HEAD dorsum with various sutures, between which there are as many areolae: namely, five between the eyes, of which the middle one is larger, and nine arranged in a circle in front of these, between which two smaller ones are inserted. Behind the eyes, there are also two larger areolae.

The teeth in the upper jaw are large, curved, and solitary on each side, hidden within a sac.

The *upper lip margin* is white on the sides.

---

[126]A Phyllanthaceæ. Several species are used today against snake bites.

[127]By Lucano's time, Lybia was the name given to the whole North of Africa.

[128]Agathocles, tyrant of Syracuse, was killed in 289 BC by Menon, who gave him a poisoned toothpick.

[129]Cited by Bœrhaave in his Praxis medica... In the third edition (Bœrhaave, 1738), the comment is in p. 143: "...*Legimus Agathoclem, qui terror erat vicinis, porrecto dentiscalpio, quo post prandium gingivas scalperet, cujus in cuspide erat venenum toto capite & gingivis computruisse & contabuisse...*". [We read about Agathocles, who was a terror to his neighbors, extending a toothpick, with which he used to scrape his gums after lunch, the tip of which had poison, which caused his whole head and gums to decay and rot.].

[130]An American rattlesnake.

[131]*Polygala* is a genus of Polygalaceæ (Fabales). The use of its roots as antivenin was first reported by Tennent (1738).

[132]Undetermined species of the genus *Naja*

[133]Seba (1735). The figures refers to *Serpens, ex Regno Peru, conspicillom insignita; Serpens, Noja* (sic) *cum Conspicillo, seu Cobra de Capello vel Cabelo ditus; Serpens, cum conspicllo, minor; Serpens, Indicus, cum Conspicillo, lepide circulatus; Serpens, Brasiliensis, cum conspicillo, Cordis oculati formam habente; Serpens, Indicus, coronatus diademate, seu conspicillo insignitus, Lusitanis Cobras de Capello dictus; Cobra de Capello, fœmella, sine perspicillo, seu Tripudia; Serpens, Malabarica, Diademate coronata, sive conspicillo insignita, & Cobra de Capello dicta; Serpens, Indicus, Noja* (sic); *seu Lusitanis Cobra de Capello dictus, maximus, conspicillo notatus; mas; Serpens, Indicus, seu Cobra de Capello; fœmella, sine perspicillo; Serpens, Ceilonica, Conspicillo notata; seu Cobra de Capello*, and *Serpens, Conspicillo ornata, ex Novâ Hispaniâ.*

[134]Hast used the noun *scuta* for both, scutes and scales. Note that in 1752 Linnæus published this method with his signature, without mentioning this dissertation or its re-edition in *Amœnitates Academicæ* (Hast, 1749).

[135]*Coluber berus* in Linnæus, 1758. Henceforth SN 10.

TRUNK of the snake covered by seventeen rows of scales, which are oval and keeled, except for the lowest row of lateral scales, which are smooth and not keeled.

Abdominal scutes 144.

TAIL covered underneath by 39 pairs of broad scales, except for the last one which is pointed.

COLOR of the entire snake black.

The poison lies dormant in its big, sharp teeth, only one of which, on each side of the jaw, is the killer. They are curved, hollow and have an opening at the end.

These teeth are attached to vesicles full of poisonous liquor, which is injected into the wound through the cavity of the tooth. The poison comes out by compression of the vesicle.

Rhedi[136] observed in experiments he conducted with vipers that if a piece of bread is pressed against a viper's fang, it completely soaks up the venom, and the viper's bite produces no other effect than a wound from a needle. He also found that giving this venomous fluid to an animal had no harmful effect. But when an animal was wounded with a needle infected with this venom, it produced the same effect as a viper bite. [p. 6]

β. *Anguis cinerea, macula dorsi fusca longitudinaliter dentata*. Act. Ups. 1736. pag. 11. n. 4.

Vipera anglica fusca, dorso linea undata nigricante conspicua. Petiv. mus. pag. 17. n. 103.

Vipera Raj. quadr. pag. 285.

Vipera vera indiae orientalis. Seb. thesaur. t. II. p. 9. tab. 8.

This variety has the same number of abdominal and caudal scutes, and the dorsal scales and the head have the same structure as the preceding, but the color is different.

HEAD with upper jaw variegated with black and white spots.

DORSUM gray with a black, longitudinal, dentate and sinuous band on each side; laterally, with black spots corresponding to each sinus.

ABDOMEN black, as in the preceding.

Penis subround and spiny, with many strong spines prominently pointing backward.

Obs. The black viper is rare among us, but the gray one, which we have included under the variety β), is quite common.

Those who walk in bare feet through fields, meadows, and woods during the summer are not uncommonly bitten by this snake, from which inflammation, pain, swelling, fever, sleeplessness, and jaundice result, although death rarely occurs.

This snake has the same ability as the rattlesnake; upon seeing a mouse or shrew, it immediately compels the animal, even against its will, to rush into its jaws. To confirm this remarkable phenomenon, an Italian[137] placed a mouse under a jar or pot where they had previously hidden a female snake, and they observed the rodent rushing into the snake's jaws. See what the illustrious President reports about the rattlesnake in the dissertation of April 9, 1743. [p. 7].

2. ANGVIS *scutis abdominalibus CLXII, caudalibus CL*. [Snake with 162 abdominal and 150 tail scutes].[138]

Serpens ceylonica lineis subfuscis. Seb. thesaur. tom. II. pag. 14. tab. 12. fig. 3.

Anguis borneocus, viridis, familiaris, pertenuis Petiv. mus. pag. 18. n. 105.

Serpens indicus gracilis viridis in arboretis & fruticetis degens, Ahaetulla ceilonensibus, i. e. oculis infestus. Raj. quadrup. pag. 331.

HEAD covered anteriorly by two scales, followed by another two larger scales, and finally three scales between the eyes and two even larger scales.

Nostrils very small and sub-rounded.

Upper jaw white.

Color of the top of the head cerulean, with a black line between eyes, and with white separating it from the cerulean.

TRUNK one-and-a-half times longer than the tail. It is barely thicker than a goose feather towards the head, and only slightly thicker than a swan at its widest point.

Scales always smooth, without a raised ridge and with an entire apex.

Abdominal shields number 162; abdomen is only slightly angular.

Color on the back, especially at the beginning, a shining greenish-golden, with a white longitudinal median line. The margins of the scales are black, especially towards the tips, and the body is patterned with a network of lines and greenish rhomboid spots.

The abdomen is white.

TAIL remarkably thin and covered below by 150 pairs of scales arranged lengthwise.

This snake shines in a greenish-golden COLOR and is rightly considered the most beautiful in its country. [p. 8]

3. *ANGUIS scutis abdominalibus CLXXVII caudalibus LXXXV*. [Snake with 177 abdominal and 85 tail scutes].[139]

Anguis vulgaris fuscus, collo flavescente, ventre albis maculis distincto. Petiv. mus. pag. 17. n. 101.

Anguis Gesner. serpent. pag. 43.

Serpens Anguis Schwenck. rept. pag. 157.

Anguis Coluber Merret. pin. pag. 204.

Natrix torquata Aldrovand. serp. pag. 287. Jonst. serp. pag. 89. Raj. quadrup. pag. 334. Charlet. excercit. pag. 35.

Coluber it. Oeland. 6. 154.

In Scania, *Snok*; in Småland, *Ringorm*.

HEAD with two angular scales, then two round ones, then three scales between the eyes, and finally two larger oval ones.

Nostrils subrounded.

Teeth not large or retractable, like those of a viper; the smaller are equal.

Upper jaw white, with black transverse lines.

Lower jaw and throat white.

There is a large bluish-black spot on both sides near the

[136]Redi (1675).

[137]Refers to Redi (1664) or its Latin translation (Redi, 1670a).

[138]*Coluber ahætulla* in SN 10. The species is currently included in the Neotropical genus *Leptophis*, in spite of its provenance from Ceylon, as noted by Hast. In SN 10 Linnæus expanded the geographic range to "America". See Stejneger (1933).

[139]*Coluber natrix* in SN 10.

region of the ears.

TRUNK covered on the back by nineteen rows of imbricated scales. Scales are ovate, obtuse, entire, with a raised keel, except for the two rows closest to each side of the belly, whose scales are smooth and without any marked keel.

Abdominal scutes 177.

TAIL covered below by eighty-five pairs of scales.

COLOR of the whole animal black.

This harmless snake is common throughout Sweden, especially in rural Scania, where it is not only found in stables but also sometimes in the high-temperature ovens used by the locals, called 'rokstugor'.[140] *It commonly lays its eggs in the fields.* [p. 9]

4. *ANGUIS scutis abdominalibus CL, caudalibus LIV.* [Snake with 150 abdominal and 54 tail scutes][141].

Serpens Cobella dicta americana. Seb. thesaur. Tom. 2. pag. 4. tab. 2. fig. 6.

HEAD sutures as in species 2 and 3.

Throat and *jaw* white, with transverse, dark lines.

TRUNK grayish dorsally, mottled with whitish transverse bands, within which dark spots are can be seen. There are also many thin, longitudinal lines, most of which are divided above and below

There are 150 *abdominal scutes.*

Abdomen white, with blackish, wider, uneven, larger, and numerous bands.

TAIL very sharp, with 54 pairs of scales. [p. 10]

5. *ANGUIS scutis abdominalibus CXVII, caudalibus LXI.* [Snake with 117 abdominal and 61 tail scutes].[142]

Anguis bicolor elegantissimus malabaricus. Seb. thes. 2. pag. & tab. 54. fig. I.

HEAD as in the previous species 2, 3, 4, with a grayish-blue color.

Jaw and *throat* white, marked with dark, thin and transverse lines.

TRUNK white, with broad blackish bands, wider in the middle of the dorsum, narrower on the sides, and intertwined below.

Abdomen covered by 118 *scutes.*

TAIL sharp, with 61 pairs of scales underneath.

SIZE half-foot and one spithama.[143]

This small creature is so beautiful and pleasant that young ladies in the East Indies not only enjoy looking at it but also use it to cool themselves in the hot weather by placing it on their bosom. Seb.[144]. [p. 11]

6. *ANGUIS scutis abdominalibus CCL, caudalibus XXXV.* [Snake with 250 abdominal and 35 caudal scutes].[145]

Serpentula ex ins. Ceilon. Seb. thesaur. 1. pag. 19. tab. 10. fig. ult.

Serpens Ceilonica lemniscis latis. Seb. th. T. 2. pag. 76. tab. 76. fig. 3.

Serpens americanus niger, annulis latis albis, ac inter ipsos singulos duo annuli albi tenuiores, ore squamularum albarum nigricante. Mus. Petropol. p. 464. n. 218.

HEAD no longer than the body itself.

Nostrils very small, between which there are two rhomboidal scales, followed by two slightly larger pentagonal scales, then three scales between the eyes, of which the central is the largest and, finally, behind these are two very large scales.

Tip of the *snout* black, with a black transverse band in the middle of the head and a round spot in the occipital region.

Lower jaw grey on the sides.

The TRUNK has the thickness of a swan's feather and is covered with rhomboidal, somewhat blunt, smooth and not keeled scales, arranged in fifteen rows.

There are 250 *abdominal scutes.*

TAIL somewhat obtuse and covered on the underside with 35 pairs of scales.

COLOR of the entire body elegantly variegated with black and white; black segments, except on the head, delimited by 43 white rings; of these, each third band being four times wider than the others; they are always wider on the belly than on the back. All the white scales on the back have the tip and margins rust-colored.

LENGTH of the whole snake one and a half feet. [p. 12]

7. *ANGUIS scutis abdominalibus CXXIV, caudalibus LX.* [Snake with 124 abdominal and 60 caudal scutes].[146]

Serpens americana vipera aemula. Seb. thes. T. 2. p. 15.

---

[140]A *rökstuga* is a type of house in Sweden that has no chimney, but instead has an open pit or oven where the smoke is contained under the roof and regulated by ventilation. They were historically associated with the Finnish population and were sometimes used for temporary shelter during work in the forest. Similar house designs existed in other parts of Scandinavia and in Sami areas. Linnæus describes, as *Rökstufwor*, one of these buildings in eastern Skåne (Ravlunda parish). See entrance of May 29 in Linnæus (1751).

[141]*Coluber cobella* in SN 10.

[142]*Coluber domicella* in SN 10.

[143]*Spithama* (from G. σπιθαμή), is Greek measure of length, measured by a human hand, equivalent to twelve fingers, or the distance from the tip of the middle finger to the tip of the thumb when the hand is outstretched. A spithama was approximately equivalent to 19 centimeters or 7.5 inches.

[144]Seba's (1735) comments: "*.... Bestiola adeò usquequaque venista & amœna est, ut Domicellæ in Indiis Orientalibus ejus conspectu non solum delectentur, sed & in refrigerium sinui suo, æstuante cœlo, applicent...*". ["...This little creature is so entirely charming and *pleasant* that young ladies in the East Indies not only delight in its sight but also use it to cool themselves by applying it to their bosom in the heat of the summer...."].

[145]*Coluber lemniscatus* in SN 10.

[146]*Coluber angulatus* in SN 10.

tab. 12. fig. 3.

HEAD covered by numerous unequal scales, two of which are perforated by a small nostril on each side. Then a certain triangular one is placed in between, followed by two round ones, then three larger ones placed between the eyes. Finally, two larger scales are placed between the eyes, which are ovate-triangular in shape.

Head color ashen-livid.

TRUNK covered with nineteen longitudinal rows of scales.

All *scales* are notched, oval and keeled, except for the lowest lateral row on both sides, which are not keeled.

Color whitish, with broad, angular, and brown bands with black edges; they are wider at the back.

Abdomen with 124 scutes, painted with alternating, quadrangular, rust-colored spots.

TAIL strongly acuminated, with 60 pairs of ventral scales.

LENGTH of one or one and a half feet, with the thickness of a musket barrel[147]. [p. 13]

8. *ANGUIS scutis abdominalibus CCVII, caudalibus LXXXV.* [*Snake with 207* abdominal and 85 caudal scutes].[148]

Serpens americana Petola dicta. Seb. thes. T. 1. pag. 89. tab. 54. fig. 4.

HEAD as in the previous species 2, 3 and 4.

Upper and lower jaws with white margins.

Teeth are numerous, small, sharp and curved backward.

Color in the upper body dark-livid.

TRUNK covered with nineteen rows of oval, smooth, entire and not keeled scales.

Abdominal scutes count 207.

Dorsal color pale, with numerous broad transverse bands.

Abdomen yellowish-white, immaculate.

TAIL tapering, thin, scarcely pointed; scales not keeled; ventrally with 85 pairs of scales.

LENGTH of about two feet; at mid-body has the thickness of a musket barrel, while towards the head the thickness is of a finger. [p. 14]

9. *ANGUIS scutis abdominalibus CLXXXVI, caudalibus LXIV.* [Snake with 186 abdominal and 64 caudal scutes].[149]

Serpens americana annulata. Seb. thesaur. t. II. p. 38. tab. 38. fig. 2.

HEAD with two large scales at the beginning, then many very small scales.

Dorsal color whitish, with a dark spot on each side behind the eyes.

TRUNK four times longer than the tail; dorsally with 19 longitudinal rows of oval-oblong, blunt and smooth scales without raised furrows.

Abdomen with 186 scutes, white in color.

Dorsal color white, with dark, rounded scales alternating in the mid-dorsal region; sometimes two or three converge, creating a longitudinal, curved spot. The spot closest to the head is bifurcated anteriorly.

TAIL blunt, covered with 64 pairs of scales.

LENGTH of a foot and a half; width near head like a swan feather, while towards mid-body it is as thin as a finger. [p. 15]

CHAP. II.

LIZARDS.

There are numerous species of lizards, and zoologist's descriptions and travel narratives confirm it, while our country is home to at least four or five of them.

They differ in many ways; they all have tails, which can be pressed flat and perpendicular, as in crocodiles and many others that live in water, while terrestrial lizards, on the other hand, have tails that are either cylindrical, angular, or articulated, and so, continuing in an extensive series.

As for the body itself, lizards also vary. Some are naked, as *Salamandra*, while others, such as almost all lizards found in warmer regions, have a scaly body resembling that of fish. Furthermore, some lizards are covered in callosities or bumps dispersed throughout their bodies, like the *Crocodilus* and *Gecko*, while others, like *Draco volans*, are crested on their backs, under their throats, or along their sides.

In those that retain legs, they can have separated or webbed digits, and its number on their front and rear feet can differ. The presence of claws is also variable, with some lizards having them on all digits, while others have them only on a few or none at all.

From these facts, it is easy to see which characteristics are most useful in identifying specific differences and which traits remain constant and never vary. [p. 16]

10. *LACERTA cauda ancipiti, pedibus triungulatis, palmis pentadactylis, plantis palmatis tetradactylis.* [Lizard with laterally compressed tail, feet with three claws; hands with five fingers; feet with four toes].[150]

Crocodilus Seb. thesaur. tom. I. p. 160. tab. 103, 104. fig. omnes. Jonst. quadr. p. 200. tab. 79. fig. 3. Schwenck. rept. p. 145. Aldrov. quadr. p. 677. Charlet. exercit. p. 29. Gesner. quadr. p. 9. Rondel. pisc. p. 232. Bellon. aquat. p. 41. Sloan. hist. II. p. 332. Olear. mus. p. 8. tab. 7. fig. 3. Mathiol. dioscorid. p. 278. tab. 278.

Crocodilus, qui per totam indiam, Cayman audit. Bont. Jav. p. 55. tab. 55.

Crocodilus Leviathan Jobi. Bochart. hieroz. lib. I. c. 7.

Jacove seu Crocodilus Pison. americ. p. & tab. 288.

Lacertus omnium maximus, Crocodilus dictus. Raj. quadr. p. 261.

Crocodilus, lacertorum omnium maximus, africanus. Mus. Petropol. I. p. 433. n. 1.

HEAD elongated, depressed, marked by two sutures extending backwards from the eye; behind the eyes, it is covered by a slightly flattened square bone; in the preocular region, there are two raised, confluent, wrinkles. Ventrally smooth, with net-like wrinkles and quadrangular spots. The occipital region with longitudinally acute tubercles ar-

---

[147]In the sense of a portable, long-barrelled firearm.

[148]*Coluber petola* in SN 10.

[149]*Coluber annulatus* in SN 10.

[150]*Lacerta crocodilus* in SN 10.

ranged in two transverse series. The occiput is fortified with callous tubercles arranged longitudinally and sharply into two transverse rows.

Nostrils elevated, with crescent-shaped openings and with small "horns" pointing backward.

Maxillary teeth unevenly arranged in a single marginal row.

Upper eyelids raised and highly wrinkled, giving a fierce appearance; their edges are thinner and transversely striped.

Neck is covered in its upper part by four or five pairs of tubercles, while below it is divided into eight segments.

TRUNK, after the neck tubercles, covered with twenty segments; each of them is square, with a small dorsal line raised towards the flanks.

Abdomen white, with 24 segments.

Anus as a prominent longitudinal fissure, covered with wrinkles all around.

TAIL: The anterior part consists of ten segments, and on the back, it is furnished with somewhat flattened scales and lateral ridges raised by the lateral scales, while the posterior part is compressed with a sharp upper margin and a rounded lower one

The tail is terminated by twenty-two segments, the apex of each of the uppermost ones facing backward, except for the last three segments.

Forelimbs with five, separate fingers; the first three having claws; 4 and 5 without. Hind legs with four semi-palmated digits; the outermost lacks claw.

COLOR ash gray to somewhat bluish, with wavy transverse bands. The first is on the neck, the second between the arms, third and fourth on the back, the fifth in front of the thighs, the sixth behind the thighs, and seventh, along with the rest up to the fourteenth, encircles the tail.

The animal is sluggish and cannot easily be turned on its side. Otherwise, it is harmful, as it crushes with its teeth and scatters the densest bones. Ray. Nevertheless, some country people attack and capture one or two young crocodiles (which are capable of resisting many men), grasp them with their hands, tie up their feet and snouts, put them on a donkey, and transport them to Venice, where they are slaughtered and skinned. Alpin.[151]

It lays eggs the size of a goose, numbering close to 60. Mus. Petr.

The flesh is very white, but with a rough and muddy flavor. Seb. [p. 18]

11. *LACERTA cauda tereti, pedibus pentadactylis, crista dorsi longitudinali, gulae pendula antice dentata.* [Lizard with tapering tail, pentadactyl limbs, with a longitudinal dorsal crest and pendant throat, serrated anteriorly].[152]

Lacerta seu Leguana surinamensis pectinata & strumosa caerulea foemina. Seb. thesaur. tom. I. p. 149. tab. 95. fig. 2.

Lacertus americanus pectinatus & strumosus seu Leguana. Seb. ibid. fig. 1.

Leguana pectinata & strumosa asiatica. Seb. thes. I. p. 151. tab. 96. fig. 4.

Liguana senembi in nova hispania, Tamacolin dicta, pectinata & strumosa. Seb. thes. 1. p. 152. tab. 97. fig. 3.

Lacertus amboinensis pectinatus & strumosus maximus, Senembi & Iguana, dictus, amphibius. Seb. thes. 1. p. 153. tab. 98. fig. 1.

Lacerta vulgo Leguan dicta. Bont. Jav. pag. 56. tab. 56.

Yvana. Ovied. amer. lib. 13. c. 3. Olear. mus. pag. 6. tab. 6. fig. 1. &. p. 8. tab. 7. fig. 2.

Yvana seu Iguana. Worm. mus. p. 313.

Iguana. Nieremb. nat. pag. 271. tab. 271. Roberg. diss. de lue vener. pag. 9. tab. 9. Jacob. mus. tab. 4.

Senembi seu Iguana. Marcgrav. pag. 236. tab. 236. Jonst. quadr. p. 191. tab. 77. fig. 5.

Lacertus indicus Senembi & Iguana dictus. Raj. quadr. p. 265. Sloan. hist. II. p. 333.

Lacertus indicus Senembi & Iguana item Leguan dictus. Mus. petrop. 1. p. 431.

HEAD, in the upper part, covered with round scales and tubercles.

Nostrils are slightly prominent and cylindrical.

Jaws covered with smooth warts laterally.

Ear-opening somewhat large, round and excavated.

TRUNK adorned with numerous rows of scales, of which those on the abdomen are much more distinct.

Anus as a transverse, oblong slit. [p. 19]

TAIL twice as long as the trunk, with 24 raised striae, which can be more or less numerous.

FORELIMBS with five fingers, of which the 1 is also the shortest, 2 and 5 of equal length, while 3 and 4 are almost equal to each other, but longer.

Hind-limbs with five toes, of which 1 is very short, and 2 and 5 of the same length. The 5 is more distinctly separate from the others and the 3 is longer, but the 4 is the longest.

Claws on both the hands and feet are sharp and compressed.

LONGITUDINAL CREST prominent, from the head to the tip of the tail, consisting of more than 320 compressed, double-edged, and erectly recurved denticles. There is also another membranous crest hanging under the throat, anteriorly serrated with 12 denticles, whilst the back is entire.

DORSAL COLOR blackish-blue, with white spots on the arms, thighs, abdomen, and sides. Below, 16 gray warts are arranged in a longitudinal line, and other raised, calloused warts can be seen on the sides of the neck.

This animal, found in both Eastern and Western India, is considered a food source by both the locals and some Christians who consume its eggs as well. Seb.[153] Lactant.[154] Marcgr.[155] Bochart.[156]

It swims in rivers and climbs trees; has a horrifying appearance and is silent, never moaning or screaming, and

[151]The reference to the crocodiles being captured and sent to Venice is taken from Alpini, 1735: 218-219, who wrote: "...*Rem maxime miram dicam, scilicet tam grandem, immanemque belluam, quæ numerosissimis hominibus valide obsistere apta nata est, homunculos quosdam rusticos gracillimos, unum, vel duos invadere, eamque manibus apprehendere, & ita tractare, ut fune os vinciant, atque pedes, & ita legatis pedibus, & ore, asino superponentes, vehant ipsam ad mercatores Venetos, eamque belluam vendant, occidant, pellemque decorticent...*" (I will say a most amazing thing: some very skinny country folk are capable of attacking such a large and immense beast that is capable of resisting a great number of people. They grab it with their hands and handle it in such a way that they tie it up with a rope around the bones and legs, then tie the legs and mouth and place it on a donkey to transport it to Venetian merchants, who buy the beast, kill it, and skin it).

[152]*Lacerta igvana* in SN 10.

is harmless. The young ones swim very fast, but when they grow bigger, they sink and cannot swim anymore. The harm from eating this animal is claimed by those who suffer from venereal disease, for the disease, although dormant for a long time, renews[157]. Oviedo[158]. Worm[159].

It is skillfully caught by the Brazilians; namely, it is accustomed to cling to trees, which the Brazilians, knowing this, adapt a long stick with a rope. The animal, seeing this, is amazed but does not move. It allows itself to be caught by the rope and is captured; although otherwise it runs extremely fast. Marcgr. [p. 20]

12. *LACERTA cauda tereti, pedibus pentadactylis, alis femore connexis, crista gule triplici.* [Lizard with tapering tail, pentadactyl limbs, wings attached to the femora and triple gular crest].[160]

Lacerta africana volans seu Draco volans. Seb. thesaur. 2. p. 92. tab. 86. fig. 3.

Lacertus volans seu Dracunculus alatus. Bont. Jav. p. 57. tab. 57.[161]

Lacerta volans indica. Raj. quadr. 275.

HEAD uneven, with a calloused wart protruding on both sides behind the eyes; towards the cheek, a crest of four compressed segments is elevated, with a black spot and two prominent scales on each side.

Nostrils round, convex.

Ear-opening round, concave, with a protuberance in the middle.

Throat with three crests, of which the lateral ones hanging directed backward; the medial one being larger and folded, with a lobe in the middle also backwardly oriented.

Back of the *neck* rough, with prominent angles.

TRUNK covered on both sides with overlapping, blunt scales, without any distinct segments; of a bluish color with black markings. The *anus* is a transverse slit.

TAIL one and a half times longer than the trunk, covered with keeled and overlapping scales, with more angles towards the base than towards the tip.

FORELIMBS with five separated fingers, of which 1 is short, the 2 and 5 are nearly equal in length, and 3 and 4 are equal to each other but longer than the others.

Hind-limbs with five toes, of which 1 is short, 2 and 5 are of the same length, with the fifth distinctly separated from the others; 3 and 4 are longer, with the fourth being the longest.

All digits are equipped with sharp *claws.*

WINGS on the sides of the abdomen, distinct and separate from the arms, attached to the thighs and consisting of six rays.

It has two yellow-colored bubbles on the outer throat that it inflates during flight. Bont.[162]. [p. 21]

13. *LACERTA cauda tereti, corpore duplo longiore, pedibus pentadactylis, crista nulla, scutis abdominalibus XXX.* [Lizard with tapering tail, twice longer than the body, without crest and with 30 abdominal scutes].[163]

Lacerta americana maculata. Seb. thesaur. 1. pag. 136. tab. 85. fig. 2.

---

[153]Seba (1734: 149): "... *Non a Æthiopes sole hisce vescuntur Leguanis, uti nos Gallinæceo genere; sed vel Christiani illarum regionum incolæ illas in deliciis habent, unoque omnes ore confitentur, carns earum sapore proximè accedere ad carnem Gallinarum...*" ("...Not only are blacks the ones who eat this kind of lizards with the same pleasure that we eat a chicken, but also Christians who live there consider them delicious, and almost unanimously say they have the same taste as hen...").

[154]Unidentified. This dœs not refer to the Christian author Lucius Cæcilius Firmianus Lactantius, as he lived between the 3rd and 4th centuries, well before the discovery of America and iguanas.

[155]Marcgravius (1648: 236): "...*Caro boni est saporis, cocta & diu frixa butyro, & æque bene sapit ac caro pullorum aut cuniculorum. Vigesimo occtavo Maij, 1640, mactavi & secui unum, pinguedinem habentem multam instar gallinæ...*" ("... The meat tastes good fried in butter for a long time, and has the good taste of the flesh of chickens or rabbits. On May 28, 1640, we killed and opened one, and its parts looked like a fat hen...").

[156]Bochart (1663: Col. 1072): "... *Caro ejus calidissima est, & maxime pinguefacit...*" ("...Its flesh is very nice and has a lot of fat...").

[157]Fernández de Oviedo y Valdez (1535) was the first who drew attention on the belief that the flesh of the iguana was dangerous for those suffering from syphilis (see below). A similar comment was included by Linnæus in SN 10 (Linnæus, 1758: 207), a peculiar fact, since Linnæus began his medical practice specializing in venereal diseases (ia, Blunt, 2001).

[158]Fernandez de Oviedo y Valdéz(1535) "...Solo un daño tienen /de que yo he oydo quexarse a algunos: y es: que dizen que los que han seydo tocados del mal de las buas quando comen este animal les torna a tentar aquella dolencia /aun que aya algun tiempo que esten sanos...". ["...The only complain I've heard was that if someone that had the evil bubœs eat this animal eat this animal, the disease will return with greater force, although he was cured since long time..."].

[159]Worm (1655: 314), "...*carnes hujus animalis noxias essee illis, qui aliquando lue Venerea laborarunt, recrudescere enim malum dudum sanatum...*" ("...the flesh of these animals is harmful to those who have suffered in past venereal disease, because it can be intensified...")

[160]*Draco volans* in SN 10.

[161]The reference to *Lacertus volans* in Bontius (1658) is on page 59, not 57 as noted.

[162]Bontius (1658) wrote: "...*Belonius effigiem hujus Insecti adsert bipedem, quum quadrupes sit, prœterea duas bullas flavi coloris circa guttur externe gerit, quas inter volandum inflat & veli loco utitur...*" ("...Belon's illustration depicts this insect as bipedal, when in fact it is quadrupedal. In addition, it carries two yellow bubbles on the outer part of its throat that it inflates during flight, using them like sails..."). The referred illustration of the "biped insect" mentioned by Bontius corresponds to the winged serpent from Arabia, depicted by Belon (1554: 292).

[163]*Lacerta ameiva* in SN 10.

Lacerta americana minor caerulea, Argus dicta. Seb. ibid. fig. 3.

Lacerta Surinamensis major, ameira dicta. Seb. thesaur. 1. p. 140. tab. 88. fig. 2.

Lacertus indicus. Clus. exot. p. 115. tab. 115. Raj. quadr. 2. 270. Worm. mus. p. 313. tab. 313.

Lacerta ceilonica crocodili capite, maxima & ex griseo & nigro varia stellulisque albis transversim sitis notata, cauda longa. Mus. Petrop. 1. p. 436.

HEAD somewhat flattened on the upper part and compressed on the sides, with a calloused apex.

Ear-opening round and excavated.

Teeth very small.

TRUNK covered dorsally with tiny and scarcely noticeable scales.

Abdomen divided into 30 transverse segments made up of quadrangular scales.

Anus as a transverse slit.

TAIL twice as long as the body, tapering, with striations and furrows and composed of over 100 segments.

FORELIMBS with five separate fingers, of which 1 and 5 are subequal, 2 and 4 are longer, and 3 is longer still, all with straight claws.

Hind-limbs also have five separate digits, of which 1 and 5 are approximately equally short, 2 is longer, 3 is twice as long as the second, and 4 is double the length of the third; claws are slightly elevated.

Dorsally, the COLOR is cerulean with black spots, while ventrally it is white with scattered small, round, black spots. [p. 22]

14. *LACERTA cauda tereti, corpore duplo longiore, pedibus pentadactylis, crista nulla, hypochondriis plicatis*. [Lizard with tapering tail, twice longer than the body, pentadactyl limbs, without crest and with folded sides].[164]

Lacerta teguixin seu Tejuguacu altera. Seb. thesaur. 1. p. 150. tab. 96. fig. 1.

HEAD oblong, acuminate, depressed, slightly convex above and marked with various sutures; behind the frontal plate are two additional small plates, then a larger one in the middle, with four smaller ones above the eyes on each side, followed by two oblong ones and finally three angular scales. Thus, the head is almost serpent-like.

The skin beneath the *neck* is loose and adorned with black spots that are darker than those on the rest of the body.

TRUNK covered with numerous circular striae, moniliform or filiform and articulated. On each side, from the head to the hind legs, there are numerous blunt, beaded folds.

TAIL with approximately 200 segments, which alternate dorsally. Unlike the previous lizard, the tail is not crossed but is acuminate and one and a half times longer than the body.

FORELIMBS with five fingers, with 1 and 5 being subequal in length, 2 being longer, and 3 and 4 being almost equal and slightly longer than the second.

Hind-limbs also with five digits, with the internal toe being minute, 2 being longer than 3, 4 being the longest, and 5 hardly longer than the first and set apart from the others by some distance.

Claws very sharp and moderately curved.

COLOR entirely off-white with a slight tinge of cerulean, and there are dark gray bands present. Oval white spots are scattered over the back (though less so in the tail) and the outer parts of the femora.

Therefore, this lizard differs little from the previous one; however, it is a distinct species. [p. 23]

15. *LACERTA cauda tereti, corpore triplo longiore, pedibus pentadactylis, gula subcristata*. [Lizard with tapering tail, three times longer than the body; pentadactyl limbs; sub crested throat].[165]

Lacerta americana, cum cauda longissima, Temapara dicta. Seb. thesaur. I. p. 140. tab. 88. fig. 4.

Lacerta Chalcitica marmorata ex gallecia Seb. thesaur. 2. p. 79. tab. 76. fig. 4.

HEAD unequal in the upper part, with a greenish color.

Nostrils with round openings, facing backward.

Ear openings are also round.

Teeth small and somewhat blunt.

Throat covered underneath with less sharp scales, but still somewhat humpy, and the intermediate row, extending from the tip of the lower jaw towards the chest, almost forms a crest with longer tooth-like scales, but not at all membranous.

TRUNK covered with rounded scales on the back.

Abdomen covered with ovate and acuminate scales.

Illiac bones on both sides prominent.

Anus as a transverse fissure.

TAIL three times longer than the trunk, tapering, nine-sided, with octagonal scales, pale, and marked with rust-colored spots marked by a white center.

FORELIMBS with five separated fingers, with the first digit shorter, 2 and 5 almost equal, and 3 and 4 longer.

Hind-limbs toes similar to the front ones, but the third and fourth are twice as long as the others.

All the nails are more curved and pointed in this species than in the others; they are compressed into a crescent shape, black at the back and bluish at the base.

COLOR of the animal on the upper part whitish-gray, with reddish-brown thighs, arms, and sides of the abdomen; also, with white and dark spots. [p. 24]

16. *LACERTA cauda tereti corpore sesqui longiore; pedibus pentadactylis, dorso lineis longitudinalibus striato*. [Lizard with tapering tail, one and a half times longer than the body; pentadactyl limbs; dorsum striped with longitudinal lines].[166]

Lacertus americanus lemmiscatus. Seb. thesaur. 1. p. 88. tab. 53. fig. 9.

Lacerta africana guineensis. Seb. thesaur. 1. p. 145. tab. 92. fig. 4.

Lacerta amboinensis taeniolis fimbriatis. Seb. thesaur. 2. p. 11. tab. 9. fig. 5.6.

HEAD moderately convex and smooth on the upper part, compressed on the sides, rough underneath.

Nostrils very small.

Ear openings round.

TRUNK covered with small, barely noticeable, and

[164] *Lacerta teguixin* in SN 10.

[165] *Lacerta marmorata* in SN 10.

[166] *Lacerta lemniscata* in SN 10.

round scales on the back.

Abdomen divided by about 30 transverse segments and 7 longitudinal ones.

TAIL longer than the trunk, distinguished by numerous segments and numerous angles.

Obs. The number of segments in the tail cannot be determined, as it is mutilated in our specimens.

FORELIMBS with five separated fingers, with digit 1 being the shortest, and 2 and 5 being of equal length, while 3 and 4 are longer and equal.

Hind-limbs also with five digits, with 1 being the shortest, 2 and 5 being almost equal, and 5 being far away from the rest. 3 is still longer, and 4 is the longest.

COLOR bluish, with a blackish back, eight longitudinal white lines, and white round spots scattered throughout on the thighs and arms. [p. 25]

17. LACERTA *cauda tereti, pedibus inermibus, palmis tetradactylis, plantis pentadactylis, corpore nudo punctis perforato.* [Lizard with tapering tail, limbs without claws, forelimbs with four fingers, hind-limbs with five toes and body naked, pierced].[167]

Salamandra maculosa nostras. Seb. thesaur. 2. p. 15. tab. 12. fig. 5.

Salamandra. Gesner quadr. p. 8o. Aldrov. quadr. p. 639. Schwenck. reptil. p. 163. Olear. mus. p. 9. tab. 8. fig. 4. Mathiol. dioscor. p. 274. tab. 274.

Salamandra terrestris maculis lutaeis distincta. Charlet. excert. p. 28.

Salamandra terrestris. Raj. quadr. p. 273. Jonst. quadr. p. 194. tab. 77. fig. 10.

HEAD smooth and slightly flat on the upper part.

Eyes bulging like those of frogs.

Nostril openings very small.

No trace of *ears*.

TRUNK smooth and devoid of scales.

Anus with a longitudinal, bulging cleft.

TAIL cylindrical and shorter than the body.

FORELIMBS with four, separated fingers, of which 1 is the shortest, followed by 4, 2, 3 in that order, being the third the longest.

Hind-limbs with five toes, of which 1 is the shortest and 5, 2, 4 and 3 being somewhat longer.

No *claws* on either hands or feet.

COLOR of the entire animal blackish, with yellowish oblong spots above the eyes, in the region of the ears, on the sides of the back, above the tail, above the thighs, on the feet and hands, under the throat, and on the sides of the abdomen.

This lizard is remarkable for its tiny dots or pores scattered throughout its upper body, such as on the sides of the head, behind the eyes, on the back and sides of the abdomen, above the tail, and behind the thighs.

It is fabled that this lizard can live in fire; the copious moisture secreted through these wide pores are at the base of this myth. [p. 26]

18. LACERTA *cauda verticillata, squamis denticulatis, pedibus pentadactylis.* [Lizard with a whorled tail, denticulate scales, and pentadactyl limbs].[168]

Lacertus africanus, cauda spinosa, mas & foemina. Seb. thesaur. 1. p. 136. tab. 84. fig. 3.4.

HEAD formed by four sutures intersecting each other at the vertex, with a quadrangular bone in the middle, adjacent to smaller segments with distinct sutures. Anterior part of the head highly rugose.

Nostrils very small and round.

Lower jaw covered with various wrinkles.

Throat covered with imbricated scales.

TRUNK composed of 26 almost equal, transverse, imbricating segments on the back. The first one made up of squarer and pointed tubercles. Segments 2, 3, and 4 interrupted in the middle.

The *back* is flat, and the sides are perpendicular, resulting in a nearly square abdomen, with segments on the sides, similar to those on the back.

Abdomen divided into 20 transverse segments between the legs and arms.

Anus as a small, round opening.

TAIL tapering, shorter than the body, and distinguished by 20 verticillate incisions with scales angled and keeled towards the apex, which extends into a spine, armed on one or both sides with a small tooth.

FORELIMBS with five fingers, with 1 being the smallest, 2 and 5 being almost equal, while 3 and 4 are still longer, all with claws.

Hind-limbs also pentadactyl, with 1 being the shortest, 2 and 5 of equal length, while 3 and 4 are also equal but longer, with the fourth being slightly longer.

Femurs and *tibiae* covered with acuminated scales.

COLOR of the entire animal grayish-lilac. [p. 27]

19. LACERTA *cauda tereti, pedibus pentadactylis, digitis utrinque cristatis subtus lunulato-imbricatis, corpore verrucoso.* [Lizard with tapering tail, pentadactyl limbs, digits crested on each side and lunulate-imbricate below; body warty].[169]

Lacerta indica squamis & verrucis rotundis, digitis latus interne rugosis. Petiv. mus. p. 19. n. 118.

Lacerta indica, seu Salamandra bufonis capite, grisea, tuberculis albis notata. Mus. petrop. I. p. 444. n. 171.

Salamandra vera seu Gecko Ceilonicus maximus, brevi cauda, amphibius. Seb. thesaur. 1. p. 170. &. 171. tab. 168. fig. 1.2. 3. 4. 5. 6. 7. 8. In figure 1 the tail is mutilated[170].

HEAD thicker than the body, with inflated cheeks.

Nostrils subrounded and directed backward.

Eyes large.

Teeth diminute.

Ear openings large, oval, and hollow.

TRUNK covered with smooth warts scattered here and there.

TAIL as long as the body and divided into 25 seg-

[167]*Lacerta salamandra* in SN 10.

[168]*Lacerta cordylus* in SN 10.

[169]*Lacerta gecko* in SN 10.

[170]There is no plate 168 in Seba (1734). Linnæus references correspond to figures 1-8 of plate 108, Gekko Ceilonicus. Figure 5 is misnumbered, as N° 4.

ments, each consisting of about 5 rows of transverse scales. On these segments, small callous, humped, and prominent points can be seen.

The FEET have five digits each in the hands and feet, and each of them is singular. They are broadened from the sides by a membrane on each side, and are nearly equal in length, although the fifth digit on the feet is situated deeper in a sinus. All are neatly covered with imbricated segments underneath.

Nails very sharp, very short, highly curved, and the first digit is devoid of a nail in the feet and hands.

Femora, tibiae, and *arms* with scattered, smooth warts above.

COLOR of the entire body is white.

This lizard was named after the sound it makes before the rainy season; then it exclaims "Gekco" out of joy. Seb.[171]. [p. 28]

Chap. III.

Frogs.

The differences of frogs can be taken from various parts of their body. For instance, the feet have different structures in different species. Some have more or fewer digits, some are equipped with claws, some lack them, and finally, some have webbed feet.

Furthermore, some frogs have an elongated and slender body, while others have a gibbous and rounded one. Some have a smooth trunk, while others have it bristled with warts and tubercles.

Modern authors have explained the unique way of conceiving frogs in various ways, but this hypothesis is the most celebrated among them: During the springtime, a small wart grows on the thumb of the male frog in each hand, which, resembling a penis, is inserted into the female's armpit, and fertilization takes place in this way.

However, the very distinguished Reaumur[172], who was engrossed in the study of nature's marvels, reported to our Celebrated Master Director[173] that he had discovered the mating of frogs through a method far different from what was commonly known, and one that was more in accordance with nature. He will soon share it with the public[174]. And if we only consider the affinity between frogs and fish, such as chondropterygian fish like sharks and rays, perhaps we could investigate these nuptials through this analogy[175].

Furthermore, frog eggs lack a shell. First, a snake-like creature emerges without legs, then it is equipped with four legs and a tail like a lizard, and finally, retaining all four legs, it loses its tail and becomes a perfect frog. Therefore, it goes from an egg to a snake-like creature to a lizard-like creature and finally to a frog.

Egypt would be flooded with frogs if storks did not come every year and devastate them. Alp.[176] [p. 29].

20. *RANA pedibus fissis, unguibus subrotundis, corpore laevi, pone angustato.*[177]

Rana brasiliensis gracilis. Seb. thesar. 1. p. 177. tab. 73. fig. 3.

Ranula americana rubra. Seb. thesaur. 2. p. 76. tab. 68. fig. 5.

Rana surinamensis, prone caerulescentis, supine albi coloris, ad latera utrinque maculis nigris notata. Mus. Petrop. I. p. 427. n. 47.

HEAD smooth and convex on the upper part.

Nostrils very small, round.

Eyes oblong and covered beneath with a nictitating

---

[171]Seba (1734: 172) wrote: "...Nomen Gekko ipsis datum est ob sonum, quem edunt instante pluvia: tum enim, præ gaudio quasi, quinquies vel sexies altum exclamant Gekko, præcipue vespera appropinquante. Unde ruricola: certissimo semper augurio imminentes pluvias prognosticantur..." (The name Gecko was given to them because of the sound they make when rain is imminent: then, as if out of joy, they loudly call out Gecko five or six times, especially as evening approaches. Hence, the country folk always predict impending rains with utmost certainty from this sign).

[172]This is a refrence to a series of experiments performed by Réaumur at the end of March, 1736. Unable to get the original report, I analyzed its transcription in the "*Morceaux Choisies*" edited by Jean Torlais (Réaumur, 1939).

[173]*Cel. D:no Præsidi* in the original.

[174]No record exists of a public presentation referred to the reproduction of frogs by Linnæus, but the hypothesis outlined by Hast was included in the 10th and 12th editions of the *Systema Naturæ*.

[175]Perhaps, a reference to the claspers or pterygopods of chondrichthyans.

[176]Alpin (1735: 199) comments that storks and ibis kill snakes: "...*Ciconiis, atque Gruibus ea regio non est destituta, quæ aliundè ex longis provinciis hyeme in Ægyptum volant, serpentesque occidunt, ciconia præfertim: unde apud eas gentes vel etiam nunc in magno honore habentur, non minus quam & Ibis vocata avis, quæ & ipsa serpentes oppugnat, hæcque tota nigra est, gruinis cruribus, rostro adunco, ciconiæ magnitudine, atque figura...*" (That region is not deprived of storks and cranes, which fly from far-off provinces to Egypt in the winter and kill snakes, especially the stork. Hence, among these people, they are still held in great honor, no less than the bird called the ibis, which also attacks snakes. This bird is entirely black, with crane-like legs, a hooked beak, and the size and shape of a stork), but in that work I did not find references to them controlling frogs or toads. I think Hast confused Alpin with Belon, who in 1605 wrote (p. 102): "...*At quæ ad sinistram est, quoniam humilis & plana, atque Nilo irrigatur, fertilior est & magis herbida; in quam aquaticæ aves hieme commeant tantâ copiâ ut agri & prata iis albescant, præfertim ciconiis, quas Ægyptii non ab re magni faciunt, cum ranæ tantâ abundantiâ istic generentur, ut nisi ciconiæ eas vorarent, nihil illis effet frequentius; tum etiam quòd serpentes capiant, & devorent...*" (But the region on the left, being low and flat, and irrigated by the Nile, is more fertile and verdant. In the winter, waterfowl migrate there in such abundance that the fields and meadows turn white with them, especially with storks, which the Egyptians rightly hold in high esteem. This is because frogs breed there in such abundance that, if the storks did not devour them, nothing would be more common than frogs; and also because the storks capture and devour snakes.).

[177]*Rana arborea* in SN 10.

membrane.

Ears scarcely visible.

TRUNK covered on all sides with papillae, but is smooth and slippery on the back, and tapers behind.

HANDS with four separated fingers, 1 being very short, 2, 4, and 3 longer.

Feet with five, separated toes, being 1 and 2 equal, 5 longer, and 3 the longest of all, but 4 is the shortest.

Nails on both the hands and feet are roundish and similar to those of humans.[178]

COLOR of the upper part of the animal bluish-grey, while the entire underside of the animal, along with the margin of the upper jaw, is white. Rust-colored transverse lines mark both sides of the thighs and the hypochondria. [p. 30]

Chap. IV.

Turtles.

We refer to turtles as creatures that have four legs and are covered by a shell, under which they can hide their head and legs. This shell is nothing but an enlarged vertebra of the dorsal spine covered by several scutes that form different patterns on the back. This is where the common term *chelonium*[179] comes from, which artisans use to make boxes and other items.

FEET with small scales or warts scattered all over. These creatures are not covered in fur like mammals[180], feathers like birds, scales like fish, or bony skin like insects. Instead, they are protected by these horny warts, which serve them well when moving on rocky banks mixed with coral and shell fragments.

Some turtles mostly live underwater; some can also live outside of it; hence authors divided them into aquatic and terrestrial. Therefore, when we observe that the structure of the feet in aquatic turtles differs from that of their terrestrial counterparts, a very consistent specific characteristic can be derived from this difference. This characteristic should be considered as the primary one in distinguishing true species differences. However, since the head and feet are often missing in museum specimens, we are compelled to rely on the characteristics provided by our own specimens, to avoid being deceived by ambiguous descriptions provided by others. [p. 31]

21. *TESTUDO major terrestris americana, mydas dicta.* [Turtle large, terrestrial, American, called *mydas*]. Seb. thesaur. 1. p. 127. tab. 80. fig. 1.[181]

Jurucua Brasiliensibus. Marcgr. bras. 241. Raj. quadr. 256.

From this, only the upper part of this shell can be seen in the Gyllenborgian collection.

The MARGIN of the carapace is cartilaginous and thick, to which are fixed the ends of the ribs, which cannot bend; it is continuous, not serrated.

13 SCUTES[182] cover the entire shell, of which 5 are dorsal and longitudinal, almost hexagonal, smooth and without any hump, not even the last one; moreover, the first and the last are twice wider than the others; additionally, there are 4 oblong scutes on each side.

The SHAPE of the shell is oval and moderately convex, almost no keeled.

The external COLOR is bluish with black spots.

When viewed from below, the shell consists of 10 vertebrae, from which 9 ribs protrude on each side, which are as broad as those of a bull; however, the joints of the vertebrae are uncertain, as they are arranged so that each rib emerges from two vertebrae, making the spine unable to bend.

Travel writers report that in Mauritius island, 5 or 6 people can ride on its back and move it forward, and more people could do so if the curved surface of the shell provided more stable support. Worm.[183]

In all of Greece, men do not dare to eat tortoises, and the Patriarch of Alexandria excommunicated all who did.[184] Bell. itiner. 131.[185]. [p. 32]

22. *TESTUDO unguibus acuminatis, palmarum duobus, plantarum unico.* [Turtle with two acuminate claws on fore-limbs and one on hind-limbs].[186]

Testudo marina americana. Seb. thesaur. 1. p. 130. tab. 80. fig. 9.

Testudo marina. Gesner. quadr. p. 78. Aldrov. quadr. p. 712. tab. 714. Grew. mus. p. 38. tab. 3. fig. 4.

Testudo. Olear. mus. p. 27. tab. tab. 17. fig. 1.

HEAD with a sharp-edged beak, lacking teeth, almost like a hawk, but blunter.".

TRUNK dorsally covered by an oval shell, surrounded

---

[178]Here Hast confused the dilated end of the digits with nails.

[179]From Greek χέλειον, noun originally designating the crab's shell.

[180]*Quadrupedia* in the original.

[181]*Testudo mydas* var. β. in SN 10. In the synonymy of *Testudo græca* fide Duméril & Bibron, 1837.

[182]*Bracteæ* in the original, being *bractea* a thin plate of metal, gold-leaf (thicker plates of metal are called *laminæ*). Translated as scutes following current use.

[183]Reference not located.

[184]*Anathematis fulmine* in the original, used to denote a severe punishment for those who have committed grave sins or heresies; is derived from the idea that the censure is like a thunderbolt from God, a divine punishment for disobedience.

[185]Belon (1605: 131): "...*Ingentes marinas testudines capiunt, quarum integumentum sive corium non minus amplum est quam eadificis alicuius ianua. Olim Christiani iis vesci non audebant, quonian Patriarcha Alexandrinus omnes qui vescerentur, anathematis fulmine petiit: sed postea ab eo soluti, iis non vescuntur...*" ("...Enormous sea turtles were caught, with shells or covers of the size of building doors. Formerly Christians did not dare to eat them, because the Patriarch of Alexandria had anathematized, and the one to do so would fall thunderstruck. Although later the ban was lifted, they do not eat...").

[186]*Testudo mydas* in SN 10.

by a margin of 27 small scutes, arranged in such a way that the posterior ones, which are larger, protrude into an angle, so that the margin has a serrated appearance. The shell is more convex than the previous one and is covered by 15 scutes, six of which are angular; the first is the smallest and the last is the most protuberant, while the lateral ones are oblong and smooth.

The *abdomen* is covered with a less hard scute than the others, in which this singular observation is made, that it is sprinkled here and there with marine barnacles, which adhere tightly to the shell and the feet below, providing a certain and clear indication that this was a marine turtle.

FEET covered with scaly calluses.

Forelimbs with 5 fingers, tightly united, of which 1 is the shortest, 2 is twice as long, 3 is the longest, 4 is longer than the second, and 5 is slightly longer than the first. The claws are pointed in the first and second digits, and membranous and barely visible in the others.

Hind-limbs shorter and blunter, pentadactyl, with tightly united digits, of which 1, which is shorter than the others, is armed with a large and pointed claw, 2 with an oval claw, and 3, 4, and 5 with membranous claws that are barely distinguishable.

COLOR on the upper side is reddish-brown.

In the past, infants were bathed in such shells by the Turks, who also used them as scutes. Worm.[187] [p. 33]

23. TESTUDO *unguibus acuminatis palmarum plantarumque quaternis.* [Turtle with four acuminated claws on hands and feet][188].

Testudo virginica. Grew. mus. p. 38. tab. 3. fig 3.

Testuda terrestris pusilla ex india orientali. Worm. mus. p. 313. Raj. quadr. p. 259.

TRUNK covered by a shell more convex backward, the margin of which consists of 14 scutes, the first of which is the smallest and the other are arranged at right angles.

The *carapace* is protected by 5 longitudinal scutes and 4 lateral ones on each side, all of which are angular and marked by angled striations, but the center is rough with dotted markings.

The *color* at the edges of the scutes is black or rust-colored, while in the middle it is pale.

The *sternum*, or lower shell, is forked and blunt; is divided by 5 transverse sutures and a single longitudinal one.

LIMBS are covered with ovate, broad and round scales, not dilated on the fingers.

Forelimbs always with four claws.

Hind-limbs are also armed with four claws, each straight, pointed, and almost equal. [p. 34]

24. *TESTUDO picta vel stellata.* [Turtle painted or stellate][189]

Worm. mus. 317.

Testudo testa tessellata major, e madagascar. Grew. mus. p. 36. tab. 3. fig 1. 2.

Testudo minor amboinensis. Seb. thesaur. 1. p. 130. tab. 80. fig. 8.

Testudo tessellata minor. Raj. quadr. p. 259.

Jaboti seu Testudo nigricantibus & flavescentibus figueris geometricis. Pis. amer. 1658. p. 106. tab. 105.

UPPER SHELL black, painted in a wonderful manner with a yellow net[190], three times larger and more convex than the previous one, consisting of the same number of scutes both on the margin and the disc.

The *carapace* is highly convex, consisting of distinct scutes with deeper cavities and grooves than the previous one. Each scute is highly convex, with a yellow, flattened tip, a faintly dotted center, and angulated, striated, dark sides, variegated with nine or ten yellow lines running down the center.

Sternum as the previous species, but largely forked and acute posteriorly.

The COLOR of each scute is yellow, emitting 9 or 10 wide, yellow lines from the edges that uniquely blend with the others.

THAT'S ALL.

[187]Worm (1655: 316). Among various properties of the shell, mentions that: "...*In ejusmodi testis infantes olim lavabantur apud Turcas, scutorum etiam loco utuntur, quia ferris Ictibus resistunt...*" ("...Turkish bath their children in this kind of containers, and also use them instead of scutes, as they resist the blows of the irons...").

[188]*Testudo pusilla* in SN 10.

[189]*Testudo geometrica* in SN 10.

[190]*Rete mirabilem* in the original.

[s. n.]

To The Author

By Säla River's verdant shore,
Where nymphs reside, forevermore,
Let them bear witness, friend so dear,
To your prowess, both far and near.
From head to toe, you've never tired,
On paths of wishes, ever inspired.
For what you learned in halls of lore,
Must valleys, mountains, shores explore,
Lakes and lands, all I can name,
Willingly or forced, they acclaim,
From their essence, in your hand they claim.
Not even the tiniest blade of grass,
Escapes your gaze, nor creatures pass.
Birds of fine lineage, so grand,
Fish of the sea, a silent band,
All that binds them in city's hand,
Serves you in its own ordained strand.
But what might be your thought today,
With this strange star, you fill the way?
Do you seek to stretch, or serve Pindus' call,
I cannot fully discern it all.

Yet I know what you've embraced so true,
The art of Aesculapius, you pursue,
Which sees in every ailment's pain,
A chance to craft, from simple strain,
The finest cure, to ease all bane. [s. n.]

Follow the wise leader's guiding light,
My friend, let not fear cause you fright,
Though Fortune threatens, do not despair,
When you've learned to conquer death's snare,
She'll soon unveil her gracious stare,
And look upon you with tender care.

May heaven guide what you begin,
So that your hard-earned success won't dim,
For heaven rewards virtue's embrace,
I foresee you'll find boundless grace,
In measure infinite, time and space.[191]

Carl Krook

Sir[192],

No one could adequately praise and thank those who not only love, but also protect the sciences. What title should be given, then, to the particular grace and magnificence of His Excellency Count Gyllenborg, who not only supports our Academy but also wishes to honor it with everything that can embellish it, even providing the means to sharpen the minds of those who are there. I am delighted to see you, Sir, demonstrate such a beautiful subject with such accuracy in the wonderful dissertation you have just given to the public. It seems to me that you could never have chosen a better one, and as the curiosities you address deserve everyone's attention, the words that express your thoughts also charm the reader with a skill that is unmatched. Our friendship prevents me from saying more, asking only that you always believe me to be one of your most sincere friends. I am

SIR,
Your very humble servant
FREDRICH BÖCKELMAN

[191] In Swedish in the original.

[192] In French in the original.

I. N. J.

MUSEUM ADOLPHO-FRIDERICIANUM,

QUOD,

CUM CONSENSU AMPLISS. FAC. MEDICÆ IN REGIA ACAD. UPSALIENSI,

Sub PRÆSIDIO

VIRI CELEBERRIMI,

D. D. CAROLI LINNÆI

MEDIC. ET BOTAN. PROFESS. REG. ET. ORDIN. ACAD. IMPER. REGG. MONS. STOCKH. ET UPSAL. SOCII HUJUSQ. SECRETARII.

SPECIMINIS ACADEMICI

LOCO

PUBLICO BONORUM EXAMINI SUBMITTIT

LAURENTIUS BALK FIL.

GEVALIA-GESTRICIUS.

IN AUD: CAROL. MAJORI, DIE XXXI MAJI, ANNO MDCCXLVI.

HORIS ANT. MERIDIEM, CONSUETIS.

HOLMIÆ,

Typis LAURENTII SALVII.

Balk, L. 1746. Museum Adolpho-Fridericianum. Holmiae. Laurentii Salvii. [8] + 2-48 + 2 pl. + [2].

The donation of Crown-Prince Adolph-Fredrik studied by Balk (1746) consisted of duplicate materials of the main collection housed at the castle of Ulriksdal (Wallin, 1992). Balk described one turtle, three frogs, 12 lizards and salamanders (one in a footnote), and 15 snakes and amphisbaenians. As was customary, Linnaeus re-edited this dissertation in the first volume of the *Amoenitates Academicae* (Balk, 1749) and, as in the previous case, was frequently cited from that source in the tenth edition of the *Systema Naturae* (Linnaeus, 1758). Under Thunberg's direction, Radloff (1787) reported one turtle, five frogs, 19 lizards and 19 "snakes" (including true ophidians plus caecilians and amphisbaenians). See Wallin, 1992, for a probable explanation on the origin of these additional specimens. Lönnberg (1896), in turn, noted that the specimens 7; 10; 12; 18; 23; 24; 27; 29; 31; 32; 34 and 35 described by Balk (1746; see below) were lost by the end of the XIX century. Today, according to Wallin (2001), Uppsala Museum houses *Rana arborea, Rana boans, Rana gibbosa, Amphisbaena fuliginosa, Coluber agilis, C. annulatus, C. cenchoa, C. petola, C. plicatilis, C. pullatus, C. sibilans, Lacerta agama, L. agilis, L. ameiva, L. aurata, L. calotes, L. chamaeleon, L. gecko, L. marmorata, L. principalis, L. seps, L. umbra,* and *Testudo mydas,* among the species described in Balk (1746), plus *Rana paradoxa, Rana pipa, Caecilia glutinosa, Boa canina, Boa constrictor, Boa hortulana, C. laticaudatus, L. bicarinata, L. crocodilus, L. orbicularis, L. palustris, L. sexlineatus,* and *L. superciliosa,* among those referred by Radloff (1787) to this donation.

I. N. J.[193] Adolpho-Fridericianum Museum, which, with the consent of the most distinguished Faculty of Medicine in the Royal Academy of Uppsala, under the Presidency of the celebrated man, Mr. Carl Linnaeus, Professor of Medicine and Botany in the Royal and Ordained Academy of the Imperial Royal Monarchies of Stockholm and Uppsala, and Secretary of this Society, submits academic specimen to the public examination, in the place of academic specimens, Laurentius Balk, son. of Gevalia-Gestricius. in the Great Hall of Charles, on may 31st, 1746 in the morning, as usual. *Stockholm*, by the press of Laurentius Salvii.

[s. n.]

His Royal Highness Duke Adolph Frederick, Prince of Sweden, Goths, and Vandals, the heir to the throne.

[s. n.]

Her Royal Highness Princess Louisa Ulrica, Crown Princess of Sweden, Goths, and Vandals.

[s. n.]

Most noble Prince and Heir to the Throne!

Among the offerings that the Royal Academy of Uppsala has made in recent years, the most important is the Royal gift that it has received from YOUR ROYAL HIGHNESS's gracious hand. YOUR ROYAL HIGHNESS has thus put our *Medical Garden* on par with the best foreign ones.

We can now see with our own eyes in this valuable collection of Indian animals that YOUR ROYAL HIGHNESS has bestowed upon us what we have had to grasp from others' imperfect descriptions until now, [s. n.] and we can also learn from our own resources how to identify the animals that were unknown to us and to them.

But forgive me, great Prince, for daring to undertake such a subject, from which both the donor's splendor and the gift's magnificence had intimidated me, had not the Grace, which shines so kindly from YOUR ROYAL HIGHNESS's eyes, encouraged me to continue, and also to adorn my humble work with YOUR ROYAL HIGHNESS's precious name.

I am with deepest respect and constant loyalty,

YOUR ROYAL HIGHNESS's

Most obedient servant

Lars Balk.

[193] *I.N.J.* = *In Nomine Jesu* in the original. Translated as "In the name of Jesus".

[s. n.]

Most gracious Crown Princess,

His Royal Highness, our gracious hereditary prince, has deigned to present to our Academy an immense collection of the rarest animals produced by both the Indies, about which we previously had only inadequate knowledge from those who have spent their whole lives on long expeditions in those far-flung lands. Now we have the privilege of seeing this collection with our own eyes: the rattlesnake, which charms the squirrel; the electric fish, which bewitcheth men; the cobra, *itself charmed by* the Indians; the amphisbaena, *which moves both forward and backward;* the scolopendra, *against which* the natives *must keep constant fire*; the golden fish, *which glitters in* China; the spider *that catches birds*; the scorpion, *which delivers deadly stings*; the didelphis, [s. n.] *which nurses its young inside its own body; the fish* Callichthys, *which moves on land from one stream to another*; the aspredo, *which grinds its food with the movement of its fins*; the inkfish, *which clouds the water like black ink, and many other such fascinating creatures.*

Among this collection, I have presented the Chinese Pheasant, *which Your Royal Highness has graciously bestowed upon us, and which previously had never been accurately described. Your Royal Highness's gift, as well as the many wonders of nature that shine therein, have led me to offer Your Royal Highness these pages, which give so much attention to everything the Creator has created.*

I am with the utmost respect and unceasing loyalty,

YOUR ROYAL HIGHNESS'S

Most obedient servant
Lars Balk.

[s. n.]

L. B[194].

With the exceptional and unique kindness of our Hereditary Prince, the Most Serene and August ADOLPH FREDERICK, *it has befallen our Uppsala Academic Museum to be enriched by a splendid donation, exceeding all other possessions in both value and rarity, as much as the Royal Majesty itself surpasses all others. The exotic and rare animals observed here among others have offered the opportunity to celebrate the royal favor with the utmost devotion, and therefore I believed that I could offer you, Carolus Linnaeus[195], nothing more pleasing or agreeable than to share with you a sample of the academic collection as I present this particular section to you. Especially since you well know how incomplete and imperfect all that has been written on this topic until this day. I have given concise descriptions that are sufficient for the knowledge of the species.*

The study of snakes, for which the illustrious Chancellor of our Academy, the King and Senator of the Kingdom, and the excellent President of the Royal Chancellery, Lord Count CARL GYLLENBORG, *laid the foundations with his splendid donation, has taken a great leap with this kind donation. The rare study of ichthyology, long shrouded in the darkness of the unknown, has also made great strides in terms of both genera and species, and no other parts of Zoology have been left behind.*

I have provided the synonyms of authors where I found them to be certain and indubitable, while I have gladly omitted others altogether.

I have given the figures of the animals described here in their natural size, except for the Phasianus, which is twice as large as it appears in Plate I. I have not drawn any others, except for those for which we previously had no figures or which were so erroneous as to be entirely useless.

[s. n.]

[194] L. B. is the abbreviation of Latin *Lectori bono*, translated as "To the good (or benevolent) reader".

[195] C. L. in the original. The abbreviation, translated here as *Carolus Linnaeus*, the supervisor of this dissertation, could also refer to the potential readers of the thesis (*carus lector* or *clarus lector*). The subsequent content of the paragraph suggests that the author was addressing someone with authority and an understanding of the problems that the absence of collections posed for the development and study of zoology at the time, rather than a layperson approaching the subject for the first time.

Class I.

QUADRUPEDIA

S 1. *SIMIA cauda abrupta, ungvivus compressis obtusiusculis, pollice palmarum digitis adhaerente.* [Simia with broken tail, nails compressed and somewhat blunt, with the thumb close to the fingers on the palm].[196]

Simia cynocephalos parva, quae caudam non habet. [Small dog-faced (cynocephalus) monkey, which has no tail]. Alpin. Aegypt. 142. t. 20. f. 1. (an?)

HEAD somewhat round. *Snout* prominent and slightly blunt. *Ears* round and nude.

BODY completely covered with slightly yellowish woolly fur; hair tips blackish.

TAIL very short, barely two finger-widths long, blunt, and hairy.

Digits of hands elongated, with the thumb adhering to the other fingers as in human feet. *Digits of feet* longer, with the thumb more separated, as in human hands. *Nails* long, curved, convex, compressed, linear, and blunt.

This monkey is very playful. It is the size of a cat.

2. *SIMIA a cauda, digitorum indicum ungue subulato.* [Simia with a tail; pointed nail on its index finger].[197]

Simia Charletoni a Robino visa. [Charlton's monkey seen by Robinson][198]. Raj. quadrup. 161. [p. 2]

Simia parva ex cinereo-fusca, naso productiore, brachiis manibus pedibusque longis tenuibus. [A small monkey, grayish-brown in color, with a relatively long nose and thin arms, hands, and feet]; Mus. Petrop. 339. N. 38. 39. 40. 41. 42. 43. 44. *Animalculum cynocephalum ceilanicum tardigradum.* [Small, slow-moving, Ceylonese dog-faced (cynocephalus) animal], Seb. thes. 1. p. 55. t. 35. f. 1. 2. *Cercopithecus ceilonicus seu Tardigradus dictus major.* [Larger Ceylonese Cercopithecus, also known as Tardigradus]. Seb. thes. 1. p. 75. t. 47. f. 1. *Nails of hands and feet somewhat round, like those of humans, except for the pointed or cylindrical ones on the index fingers, resembling those of birds. female with four breasts on chest; she arouses the torpid male for mating. Male tastes the fruits of trees and then offers them to the female. It inhabits Ceylon. Only its skull is held in the museum.*

3. *DIDELPHIS mammis intra abdomen.* [Didelphis with nipples inside the abdomen][199] (a)[200]. *Philander maximus orientalis* Seb. thes. 1. p. 64. t. 39. *Philander amer.* Seb. thes. 1. p. 57. t. 36. f. 2. foemina. *Philander, Opossum* Seb. thes. 1. p. 56. t. 36. f. 1. mas. *Marsupialis amer.* Thyson. act. Carigue. Laet. Amer. 551. *Rapoza Lusitanis* Schott. Phys. 953. t. 32. *Carigurija brasiliensibus* Marcgr. bras. 222. t. 222. Pison. bras. 323. Jonst. quadrup. 136. t. 63. *Maritacaca.* Raj. quadrup. 182. *Taij-ibi brasiliensibus.* Raj. quadrup. 185. *Tlaquatzien* Hern. mex. 330.

HEAD acuminate. *Ears* large, round, membranous, naked and transparent. *Snout* prominent. *Nostrils* in a crescent shape. *Teeth in the lower jaw* eight small, wedge-shaped at the front, followed by three canines on each side, and some sharp and lobed molars. In the *upper jaw*, two primary molars, followed by three small conical incisors on each side, then [p. 3] three canines and several molar teeth with cusps. *Tongue* with papillae and cilia at the tip.

BODY of a fulvous (reddish-yellow) color above, slightly blackish along the length of the back, and whitish on the abdomen and front.

TAIL barely the length of the body, hairy at the base but otherwise naked like a mouse. Thus, it has a fox's snout, a mouse's tail, bat-like ears, and monkey-like feet.

Front FEET relatively wide, like those of monkeys, but the thumb is not separate; all nails are claw-shaped. *Hind feet* are also pentadactyl and split like the front ones, but the thumb is more separated like in monkeys; all nails are claw-shaped except for the thumb, which is flat and rounded like a human's.

Obs. The one we are describing is a young offspring, so we are uncertain whether it will change its teeth into other forms. A peculiar thing about this animal is that when young ones are suckling or frightened, they enter the mother's womb and hide there, even refusing the teats. It lives among trees and feeds on sugar cane and birds.

4. *ERINACEUS loricatus, cingulis septenis, palmis tetradactylis, plantis pentadactylis.* [Armored Erinaceus, with seven bands, four-toed front paws, and five-toed hind paws][201].

Dasypus, Hernand. mex. 314. *Tatuete brasiliensibus*, Marcgr. bras. 231. *Tatou seu Armodillo amer.* Seb. thes. 1. p. 45. t. 29. f. 1. *Tatou juvenis seu Armodillus minor Amer.* Seb. thes. 1. p. 87. t. 53. f. 6.

HEAD ovate, smooth, reticulated with numerous sutures, with a prominent snout. *Nostrils* crescent-shaped. *Ears* sub-ovate, covered with dot-shaped papillae.

BODY covered with a bony shell, whose anterior part is wide, bordered, and covered with slightly raised, rounded dots, and transversely striated; seven narrow belt-like segments, ovate-lanceolate in shape, covered with raised dots;.

---

[196] *Simia apedia* in SN 10.

[197] *Lemur tardigradus* in SN 10.

[198] Ray (1693) wrote: "...*D. Tancred Robinson observavit elegantissimum Animal in Museo clarissimi & generosissimi viri D. Charleton...*".

[199] *Didelphis marsupialis* in SN 10.

[200] Footnote on p. 2: (a) *Didelphis mammis extra abdomen. Mus. f. sorex sylv.* Americ. Seb. thes. 1. p. 49. t. 31 f. 1. 2. 4. 5. 6. 8. *Mus african. Hayopolin dictus*, Seb. thes. 1. p. 49 t. 31 f. 3. *Hujus generis ejam est animal.*

[201] Name and synonyms shared by *Dasypus septencinctus* and *Dasypus novemcinctus* in SN 10.

[p. 4] posterior part of the same size as the anterior and marked with elevated calloused dots. *Penis* exposed, keeled underneath.

TAIL whorled with numerous bands, bare, slightly shorter than the body.

Anterior FEET with four digits, deeply divided, somewhat cylindrical, the inner two longer and nearly equal; the outer two shorter and nearly equal to each other. *Claws* are large, erect, awl-shaped, and very sharp. *Posterior feet* with five digits, the lateral ones more deeply separated and shorter than the middle ones, which are less divided and longer. *Claws* blunter than those of the anterior feet.

The specimen is a fetus recently expelled, and as such, no hairs or teeth are visible yet."

5. *SCIURUS cauda tereti: pilis brevibus, auribus subrotundis.* [Sciurus with cylindrical tail, short hairs, and slightly rounded ears].[202]

APPEARANCE of a common squirrel, but smaller. *Ears* somewhat naked and round, resembling those of humans, not terminated by long hairs. *Bristles* abundant on the sides of the upper lip and some on the cheeks.

TAIL hairy, cylindrical, and not divided in two by long hairs arranged in a comb-like fashion as in our [squirrels].

Hands with five digits, but with a very short and almost absent thumb, armed with an even shorter claw.

Feet also with five digits, with a longer thumb, armed with a more acute claw than in the hands.

COLOR yellowish, with the tips of the hairs whitish. [p. 5]

Class II.

AVES.

6. *PHASIANUS crista flava, pectore coccineo, remigibus secundariis caeruleis.* [Phasianus with a yellow crest, scarlet breast, and blue secondary remiges].[203]

HEAD covered with a feathered crest, yellow, hanging down behind. *Beak* short, convex, somewhat oblong, entirely like that of a hen. *Area* or circle around the eyes naked, flesh-colored. *Neck* covered with yellowish or more saturated yellow feathers, with black margins; on this side yellow with black transverse lines. *Throat* grey.

BREAST, abdomen & front part of the neck blood-red or crimson. *Shoulders* with green, shiny feathers with black margins; hence every feather is green with transverse black lines. *Back* yellow, the *upper tail* feathers are long and scarlet, waving on both sides towards the tail. *Thighs* rust-red.

TAIL wedge-shaped, bent downwards and curved on both sides, as long as the body; rectrices dark-tipped and pale-spotted, and the intermediate ones are squashed by those that surround them.

FEET bare, with four toes, just like in a hen. *Tibia* short, with a spur like in hens. *Wings* rusty-brown: primary rectrices brown, secondary blue on the outer part. Size almost that of a hen.

I establish a new genus of pheasant, whose characteristic is the naked area around the eyes that lacks feathers.

Her sacred and royal highness, the most serene Princess LOVISA ULRICA, who received this bird from China and nourished it for a long time, sent its stuffed remains to the illustrious president recently, which is now kept in the museum of His highness, the sacred and royal prince. See Pl. I. [p. 6]

Class III.

AMPHIBIA.

The CLASS Amphibia consists of four genera: turtles, frogs, lizards, and snakes[204]. The snake genus stands out with the largest number of species, closely followed by the lizard family, which also has quite a few species. Frogs, on the other hand, are fewer in number, and turtles are the fewest.

The differences between the snake species are difficult to determine because this group varies in size and color and lacks limbs. Differences among other animals are easily determined by their limbs. Therefore, the knowledge of snakes has been difficult and very obscure since ancient times. However, in a dissertation presented here a year ago about Gyllenborgian Amphibians[205], a most remarkable discovery was made regarding the specific character of snakes, which consists solely and uniquely in the number of abdominal scutes and tail scales, by which the names and most distinct species of snakes are differentiated.

In this truly vast swarm of snakes that we have undertaken to examine, many species have been found that were not present in the Gyllenborgian collection, and which provide new observations and, above all, illustrate the history of snakes. Here there are snakes that have scutes under their tails, not scales; there are those whose abdomen is covered in scales; finally, there are others whose body is surrounded by rings, so that scutes cover the whole body and exclude all scales. Hence, a fourfold division of the group emerges: [p. 7]

1. *Snakes with abdominal and caudal rings* like the Amphisbaena.

2. *Snakes with abdominal and caudal scales* like Caecilians[206].

3. *Snakes with abdominal and caudal scutes* like the Rattlesnake.

---

[202]*Sciurus flavus* in SN 10.

[203]*Phasianus pictus* in SN 10.

[204]*Testudines, Rana, Lacerta, Angue* in the original.

[205]Hast (1745), see above.

4. *Snakes with abdominal scutes and caudal scales,* like all Gyllenborgian snakes and most of the snake family.

7. *TESTUDO ungvibus acuminatis, palmarum plantarumque solitariis.* [Testudo with pointed claws, single in hands and feet].[207]

Testudo marina americana, Seb. thes. 1. p. 127. t. 79. f. 4. 5. 6

HEAD with an acuminate beak, equipped with nostrils; [mandibular] edge sharp, toothless, not bent as in hawks. Lower eyelid of the eyes divided into twelve deeper notches or tubercles, while the upper eyelid is striated.

TRUNK covered above. Shell oval, surrounded by a margin divided into twenty-five parts, behind which are little protrusions, almost serrated. Disc convex, slightly keeled, covered by fifteen hexagonal scales in the middle and numerous square ones at the sides. The lower part of the shell is composed of thirteen scales in the disc, excluding the lateral ones.

TAIL short, soft.

FEET, both front and back, with the posterior margin wider, sharper, and serrated; the anterior margin is thicker, armed with a single large claw in the middle and another claw barely noticeable, located outside like a sharp scale.

8. *RANA palmis tetradactylis fissis, plantis pentadactylis palmatis, apicibus digitorum subrotundis.* [RANA with four, separate digits on hands, and five palmate digits on feet; tips of digits rounded][208]. [p. 8]

Rana surinamensis, Seb. thes. 1. p. 114. t. 71 f. 3. 4.

MOUTH very large, round, toothless. *Eyes* prominent, with a nictitating membrane underneath. *Ear* openings closed by a membrane and located behind the eyes.

BODY oblong, narrowed towards the back, relatively smooth, and covered underneath with small bumps (papillae).

Palm with five, separate fingers, not connected by a membrane. *Arms* short. *Nails* round, soft, wider than the digits. The lateral fingers are sharp, with the first shorter, the second and fourth slightly longer, and the third longer still, with round notches on both sides. *Plant* with five connected toes, the innermost being the shortest, the second slightly longer, the third and fifth even longer, and the fourth the longest. *Thighs* and *shins* longer and narrower than the whole body. *Nails* also round, soft and wider."

COLOR livid, consisting of faint greyish bands on the thighs and shins, as well as on the hypochondria.

Singularities in this species. The color is snowy or milky, speckled on the arms, thighs, and back without a definite order or determined pattern of spots, hence variously diverse.

9. *RANA*[209] *pedibus fissis, palmis tetradactylis, plantis pentadactylis, geniculis subtus tuberosis.* [RANA with split feet, with four fingers on the hands and five on the feet; knees rough underneath][210].

HEAD small, slightly pointed, smooth. *Eyes* small, equipped with a nictitating membrane.

BODY somewhat round, smooth on both sides, dark.

Palm with four fingers, separate; first finger the smallest, second and fourth slightly longer, third the longest, all blunt, with several tubercles on the underside of the carpus and a tubercle under each knuckle of the fingers. *Feet* with five digits, separate: first digit [p. 9] the shortest, second and fifth digits longer, third even longer, fourth the longest, all blunt, and with protuberances under the knuckles, though less pronounced than in the hands.

These tubercles are also found under each knuckle of the fingers in Rana maxima, virginiana, eximia, rara, Seb. thes. 1. p. 119. t. 75. f. 1. The appearance is the same, but the number of digits is different; our is smaller, not variegated with any colors and has a smaller head than Seba's frog, which is variegated and has five digits on the hands (if properly depicted); however, this difference may be only due to sex or age.

10. *RANA palmis tetradactylis, fissis, plantis hexadactylis, subpalmatis: pollice latiusculo brevissimo.* [RANA with four separate fingers in hands, six toes, somewhat palmate, in feet: thumb (pollice in the original)) broad and very short].[211]

HEAD small, blunt, sunken into the thorax.

BODY subglobose, highly convex, rough, not truly warty, cloudy, with a pale longitudinally sinuate band on each side

Hands with four separate, unarmed fingers, slightly tuberous underneath at the knees, with the first finger shorter than the fourth, and the second and third fingers longer. *Feet* with six toes, of which the pollex (sic) is wider, shorter and [placed] farther back than the others, the second, sixth, third, fourth, and fifth toes gradually become longer.

11. *LACERTA cauda tereti, corpore duplo longiore, pedibus pentadactylis, crista gulae integerrima, dorso levi.* [Lizard with a cylindrical tail, twice the length of the body, five-fingered feet, a complete crest on the throat, and a smooth back][212]. *vid.* Plate II. Fig. II. [p. 10]

---

[206]The attribution of this character state to cæcilians is wrong; Linnæus used the presence of scales in ventral abdomen and tail to characterize *Anguis* in SN 10.

[207]*Testudo mydas* var. γ. in SN 10.

[208]*Rana boans* in SN 10.

[209]*Bana* in the original.

[210]*Rana arborea var. β.* in SN 10.

[211]*Rana gibbosa* in SN 10.

~~*Lacerta volans indica*, Raj. quadrup. 275.~~[213]

HEAD depressed, slightly pointed, somewhat wide, barely compressed on the sides, with various very thin sutures. *Nostril* openings above the snout, small and not raised. *Ear openings* small. *Gular* [p. 11] *crest* rounded, minimally serrated, with a thicker margin directed forward, with a ray and, in the middle, it is marked with a black spot and white stripes on each side.

BODY covered in very small scales, with the abdomen divided into scutes.

TAIL twice as long as the body, covered with very small scales, becoming very thin towards the tip, slightly keeled above and more striated below, finally divided into various segments with barely noticeable transverse indentations indicating articulation.

FOREFEET with five digits, the third and fourth slightly longer. *Hindfeet* with five digits, the fifth separated more deeply, then the first, second, third, fourth, and finally the fifth, which is the most separated. The first is the shortest, the second and fifth are longer, the fourth even longer, and the fifth is the longest. The penultimate joint of all the toes is wider below than the others. *Claws* on all the feet small and sharp.

COLOR of the body leaden (grayish-blue), paler on the tail, with brownish spots.

12. *LACERTA cauda tereti, pedibus pentadactylis, crista gulae pendula antice dentata, dorsi sutura denticulata.* [Lizard with cylindrical tail, pentadactyl feet, a pendulous crest with tooth-like projections on the front of the throat, and a denticulated suture on the back].[214]

Lacerta cauda tereti, pedibus pentadactylis, crista dorsi longitudinali, gulae pendula antice dentata. Amphib. Gyll. N. II.

HEAD less pointed. *Ear openings* closed by a membrane. Posterior *scales* (callum) covering the lower jaw larger, with the last one being the largest and situated just beneath the ear.

TRUNK covered with very small, quadrangular scales. No *abdominal scutes*. *Abdomen* swollen. The sole of the foot with five digits, the first having only one joint, the second having two, the third and fifth having three, and the fourth having four joints.

COLOR leaden (grayish-blue) in some areas, and whitish-blue in others, with brownish spots on the tail being more evident.

13. *LACERTA cauda tereti, corpore longo, pedibus pentadactylis, gula subcristata, antice dentata, dorso levi.* [Lizard with cylindrical tail, long body, pentadactyl feet, slightly crested underside of the throat with tooth-like projections on the front, and a smooth back].[215]

Lacerta cauda tereti, corpore triplo longiore, pedibus pentadactylis, gula subcristata. Amphib. Gyllenb. N. 15.

Lacerta chalcidica marmorata ex gallaecia, Seb. thes. 2. P. 79. t. 76. F. 4.

Our specimen benefits from adding the following features.

Ear openings closed. *Males* with membranous, pendulous throat crest while females have hardly any crest. The crest is covered with blunt, humped scales and is denticulated with compressed, pointed scales along the anterior margin.

Posterior and inferior region of *femora* marked with a longitudinal row of eight or ten truncated papillae, which is however less conspicuous in females.

COLOR grayish, with transverse white and black markings on the back, facing posteriorly on the sides.

14. *LACERTA cauda tereti longa, pedibus pentadactylis, dorso antice denticulato, collo capiteque pone aculiato.* [Lizard with long, cylindrical tail, pentadactyl feet, dorsum with small teeth-like projections along the front, and a neck and head spiny behind].[216]

Salamandra Americ. Lacertae aemula altera Seb. thes. 1. p. 170. t. 107. f. 3. *Salamandra amer. amphibia* Seb. thes. p. 169. t. 107. f. 1. 2.

HEAD oval, covered with overlapping scales facing forwards; behind the ears, it is spiny with unequal prickles, which is unique in this species. *Neck* also stiff, with scattered spiny scales. *Nostril* openings elevated and faced backward. *Eyelids* toothed. No crest [p. 12] under the throat, but loose skin. *Ear openings* closed by a membrane.

TRUNK covered with round scales, keeled on the upper side and ending in a spine, while on the underside the scales are blunter and not keeled. *Suture* on the back distinctly crested by erect, lanceolate scales, especially in the neck area. *Abdomen* lacks scutes

TAIL longer than the body, slightly angled and covered with scales that are more keeled and pointed.

Forelimbs pentadactyl, with the first and fifth digits separated more deeply than the third and fourth. The first digit is the shortest, while the second and fifth are longer, and the third is longer than the fourth. *Hindlimbs* also pentadactyl, with the fifth digit separated more deeply, and the second

---

[212] *Lacerta principalis* in SN 10.

[213] Handwritten strikeouts appear in some copies (i.a., in the one analyzed here, held at the Hunt Institute; in the copy that belonged to Albrecht von Haller, now digitized in Europeana; and in those from the Harvard University Library and the Bayerische Staatsbibliothek), and are absent in others, such as the copy held at the Lyon Public Library. The citation was removed from Balk (1749) and does not appear in SN 10, but *Lacerta volans indica*, Raj. Quadruped. 275, was included by Linnaeus in the synonymy of *Draco volans* in SN 12.

[214] *Lacerta igvana* en SN 10.

[215] *Lacerta marmorata* in SN 10.

[216] *Lacerta agama* in SN 10.

less than the others. The first is the shortest, and the fifth, second, third, and fourth gradually become longer.

Male differs from female in that the dorsal crest consists of larger scales, extending from the head to almost the middle of the back. These scales are less pointed at the back and the sides of the neck are not spiny. In females, the crest consists of smaller scales, extending from the head to just the shoulders. The female is armed with very small spines on the sides, and the scales on the back and tail are more pointed

15. *LACERTA cauda longa, pedibus pentadactylis, dorso antice dentato, capite pone denticulato.* [Lizard with long, cylindrical tail, pentadactyl feet, dorsum with teeth-like projections anteriorly, and head denticulated behind].[217]

Lacerta ceilonica caerulea, Seb. thes. 1. p. 150. t. 35. f. 4. *Lacertus ceilonicus amphibius seu Leguana soa ajer dita.* Seb. thes. 1. p. 149. t. 95. f. 3. *Lacerta ceilonica lemniscata & pectinata caerulea, Kolotes & Askalotes Graecis dicta, aliis Ophiomachus f. pugnatrix coerulea*, Seb. thes. 1. p. 146. t. 93. f. 2.

HEAD ovate, square, covered above with imbricated, [p. 13] obtuse scales, slightly pointed below; behind and above the ears the scales emerge arranged as longitudinal spines; no spines on the neck as in the previous one.

TRUNK with imbricated scales turned upwards above the shoulders. *Back* keeled, with lanceolate, distinct, large, and thin longitudinal scales from the head to the middle of the trunk, forming a crest. Scales on the abdomen and throat slightly keeled, marked by a pointed tip. Numerous raised, longitudinal ridges on the abdomen.

TAIL twice as long as the body and tapered.

Hands pentadactyl, with the middle fingers one and a half times longer. *Feet* pentadactyl, with unequal toes, with denticulated scales underneath.

COLOR cerulean; the abdomen tends to be whitish-blue.

16. *LACERTA cauda tereti brevi, pedibus pentadactylis, digitis duobus tribusque coadunatis.* [Lizard with long, cylindrical tail, pentadactyl feet, dorsum dentate anteriorly and head dentate backward].[218]

Chameleo oriental. ex amboina, Seb. thes. 1. p. 133. t. 82. *Chamaeleo ceilonicus, subcrocei coloris.* Seb. thes. 1. P. 134. t. 82. f. 3. *Chameleo pullus amboinensis*, Seb. thes. 1. p. 134. t. 72. f. 4. 5. *Chameleo promontorii bone spei, caeruleo alboque colore, marmoris instar variegatus*, Seb. thes. 1. p. 135. t. 83. f. 5. *Chameleon*, Olear. mus. 9. t. 8. f. 3.

HEAD compressed laterally, flat on top but with a transverse elevation between the eyes, with a keel-like angle on each side above the eyes, extending from the snout to the occiput. The nape is distinctly separated from the trunk by a deep notch, from which the occiput extends into a sharp angle. Eyes covered by a thick, papillose membrane, with a longitudinal slit in the middle, which therefore functions as the eyelids. This membrane is covered all over with callous points. [p. 14] Nostrils small and facing downward. The mouth is very wide and toothless. Tongue long, like a worm.

TRUNK compressed with a keeled back. The body is covered not with scales but with tubercles that are swollen and elevated, often arranged in fours on the sides. The *anus* is transverse and crescent-shaped anteriorly.

TAIL slightly shorter than the body, cylindrical, moderately compressed. *Hands* pentadactyl, with the three inner fingers completely fused and covered by a common sheath, while the two outer fingers are similarly fused. *Feet* also pentadactyl, but with the three outer toes fused together and the two inner toes similarly fused, in the opposite arrangement compared to the hands.

CREST almost inexistent on the body, but a dorsal suture from the head to the middle of the tail, as well as an abdominal suture from the tip of the snout to the anus, and a longitudinal line on each side of the head marked by larger, pointed, conical scales.

The two or three digits are directed backward to grasp branches of trees more firmly, similar to what is seen in parrots, woodpeckers, and cuckoos. Hence, it is very quick in trees, but slower on flat ground, like a tortoise. It inhabits shaded, somewhat humid, warm hedgerows and bushes. The tail is curved, which it uses like a hand to grasp nearby supports. It moves its eyes in different directions so that it can often view an object placed in front with one eye and observe the area behind with the other. The body appears thin, as if the skin were stretched over a bare skeleton, but it can swell intermittently and at regular intervals, making it look very fat, then gradually subsides and deflates. It often endures hunger for a whole month without eating a single fly; it lives on insects and for this reason is easily kept in houses in India to ward off [p. 15] *harmful insects. The natural color is gray, which it retains even after death, but when alive, it assumes various colors depending on its emotions, such as red, yellow, green, and all colors except black.*

17. *LACERTA cauda tereti, pedibus pentadactylis, digitis utrinque cristatis, subtus lunulato-imbricatis corpore verrucoso.* Amphib. Gyllenb. N. 19. p. 27. [Lizard with a cylindrical tail, five-toed feet, both sides of the digits crested, underneath lunulate-imbricated, body warty].[219]

Nostril foramen pierced. *Scales* on them minute, almost absent.

Our [species] differs from Gyllenborg's because 1. it lacks calloused spikes in the body, 2. the toes have no claws, 3. that the tail is not at all jointed, but almost from the base to the tip, as if enclosed in a little sack, or dressed in a looser, thicker skin. Whether these features pertain to sex, age, or happenstance, I do not know.

[217] *Lacerta calotes* in SN 10.

[218] *Lacerta chamæleon* in SN 10.

[219] *Lacerta gecko* in SN 10.

18. *LACERTA cauda verticillata squamis denticulatis, pedibus pentadactylis*, Amphib. Gyllenb. N. 18. p. 26. [Lizard with a tail in whorls with toothed scales, five-toed feet].[220]

Lacertus african. cauda spinosa, mas & foemina, Seb. thes. 1. p. 136. t. 84. f. 3. 4. *Cordylus vulgo.*

HEAD sides covered with six or seven convex, broad, blunt scales, as are the lower jaws with somewhat large convex scales of this kind. The openings of the ears are small, located behind the head.

TRUNK covered with keeled and pointed scales, but on the back and under the abdomen, the scales are not keeled or toothed but truncated. Hence, the animal bristles with spiny scales on the sides.

TAIL as long as the body, with twenty or twenty-two whorls.

19. *LACERTA cauda tereti, corpore duplo longiore, pedibus pantadactylis, crista nulla, scutis abdominalibus XXX.* Amphib. Gyll. p. 21. N. 13. [Lizard with a cylindrical tail, twice as long as the body, five-toed feet, no crest, and thirty abdominal scutes].[221] [p. 16]

Lacerta ceilonica crocodili capite, maxima & exgriseo & nigro varia stellulisque albis transversim sitis notata, cauda longa, Mus. Petrop. 1. p. 436.

All as in the Gyllenborgian's.

20. *LACERTA cauda verticillata, pedibus subpentadactylis, squamis quadratis.* [Lizard with verticillate tail, subpentadactyl limbs and square scales].[222]

Seps sive lacerta chalcidica. Raj. quadrup. 272.

HEAD very small. Ears tiny and located behind the head.

TRUNK oblong, covered with very fine, square scales, striated both transversely and longitudinally, which is unique in this case. Abdomen, with a sharp and reflexed margin, is attached to the trunk on both sides, which is also unique in this case; this suture of the abdominal skin starts from the head itself and ends on both sides at the anus. Anus covered with scales.

TAIL as long as the body, covered with about fifty-seven whorls, consisting of more keeled scales than in the rest of the body; hence, the tail is grooved and extremely sharp. Hands very short, five-toed, very small, and the thumb of these is barely noticeable, yet all are armed with claws; due to the shortness of the feet, the trunk in this species is very long. The feet are five-toed, with very slender digits, clawed as in others, and the thighs of these have ten papillae combed from the lower angle.

COLOR grayish-livid, darker along the middle of the trunk, with a yellowish transverse band above the anus on the back, and a whitish band below behind the anus.

21. *LACERTA cauda tereti, pedibus pentadactylis, squamis rotundatis levissimis subgriseis, lateralibus subfuscis.* [Lizard with a cylindrical tail, five-toed feet, very smooth roundish grayish scales, and somewhat dusky sides].[223] [p. 17]

HEAD oval, not thicker than the neck, somewhat flat on top, with sutures. Ear openings exposed. Very small nostrils at the tip of the snout.

TRUNK and neck with same width and thickness, smooth, without a crest, covered with roundish, very blunt, tight scales.

TAIL cylindrical, gradually tapering, as long as the body. Hands very far from the feet, five-toed, separated, very small, armed with sharp claws. Feet five-toed, separate, with very unequal digits, as in others.

Color of body unique, greenish and yellowish-gray, with a gray-brown longitudinal band from the ears above the legs. The color and structure are very similar to Stincus officinalis (a).[224]

22. *ANGVIS annulis abdominalibus CC. annulis caudalibus XXX.* [Snake with 200 abdominal and 30 caudal rings].[225]

Amphisbaena, Raj. quadrup. 288. *Ibyaru.* Margr. bras. *Amphisbaena ceilanica*, Seb. thes. 2. p. 2. t. 1. f. 7. *Serpens ceilanica, Cecilia dicta*, Seb. thes. 2. p. 19. t. 18. f. 2. *Serpens amer. Caecilia dicta.* Seb. thes. 2. p. 24. t. 22. f. 3. *Serpens caecilia amer.* Seb. thes. 2. p. 25. t. 24. f. 1. x *Amphisbaena americ. ex fusco & albo variegata* Seb. thes. 2. p. 77. t. 73. f. 4. x. *Serpens Apamea cyriaca biceps*, Seb. thes. 2. p. 106. 1. 100. f. 3. *Serpens Amer. grossa versicolor*, Seb. thes. I. p. 140. t. 88. f. 3. [p. 18]

HEAD very small, smooth, blunt, marked with a longitudinal line on top and lateral wrinkles, inscribed with a triangular line between the nostrils at the tip of the snout. Nostrils very small. Eyes almost nonexistent, replaced by as

[220]*Lacerta cordylus* in SN 10.

[221]*Lacerta ameiva* in SN 10.

[222]*Lacerta seps* in SN 10.

[223]*Lacerta stincus* in SN 10.

[224]Footnote on p. 17: (a) The appearance of this species resembles that of *Stincum marinum officinarum*, from which is easily distinguished by the following characters. *Stincus* is smaller and fat, the neck has the same width and magnitude as the head, the scales are rounded and blunt, and some of which have three high ridges; color as in the preceding. Forelimbs have five fingers, flat and marginated on the sides; the first shorter than the second and fifth, and the first (sic), third and fourth a little longer. The nails of these fingers are flat, oblong and slightly curved downwards. Hind-limbs are also pentadactyl, with tœs fringed on both sides, of which the first is the shortest, the second a little longer, third, fourth and fifth similar and somewhat longer than the second. All digits have cartilaginous, wide nails, which reminds that of the humans.

[225]*Amphisbœna fuliginosa* in SN 10.

many blackish spots. Mouth with numerous small teeth.

TRUNK cylindrical, about a foot long, with around two hundred convex ring-like segments, as in earthworms and insects; marked with about forty longitudinal striations, the twelfth of which (on both sides of the intermediate dorsal striation) is crossed like the letter X in Latin, distinguishing the abdomen from the back with a longitudinal line on both sides, marked with X-shaped characters. Anus as a transverse slit. The last abdominal segment before the anus is marked with eight small papillae arranged in a transverse line.

TAIL short, consisting of thirty annular segments without the lateral marks of X-shaped characters; the tail is very blunt and thick, like the head, not tapering as in others.

COLOR entirely variegated with black and white, with black more noticeable on the back and white more conspicuous underneath.

This is the so-called Amphisbaena by the ancients, who believed it had two heads, as it presents almost the same appearance and thickness in the tail and head regions, can move equally well forward and backward, and, when captured, may present its tail in place of its head. It is very unwise to assert that a snake with a head at each end exists in nature, and too foolish to believe it. Ray[226]. *It has annular segments like those of earthworms, as it lives in the same soil and, like worms, moves both forward and backward. Therefore, it has a more robust tail compared to other snakes.* [p. 19]

They have been found in Libya, Lemnos, Brazil, and Ceylon, feeding on ants, and especially on worms.

The bite is mortal, and against it there is no possible remedy, says the Portuguese, although it seems that we face a new paradox since they lack movable fangs, which typically are those that inject venom in these cases.

23. *ANGVIS squamis abdominalibus CCXL, caudalibus XIII.* [Snake with 240 abdominal and 13 caudal scales].[227]

Scytale amer. Amphisbaena & Cecilia vocata, Seb. thes. 2. p. 3. t. 2. f.. 1. 2. *Scytale ex nova hispania*, Seb. thes. 2. p. 4. t. 2. f.. 3. 4. *Amphisbaena amboinensis, corio quasi reticulato conspicua, squamis rubicundis obducta.* Seb. thes. 2. p. 9. t. 7. f.. 4. *Serpens amphisbaena*, Seb. thes. 2. p. 21. t. 20. f.. 3. *Serpens amphisbaena orientalis*, Seb. thes. 1. p. 27. t. 53. f. 7.

HEAD oval, small, not angular. Nostrils round. Eyes very small, set in their own scales.

TRUNK cylindrical, one and a half feet long, covered with imbricated ovate, blunt, smooth scales. The abdomen is also covered with similar but wider and blunter scales, about two hundred and forty in number.

TAIL barely as long as a finger's breadth, thick, blunt, like the head.

COLOR whitish, with brown ring-like bands interrupted here and there. The whiteness between the bands is stained with rust-colored edges of the scales.

It agrees with the preceding Amphisbaena in having a thick, blunt tail, which clearly distinguishes it from other snakes, so that we have no doubt that these snakes, also mentioned by Seba, are the same as ours, although the color in ours is very different from Seba's. This one agrees with the Caecilia in lacking abdominal scutes and instead having somewhat broader scales in their place. [p. 20]

24. *ANGVIS scutis abdominalibus CLXV, caudalibus XXX.* [Snake with 165 abdominal and 30 caudal scutes].[228]

Vipera caudisona Amer. Raj. quadrup. 296.

Vipera orientalis, maxima caudisona, Seb. thes. 2. p. 101. T. 96. f. 1. *Serpens crotalophora seu Vipera caudisona Amer.* Seb. thes. 2. p. 99. t. 95. f. 1.2. *Domina serpentum* Jonst. serpent. 26. t. 7. f. 2. Nieremb. hist. 1. 12. c. 1. Olear. mus. 17. t. 11. f. 2. *Boicocininga*, Marcg. bras. 240. Pis. bras. 41. 275. *Teuhtlacet Zaubqui* Hernand. mex. 329.

HEAD extremely blunt, certainly more so than in any species known to us. The forehead from the nose to the eyes is covered with various somewhat broader scales, overlaid with two larger oval plates. Mouth opening noticeably wide, with each lip covered on both sides by fourteen scales arranged longitudinally along the margin. One canine tooth on each side of the upper jaw, very mobile and very sharp, that fold between the vesicles.

TRUNK entirely covered with ovate scales, keeled and sharper on the back, but smaller towards the head, all arranged in a quincunx pattern. The abdomen has 165 scutes longitudinally from the throat to the anus, receiving each other.

TAIL, or the space between the anus and the tip of the body, ventrally covered by thirty scutes, not divided as in other snakes. The crepitaculum that terminates the tail consists of small cartilaginous, ventricose, pellucid, ampullate corpuscles, embracing each other, somewhat broad and almost heart-shaped, dry, fragile, and resonant.

COLOR gray, with some yellow spots on the back, in our specimen, that is a young of barely two feet long.

There are many singular features in this species. [p. 21]

(a) Head extremely blunt. (b) The scutes of the tail, which in this species are whole, are halved in others. (c) The rattle by which this species alone is distinguished: Authors state that each year this serpent adds a new segment to its rattle, which does not seem foreign to us, since our specimen, which is young, has only five segments, whereas older individuals, according to authors, often have twenty or even more. Marcgrave described one with seven segments in its rattle, and Seba with forty-four. In this, the mercy of the Supreme Creator towards the human race shines forth, for He has given humans this instrument of the rattle to protect themselves from the bite of this highly lethal serpent: whenever it bites, it quickly contracts its body to protrude its head in making the strike, thereby emitting this sound from the rattle, which humans know

---

[226]Ray (1693: 288-289).

[227]*Anguis scytale* in SN 10.

[228]*Crotalus dryinas* in SN 10.

to avoid as imminent danger, a privilege denied to animals. (d) Swedish priests[229] *from Pennsylvania unanimously report this viper: upon seeing a squirrel in a tree, it leans towards its roots, its eyes glowing like those of a squirrel, and gaping like a cat does when contemplating a mouse; then, the snake, causing the tree to tremble, seeks refuge, finally springing into its mouth as if enchanted. See Lin. de telluris incremento, praefat.*[230] *(e) The bite of this serpent is very venomous, causing death within one to at most two days. (f) The antidote for this bite is the root of Polygalae caule simplici erecto, foliis ovato lanceolatis alternis integerrimis, racemo terminatrice erecto. Gronov. flor. virg. 180. Recently, Dr. Tennent discovered this very antidote of the Virginian wilderness, teaching that the true Virginian serpentaria is neither Aristolochia, nor Dorslena, nor Althea, nor Reseda, nor Prenanthes, but the aforementioned species of Polygala, which we now use as a specific remedy in* [p. 22] *inflammatory fevers. (g) This serpent often grows up to five feet in length, especially found in Brazil and Virginia, swiftly moving through rocky places; it is of various colors and emits a foul odor, so much so that humans often sense its presence from this distinctive smell alone. It has four backward-pointing thorny penises. Its two canine teeth, unlike in other venomous snakes, are not embedded in the jaws; it uses them for puncturing rather than injecting venom. It is also said to become so torpid if wrapped in even the thinnest linen that it can be carried and handled by a small person, or even a child.*

25. *ANGVIS scutis abdominalibus CCXVII, squamis caudalibus CVIII.* [Snake with 217 abdominal scutes and 108 caudal scales].[231]

Serpens amboinensis niger & albus, Apachykoatl dicta. Seb. thes. 2. p. 21. t. 20. f. 1. *Apachykoatl* Niremb. natur. 284. Jonst. serp. 29.

HEAD large, ovate, angular, black above, spotted white on the sides and underneath. Nasal openings situated at the sides of the snout, facing backward. Teeth arranged in a double row on both sides of the upper jaw, one row along the edge of the jaw and the other placed further inward. Tongue is black, with a double filament.

TRUNK very large, two or three feet in length, flattened, with a notably compressed back, almost keeled. The back is covered with smooth, blunt, ovate scales, which are brown on one side and whitish on the other. The abdomen is somewhat flat, covered with 217 scutes. Most of the abdominal scutes are white, but every third one has a brown margin, or sometimes every other one has a brown margin on one side.

TAIL darker, more cylindrical, covered underneath with 108 pairs of scales.

Obs. Easily distinguished from the other by their dorsum particularly compressed, almost as a keel. [p. 23]

26 *ANGVIS scutis abdominalibus CXXVIII, squamis caudalibus XLVI.* [Snake with 128 abdominal scutes and 46 caudal scales].[232]

HEAD oval, barely angular, smooth. Eyes small. Nostrils tiny, at the tip of the snout. Teeth small.

BODY thick, bluish on top, with a longitudinal dark stripe on the sides, consisting of dark scales with whitish tips. Abdomen white, with a lateral longitudinal row on each side made up of small yellowish spots, one spot per scale. A few scattered dark spots on the abdomen, particularly under the neck or on the first abdominal scales. Abdominal scales numbering one hundred ~~eighteen~~[233], and the body is proportionally very thick for its length, which in our specimen is over a foot long.

TAIL very short, covered underneath with 46 pairs of scales, each with a larger dark spot.

27. *ANGVIS scutellis abdominalibus CXLII, caudalibus LXXIV.* [Snake with 142 abdominal scutelles and 74 caudal scales].[234]

Head oval, marked with various sutures. Ear openings not conspicuous. Teeth very small, without movable venomous fangs. Lips white, variegated with transverse black markings.

Body dorsal scales lanceolate, marked with an elevated line, which fades in the lateral scales. Abdominal scutes numbering 142.

Tail slender, the same color as the back, covered underneath with 74 pairs of whitish scales, each with a black spot in the middle.

Color in this species very beautiful, with a narrow black line running from the head to the tail along the back. Adjacent to this on each side is a longitudinal pale gray band, sprinkled here and there with white scales. Then on each side, there is a longitudinal black band. [p. 24] Finally, a narrow white line longitudinally distinguishes the abdomen from the back. The abdomen is black, with lanceolate transverse white bands, the same number as the scutelles.

28 *ANGVIS scutis abdominalibus CL. squamis caudalibus LIV.* Amphib. Gyllenb. p. 9. N. 4. [Snake with 150 abdominal scutes and 54 caudal scales].[235]

Color, structure, and appearance in all respects match the

[229]*Sacerdotis nostri* in the original.

[230]Linnæus (1744).

[231]*Coluber pullatus* in SN 10.

[232]*Coluber plicatilis* in SN 10.

[233]Stricken through and replaced with *viginti octo*, or by the number 28, in the marginalia of various copies of the analyzed dissertation. This correction does not appear in Balk (1749).

[234]*Coluber vittaus* in SN 10.

Gyllenborgian specimen, but there is also another specimen whose abdominal scutes number 153 and caudal scales 49, so that the number removed from the tail is added to the abdominal scutes.

29. *ANGVIS scutis abdominalibus CLII, squamis caudalibus CXXXV.* [Snake with 152 abdominal scutes and 135 caudal scales].[236]

HEAD fairly thick, angular, oval-oblong, blunt. Nasal openings on the side of the snout. No visible ear openings. A small venomous tooth, movable on each side in the upper jaw. Eyes very large.

TRUNK bluish back, with oval scales having whitish edges. Belly white, covered with 152 scutes.

TAIL long, very tapered, with a bluish seam underneath, covered with 135 scales on each side.

30. *ANGVIS scutis abdominalibus CLX, squamis caudalibus C.* [Snake with 160 abdominal scutes and 100 caudal scales].[237]

Serpens Ibiboboca brasiliensibus Seb. thes. 2. p. 21. t. 20. f. 2. (an). *Serpens ceilonica sibilans, pulchre lemniscata.* Seb. thes. 2. p. 47. t. 45. f. 5. *Serpens exquisitissima Zeilanica, Malpolon dicta*, Seb. thes. 2. p. 52. t. 52. f. 4. *Serpens african. Hippo dicta*, Seb. thes. 2. p. 57. t. 56. f. 4 *Serpens Amer. rarior, lemniscis* [p. 25] *rubris & albis longitudinalibus notata*, Seb. thes. 2. p. 115. t. 197. f. 4.

HEAD oval, somewhat elongated, angular above the eyes, pale, variegated with oblong bluish spots bordered with black, and various irregular patterns. The jaw edges are marked with six different white lines. Very small nostrils near the tip of the snout. No signs of ears. Very short venomous tooth.

TRUNK most notable color is bluish, with a black longitudinal line down the middle of the back; each scale is marked with a black and white oblong spot. Then, a black longitudinal line with a white one below it; the underside is white with a black line on the sides. Scales on the back are oblong, bluish, with black borders, especially at the tips. Abdominal scutes are 160 or 159 in number.

TAIL very slender and long, colored the same as the body, i.e., bluish above with black reticulations and two longitudinal white bands on each side.

This snake is considered among the larger species.

31. *ANGVIS scutis abdominalibus CLXV, caudalibus XXIV.* [Snake with 165 abdominal and 24 caudal scutes].[238]

Angviculus African. coeruleus, Seb. thes. 2. p. 16. t. 13. f. 3.

HEAD oval-oblong, bluish above, white below. Nostrils are very small.

TRUNK long and narrow, bluish above, white below. The scales are oval, each marked with an oblong white spot, especially on the back. Abdominal scutes number 165.

TAIL slender, a darker blue and spotless, becoming very thin towards the tip.

This species should also be counted among the larger snakes. [p. 26]

32. *ANGVIS scutis abdominalibus CLXXX, squamis caudalibus LXXXV.* [Snake with 180 abdominal scutes and 85 caudal scales].[239]

Serpens Africana ab hottentottis Sibon dicta, Seb. thes. I. p. 22. t. 14. f. 4.

HEAD nearly round, very flattened. Eyes large.

TRUNK covered with smooth, rhomboid scales, underneath with 180 abdominal scutes.

TAIL shorter, narrower, covered underneath with 135 pairs of scales.

COLOR entirely a mix of brownish-rust and white, the abdomen white sprinkled with brown spots.

33. *ANGVIS scutis abdominalibus CLXXXIV, caudalibus L.* [Snake with 184 abdominal and 50 caudal scutes].[240]

HEAD oval, slightly flattened, barely angled, white with a brown transverse band running through the eyes. Nostrils open, facing backward.

TRUNK covered with smooth, slightly blunt oval scales, with 184 scutes underneath, about a foot in length, as thick as the smallest finger.

TAIL very short, covered underneath with 50 pairs of scales and rounded.

COLOR black and white bands encircle the body, with each alternate white band being twice as narrow. This beautiful pattern is illustrated in Seba's Thesaurus, Vol. 1, p. 201, plate 11, figure 3.

This snake is similar to the species described with 250 abdominal scutes and 35 caudal scutes in Amphibia Gyllenborgiana, p. 1, no. 6. In size and almost in color, yet clearly distinct not only by the abdominal and caudal scutes but also by the eyes, which in Gyllenborg's specimen are very small dots, whereas in this one they are large and proportionately significant. [p. 27]

34. *ANGVIS scutis abdominalibus CXC, caudalibus XCVIII.* [Snake with 190 abdominal and 98 caudal scutes].[241]

Anguis scutis abdominalibus CLXXXVI. caudalibus

[235]*Coluber cobella* in SN 10.

[236]*Coluber dipsas* in SN 10.

[237]*Coluber sibilans* in SN 10.

[238]*Coluber cæruleus* in SN 10.

[239]*Coluber sibon* in SN 10.

[240]*Coluber agilis* in SN 10.

[241]*Coluber annulatus* in SN 10.

LXIV. Amphib. Gyllenb. N:o 9. p. 14.

This specimen differs slightly from Gyllenborg's in its abdominal scutes, specifically having 189 or 190, whereas his had only 186. Moreover, the number of tail scales differs significantly, as the tail in Gyllenborg's snake was blunt, whereas in this one it is much longer and more pointed. This clearly indicates that the tail of the former was mutilated, resulting in a much higher number of scales in the present specimen, which requires correction.

35. *ANGVIS scutis abdominalibus CXCVI. squamis caudalibus LXVII.* [Snake with 196 abdominal scutes and 67 caudal scales].[242]

Serpentula ceilonica seu Naja altera, Seb. thes. 1. p. 77. t. 43. f. 5.

HEAD whitish, depressed behind the eyes, flat above the eyes, and greatly compressed on the sides, entirely covered with very fine scales without sutures. Large ovate scale above each eye. Nostril openings near the eyes, open. Lethal teeth on both sides of the upper jaw, very large, hidden behind the vesicle; hence, we have no doubt it is highly venomous.

BODY very slender in proportion to the head, especially the neck, covered with small ovate lanceolate, keeled scales, giving the serpent a striped appearance. One hundred and ninety-six abdominal scutes.

TAIL about a foot long in the snake, barely over two transverse inches wide, and at the same time very narrow.

COLOR ashen with various whitish markings.

Unique characteristics include the head without any apparent sutures and an extremely short and narrow tail, which easily distinguishes it from others. [p. 28]

36. *ANGVIS scutis abdominalibus CCIX, squamis caudalibus XC.* [Snake with 209 abdominal scutes and 90 caudal scales].[243]

Angvis scutis abdominalibus CCVII, caudalibus LXXXV, Amphib. Gyllenb. p. 13. N. 8.

HEAD oblong, obtuse, depressed, whitish on the sides and occiput. No lethal teeth. Nostrils fairly large, facing backward on the sides of the rostrum.

BODY as described among Gyllenborg's Amphibians. This serpent has two hundred and nine abdominal scutes, including the first and last, which slightly deviates in number from Gyllenborg, but still maintains a similar count. The number in the tail is more difficult to determine due to the very small scales towards the tip.

COLOR this species is paler, or pale brown, with a whitish abdomen featuring numerous alternating annular bands, though not all extend across the back.

In this collection, there is another variety of this species, which has two hundred abdominal scutes and approximately one hundred pairs of caudal scales, thus having more scales in the tail than it lost in the trunk.

36 (sic)[244]. *ANGVIS scutis abdominalibus CCXX, squamis caudalibus CXXIV.* [Snake with 220 abdominal and 124 caudal scutes].[245]

Anguis de cencoatel amer. venustissimus, Seb. thes. 2. p. 18. t. 16. f. 2. 3.

CAPUT subglobosum, vix angulatum, albo & nigro varium. *Oculi* magni. *Dentes* parvi.

TRUNCUS cum cauda quatuor pedum, crassitie pennae anserinae vel cygnex. *Squamae* ovato acuminatae leves, dorsum tegunt. *Scuta Abdominalia* ducenta & viginti vel summum ducenta & viginti duo. Testiculi duo globosi echinati. [p. 29]

CAUDA tenuis unicam tertiam partem longitudinis totius animalis seu trunco dimidio brevior.

COLOR fuscus maculis pallidis, in dorso fasciae circiter viginti, niveae, versus abdomen latiores & coeuntes. Hinc abdomen maxima ex parte album.

Varietas hujus datur in hac collectione, quae alba est, in cujus dorso a capite ad anum quadraginta maculae fusco-ferruginea, cum puncto ejusdem coloris singulis maculis utrinque subjecto, praesertim in postica parte corporis. Cauda hujus tenuissima maculis fuscis-ferrugineis circiter quadraginta longitudinaliter a tergo dispositis. Est hic serpens inter omnes in suo genere, ratione longitudinis tenuissimus & fere filisormis.

HEAD subglobose, barely angular, variegated with white and black. Eyes large. Teeth small.

BODY with a tail measuring four feet in length, as thick as a goose or swan feather. Smooth ovate-acuminate scales cover the back. Two hundred and twenty or at most two hundred and twenty-two abdominal scutes. Two globular echinated testicles (sic). [p. 29]

TAIL thin, one-third the length of the entire animal or half the length of the trunk.

COLOR brown with pale spots, about twenty snowy bands on the back, broader and merging towards the abdomen, which is mostly white.

In this collection, there exists a variety which is white, with forty brownish-ferruginous spots on its back from head to tail, each spot accompanied by a dot of the same color on either side, especially on the hind part of the body. The tail of this variety is very thin with about forty longitudinally arranged brownish-ferruginous spots on its back. This serpent is among the thinnest and almost thread-like in its genus due to its length.

---

[242] *Coluber naja* in SN 10.

[243] *Coluber petola* in SN 10.

[244] Corrected to 37 in Balk (1749).

[245] *Coluber cenchoa* in SN 10.

Class IV.

PISCES.

37. *TORPEDO* Arted. Gen 73. Syn. 103. Gesner. pisc. 988. Kaempf. amoen. 509. t. 510. Torpedo in Greek Ναρχη, Raj. p. 28. *Genuensibus Batte potta* Willugb. p. 81. Kramp fisk in Swedish.[246]

FISH flat, soft, without awl-shaped scales, slightly rough above and grayish-ashen, pale and smooth underneath except for the tail, which is also rough above.

HEAD very obtuse, depressed, uneven, white on top and surrounded by a sharp margin in front, which however is split at the nostrils, so that the upper part protrudes two lanceolate membranes, and the lower part a broader membrane; the head is also slightly compressed on the sides; behind the eyes on both sides there are five transverse openings for the gills [p. 30] on each side of the neck. The mouth is wide. The anus is in front of the tail base through a longitudinal slit.

The pectoral FINS are soft, large, triangular, situated at the sides, distinct from the body by an anterior and posterior sinus. The anterior angle of these fins is narrower and thicker, while the posterior is wider and thinner; the outer edge is also thin and smaller than the posterior. The *ventral fins* are located at the posterior part of the abdomen, on each side of the anus, separated from the tail by a sinus, soft and membranous. Behind them, on the inner side, they are marked by a shorter dentiform lobe. The anterior *dorsal fin* is small, erect, oval, situated above the back at the base of the tail. The posterior *dorsal fin* is similar in size and shape, situated in the middle of the tail. The *anal fin* is obliquely perpendicular (not horizontal), slightly obtuse, bifurcated with the upper lobe slightly shorter.

Obs. (a) The description of our fish is very accurate and corresponds closely to the living object as described in Kaempf. Amoenit. exotic. p. 509. (b) *This fish, by the amazing and unique providence of nature, lacks any other form of defense or protective shield, so that if anyone touches it with hand or foot, the fish itself trembles, causing an immediate and tender horror and stiffness in the body of the toucher, gripping the solid parts in such a way that you would think all the hand and arm joints were dislocated with horror at the heart, and the astonishment of all the limbs, so that no one, even for the greatest reward, would dare to approach it again, but would rather immediately discard the fish,* see Redi experiment. nat. p. 53. Bartholin. acta haffniens. v. 5. Obs. 97. Reamur acta. gallic. Ao 1714. p. 457. Haller in Boerhav. vol. 4. P. 13. Borelli demot. Anim. 2. propr. 219. Colbe cap. bon. spei. 3. p. 127. Kaemf. amoen. 514. (c) *In Venice, [p. 31] at the fish market, it is not sold under the authority of the health prefect, Behren diaet. 225. (d) It is most noteworthy how great a power lies in the retraction or retardation of the human spirit, even within those brief intervals we call moments of time. What such retraction of the spirit may effect in other cases is not relevant to discuss here; but in this particular instance, its effect is that, if the spirit is withdrawn and held back, the thing touching it will feel little or none of its stupefying power.*

38. *OSTRACION oblongus glaber, capite longo, corpore figuris variis ornato,* Arted. gen. 60. syn. 86. vid. Plate II. fig. III. [*OSTRACION* oblong, smooth, with a long head, and body adorned with various patterns].[247]

Orbis oblongus testudinis capite, Clus. exot. 14. Willugb. icht. 147. t. 1. 2. f. 3. Raj. pisc. 43. *Orbis levis oblongus, cinereis & fuscis maculis notatus.* Sloan. hist. 2. p. 279. t. 247. f. 1.

HEAD sloping, somewhat oblong, blunt, with four teeth, two on the upper jaw and two on the lower jaw, wide and converging. Nostril openings are prominent, each with a perforated sac or wart. The neck is thicker than in Clusius depiction, not narrow. The gill openings are located in front of and next to the pectoral fins.

TRUNK convex, with a rough dorsal surface, reticulated with pale and yellowish lines, and scattered dark spots on the sides. The abdomen is flat and somewhat wide, marked with scattered, excavated dots. The anus is located near the tail, in front of the anal fin.

Pectoral Fins with fourteen rays, wide and blunt. The dorsal fin has six soft and branched rays except for the first one. The anal fin is opposite the dorsal fin, with six branched rays, the first being very small. The caudal fin has nine branched rays and is entire.

39. *OSTRACION cathetoplateo-oblongus, ventre tantum a culeato & subrotundo.* Art. gen. 58. Syn. 86. [Ostracion perpendicularly flattened, oblong, with only the belly being keeled and somewhat rounded.].[248] [p. 32]

Orbis cauda productiore, dorso levi, ventre spinoso. Willugb. icht. 144. t. 3. t. 2. *Orbis lagocephalus* Grew. mus. 108. t. 7. f. 5. Raj. pisc. 43.

HEAD compressed, slightly flattened above, elevated at the eyes. Lips cover the mouth. Two teeth in the upper jaw and two in the lower, large, smooth, equal, and closing together.

BODY for the most part smooth, or rough with very rare and scarcely visible spines. Back keeled, greatly tapering towards the tail. Belly swollen, starting from the tip of the snout, sprinkled with very small, awl-shaped scales, becoming bristly here.

Pectoral FINS near the gill covers, with seventeen rays. Dorsal fin with nine rays. Anal and dorsal fins with nine rays. Tail fin whole with seven branched rays. See Plate II. fig. IV.

40. *OSTRACION conico-oblongus; aculeis undique longis teretiformibus imprimis in lateribus.* [Ostracion coni-

[246]*Raja torpedo* in SN 10.

[247]*Tetraodon testudineus* in SN 10.

[248]*Tetraodon lagocephalus* in SN 10.

cal-oblong; with long, rounded spines all over, especially on the sides][249], Art. Gen. 60. syn. 86. *Hystrix piscis*, Clus. exot. 137. t. 138. Raj. pisc. 42. Willugb. icht. 146. t. 1. f. 6.6. Sloan. hist. 2. p. 279. *Guamajacugvara*, Marcgr. bras. 158. Pis. bras. 300.

HEAD very small, with a pointed mouth, jaws toothless almost like in birds; inside a double row of blunt transverse teeth, in each row three. Eyes fairly large, above each six longer spines. Gill opening in front of and near the pectoral fins.

BODY round, covered all over with bristle-like awl-shaped spines, with a three-sided base, as long as a finger's breadth.

TAIL fin with nine rays.

COLOR mottled brown on the back, white underneath.

41. *TRACHINUS capite inermi nudo, pinnis pectoralibus coadunatis.* [Trachinus with a bare, unarmed head, and united pectoral fins].[250] [p. 33]

HEAD smooth. Mouth with small teeth in the jaws. Gill membrane with three rays.

BODY covered with broad, blunt, smooth scales, but with none between the dorsal fin and the head. A violet spot on the back above the gill covers. Anus with a soft, pointed lobe, distinguishing two openings, anterior and posterior.

Dorsal FIN anterior with six uniform rays, not sharp though somewhat rigid. Dorsal fin posterior with ten rays, except the first, branched, the posterior ones longer. Anal fin opposite and equal to the posterior dorsal fin, with nine rays, the posterior ones longer. Ventral fins with six rays, the inner ones longer, united at the base.

Since the eyes are very close together, we refer this fish to the genus Trachinus, although the head is unarmed.

42. *ASPREDO*[251].

HEAD broad, flattened, bare, uneven almost like that of a sculpin. Eyes very small and black. Mouth very wide. Teeth are two bony warts inside the upper lip, which is broad and long. Gill membrane with four small bones.

BARBEL lateral, one on each side, long, and another on each side at the upper edge of the mouth; then one on each side at the edges of the lower lip and two other small barbels under the throat. The sides of the thorax are bony, with no scales or fleshy substance on the head or thorax.

Pectoral FINS singular, located near the head, each with eight simple rays, the outermost of which is the thickest, somewhat blunt, marked on the outside with curved teeth, and on the inside with sharp, rigid, recurved teeth. Dorsal fin with five soft rays, including the first, which is simple and soft. Ventral fins far removed from the pectoral fins and almost at the sides of the anus, each with six soft rays. Anal fin extending from the anus to the tail, consisting of fifty-five simple soft rays. Tail fin oblong, bifurcated, narrow, with eleven rays.

It resembles in appearance and perhaps in genus the fishes which Marcgrave, Piso, Willughby, Jonston, and others proposed under the genus Bague.

This fish is singular for its very strong bones and the movable pectoral fins. See Plate II. fig. V.

43. *SPARUS fasciis quinque transversis, subfuscis.* [Sparus with five transverse bands, brownish].[252]

HEAD blunt. Teeth covered by lips. Gill covers scaly, with large scales. Gill membrane with six rays.

BODY oval, somewhat thick, covered with very broad scales, with a curved lateral line.

Dorsal Fin starts from the humeral region and extends almost to the tail, consisting first of thirteen strong, acute, nearly equal rays, and then thirteen longer soft rays. This fin is scaly at the base along its entire length and tapers towards the end. Anal fin consists of thirteen rays, the first and second being strong and sharp, the rest soft; the first ray is very short. This fin is also scaly at the base and more rounded towards the rear compared to the dorsal fin. Pectoral fins consist of fifteen soft and branched rays. Ventral fins have six branched rays, except the first which is short and sharp. Tail fin has fifteen highly bifurcated branched rays.

44. *LABRUS rostro reflexo, fasciis lateralibus tribus fuscis.* [LABRUS with a reflexed snout, three brown lateral bands].[253]

Jaguacaquare Marcgr. bras. 156. (an.). [p. 35]

HEAD compressed, snout pointed, turned upward, narrow. Eyes large, naked. Nostril openings paired, the anterior with a membranous covering. Teeth like a brush with very dense and equal bristles. Gill covers scaly. Gill membrane with six rather wide rays.

BODY of the fish oval, highly compressed, covered with rather large scales, with a narrow brown band extending from the shoulders through the eyes and head horizontally. Another less obscure band, wider, runs from the base of the dorsal fin towards the abdomen, behind the pectoral fins. Another similar band extends diagonally from the middle of the dorsal fin to the middle of the anal fin.

Dorsal FIN starts at the shoulders, extends to the tail, rounds off behind and becomes fleshy at the base, covered with scales, consisting initially of twelve strong, pointed, and sharp rays, each with an appendage behind a pointed membrane, followed in the same fin by twenty-one soft, jointed, and branched rays. Pectoral fins pointed, with four-

[249]*Diodon hystrix* in SN 10.

[250]*Gobius eleotris* in SN 10.

[251]*Silurus aspredo* in SN 10?

[252]*Chætodon rotundis* in SN 10?

[253]*Chætodon striatus* in SN 10?

teen unarmed rays. Ventral fins with six rays; the anterior ones pointed and sharp, the others branched and unarmed. Anal fin with twenty rays, the first three pointed and sharp, with an appendage behind a pointed membrane, the rest soft. This fin is also rounded and fleshy at the base or towards the part near the body. Tail fin small, rounded, entire, with seventeen branched rays, a transverse whitish band, and colored.

45. *LABRUS rostro reflexo, ocello purpureo iride alba juxta caudam.* [LABRUS with a reflexed snout, a purple spot with a white iris near the tail].[254]

Klip-visch seu soldalen visch, Willugb. icht. 1. t. 5. f. 4. Raj. pisc. 151. [p. 36]

Its shape is oval with a pointed snout, scaly gill covers; fins exhibit a consistent form, and both the dorsal and anal fins have a scaly base; dark band runs through the eyes, and oblique dark stripes extend from the lateral line, agrees with the previous description. However, it differs in that it lacks the second and third lateral bands, the dorsal fin has thirteen sharp rays besides eighteen soft ones, and the anal fin has seventeen rays besides the first three sharp ones. Finally, a large, round brown spot with a white circle around it is placed on each side towards the tail, marking both the edge of the dorsal and tail fins simultaneously. Thus, it has many similarities with Orphus, Sparus, Sargus, *and* Medanurus, *all of which belong to the same genus.*

It is one of the best fishes of the Indies, with white, firm, tasty, and healthy flesh.

46. *LABRUS immaculatus, pinnae dorsalis radiis spinosis decem.* [LABRUS without spots, with ten spiny rays on the dorsal fin].[255]

HEAD compressed. Mouth large. Upper jaw extended on both sides and folded back in front. Teeth small, distinct, and spaced apart, especially on the sides. Eyes naked, large. Gill covers scaly, with three teeth behind. Gill membrane with seven rays.

BODY ovate-oblong, bluish, with a curved lateral line.

Dorsal FIN extends from the shoulders almost to the tail, of uniform height throughout, consisting of twenty rays, the first ten of which are strong and sharp, the first shorter than the others. Pectoral fins with twelve soft, branched rays. Ventral fins ovate, with six rays, all branched except the first which is strong, short, and sharp. Anal fin consists of thirteen rays, the first three of which are strong and sharp. This fin ends [p. 37] in a wider space towards the tail than the dorsal fin. Tail fin with seventeen slightly bifurcated rays.

47. *LABRUS oblongus, pinnae dorsalis unico radio inermi.* [LABRUS oblong, with a single unarmed ray on the dorsal fin].[256]

HEAD compressed, smooth. Teeth small, the interior lateral ones longer. Gill membrane with six rays.

BODY compressed, white, long, scaly; lateral line running longitudinally from the head towards the tail near the back, descending near the tail and crossing the middle side to the tail.

Dorsal FIN starts near the head and extends almost to the tail with twenty-one rays, all simple and sharp except the last, which is bifid and soft. Pectoral fins with twelve lanceolate rays. Ventral fins located below the pectoral fins, lanceolate, with six rays. Anal fin with fifteen rays, the first being the shortest. Tail fin entire, with twelve rays.

48. *BLENNIUS pinnis ocularibus subulatis, pinnae ani ossiculorum XXVI.* [BLENNIUS with subulate ocular cirri[257], anal fin with twenty-six small bones].[258]

HEAD compressed, with very small tubercles on the forehead and sides. Forehead blunt. Teeth numerous, parallel, slightly touching, equal, in the front part of the mouth above and below; on each side of the lower jaw, there is a longer curved canine tooth. Lips cover the teeth. Eyes large. A long, subulate cirrus extends from the middle of the head to the upper corner of the eye on each side.

BODY oblong, covered with naked skin, sprinkled with rusty spots or patches. Anus in the middle of the body.

Anal FIN extends from the anus to the tail, with twenty-six soft and simple rays, the first and second ending in a papilla. Dorsal fin extends from the head to the tail, with thirty-four simple, unarmed rays, though the first twelve are slightly more rigid. This fin is nearly of the same width throughout and rounded at the back. Pectoral fins ovate, with fifteen simple, soft rays. Ventral fins under the throat, subulate, with two rays. Tail fin entire, with eleven branched rays. This species closely resembles *Blennius* with two small fins near the eyes, and the anal fin has twenty-three small bones. Arted. Gen. 26. syn. 44.

49. *BLENNIUS pinnis ocularibus brevissimis palmatis, linea laterali curva.* [BLENNIUS with very short palmate ocular cirri, a curved lateral line].[259]

HEAD compressed, sloping, with a very short, blunt, split fleshy lobe above each eye. Teeth numerous, small. The lateral line begins above the head and runs near the back along the length of the abdomen, then curves along the side and runs longitudinally through the middle.

Dorsal FIN starts near the head and extends to the tail,

[254]*Chætodon capistratus* in SN 10.

[255]*Perca radula* in SN 10.

[256]*Labrus linearis* in SN 10.

[257]*Pinnis ocularibus* in the original.

[258]*Blennius cornutus* in SN 10.

[259]*Blennius superciliosus* in SN 10.

with forty rays, the first thirty-four sharp, and the last six somewhat longer and soft. Pectoral fins rounded, broad, with twelve rays. Ventral fins near the head, each with two bifid rays. Anal fin with twenty-four rays. Tail fin entire, with twelve rays.

50. *CALLICHTHYS.*[260] Plate II. Fig. I.

Tamoata brasiliensibus, lusitanis Soldigo, Marcgr. bras. 150. t. 150. Willugb. icht. 211. t. 3. f. 8. Raj. pisc. 78.

HEAD round, depressed in front, with various sutures arranged in a circle around a rather large central scale. [p. 39] Small nostril openings in front of and near the eyes on the sides of the head. Each side of the mouth has a pair of cirri, with the posterior one being longer; both are longer than the head. Gill membrane with three rays.

BODY compressed, somewhat oblong, with a double row of scales on each side, one above and one below the lateral line. These scales are very broad, twenty-six in number, nearly linear, with a slightly ciliated margin, and the last ones near the tail are incised with a triple suture. Abdomen rather broad, smooth, naked.

TAIL rounded.

Anterior dorsal FIN with eight soft rays, the first being the shortest and rough in front. Posterior dorsal fin near the tail, small, with a single rough, rigid, but not sharp ray in front. The back before this fin is compressed and rough. Pectoral fins rounded, with seven rays, the anterior thicker and denticulated on both sides. Ventral fins on each side near the anus, with five jointed rays. Anal fin near the tail with five rays, the anterior one rough.

Callichthys is a genus of acanthopterygian fish, with a gill membrane with three rays, a cirrhose mouth, and only a double row of scales on the sides.

This fish, lacking water in its river during dry times, travels overland to seek water. Marcgr.

51. *GYMNOTUS*[261] Plate II. f. 6. Arted. gen. 25. Syn. 43.

Carapo. Pis. bras. 72. Marcgr. bras. 110. Willugb. icht. 115. Raj. pisc. 41.

HEAD smooth, ovate, small, sprinkled with soft warts. Teeth very tiny. Lips fleshy, the lower one longer; the upper lip extends a short cirrus or tentacle to the side. Nostrils simple [p. 40] and almost elevated above the head. Eyes small, covered with skin. Gill membrane consists of five broad bones. Gill openings fairly wide.

Body long, about a foot, narrow, back thick, rounded, and fleshy, with a keel below, gradually tapering towards the tail and becoming subulate. Thus, it has the shape of a knife. Scales orbicular, imbricate, with annular and transverse ridges.

LATERAL LINE toothed downward, with soft denticles perforated at the tip, blunt, secreting fluid. Anus located under the throat near the head.

No dorsal or ventral FINS. Pectoral fins small, lanceolate, with thirteen soft rays. Anal fin starts near the head about a thumb's breadth from the anus and extends to the tip of the tail, linear, with about two hundred soft rays. No caudal fin, hence the peculiar appearance of the fish with only the lower side of the tail being rayed, and entirely bare above.

52. *MURAENA pinnis pectoralibus carens*. [MURAENA, lacking pectoral fins].[262] Art. Gen. 26. sy. n. 41. *Muraena* Rondel. pisc. 402. Bellon. aquat. 158. t. 160. Jonst. pisc. tab. 20. f. 4. *Muraena sive Conger brasiliensibus* Seb. thes. 2. p. 72. t. 69. f. 4. 5.

EYES covered by a membrane. Nostril openings solitary on each side, cylindrical and prominent, located nearly between the eyes, emitting tentacles similar to those seen in snails; behind the head on each side is a hole for the gills. Teeth very small, arranged in a double row in the front part of the mouth, and in a single row in the inner part.

BODY eel-shaped, compressed, brown with white transverse and lateral lines, finely cancellated. The abdomen is more spotted. [p. 41]

TAIL lanceolate and tapered, without a distinct fin.

Dorsal FIN starts from the head and extends in a continuous series to the tip of the tail, being adipose. Anal fin starts from the anus and ends at the tip of the tail, similarly adipose. Both fins are of the same color as the body. No pectoral or abdominal fins.

53. *ECHENEIS*[263] Arted. Gen. 15. Syn. 28. *Remora* Rondel. pisc. 436. Bellon. aquatil. 440. Olear. mus. 42. 35. f. 2. Jonst. pisc. 16. t. 4. 3.

HEAD entirely flat above, this flatness oval with a wide membranous margin encircling it all around. This flat area is longitudinally divided by a raised line into two parts, from which transversely towards the margin on each side, eighteen or nineteen transverse membranous folds project backward, rough forward. Mouth roughened by very small teeth, blunt. Lower jaw longer. Nostril openings double, one with an elevated margin, the other wide. Eyes round at the side of the head. Gill covers smooth. Gill membrane with nine bones.

Pectoral FINS subovate, with twenty-five rays. Ventral fins more acute, connected at the base, each with five branched rays. Anal fin with twenty-one rays. Dorsal fin from the region of the anal fin, very similar in appearance and size, with twenty-four rays. Caudal fin sharp, slightly bifurcated, with sixteen rays.

The fish is covered with soft skin, rarely reaching a span in length.

Obs. Ancient sailors believed this fish attached to keels

[260] *Silurus callichthys* in SN 10.

[261] *Gymnotus* in SN 10.

[262] *Murœna helena* in SN 10.

[263] *Echeneis remora* in sn 10.

would slow down ships, hence they called it the Remora; this name was retained in the Adriatic Sea for a ship of Praetorian Antonius and later Gaius Caligula. [p. 42] Bellon. aquat. 40; but fragments.

54. *EXOCOeTUS*[264] Arted. gen. 8. syn. 18. spec. 35. *Mugil alatus* Rond. pisc. 267. *Hirundo*, Bellon. aquat. 198. *Hirundo Plini*, Jonst. pisc. 66. t. 18. *In Swedish*, Flygande-Fisk.

APPEARANCE and size similar to a herring. Gill membrane[265] with ten small bones.

Dorsal FIN positioned behind the center of balance, with eleven rays, extending across the diameter of the anal fin. Anal fin also has eleven rays. Ventral fins located in the middle of the abdomen with six rays. Caudal fin, which is highly bifurcated, has fifteen rays. Pectoral fins in this species are distinctive, with seventeen rays. The upper rays reach the length of the beginning of the tail, almost the entire length of the fish; the lower rays gradually become shorter. With these fins, the fish glides.

Obs: *A perfect description is found in Artedi's species. Authors incorrectly place two lateral fins at the base of the anal fin.*

55. *ARGENTINA pinna dorsali pinnae ani opposita.* [ARGENTINA, with the dorsal fin opposite the anal fin].[266]

HEAD oblong, compressed, flattened above. Mouth internally roughened by small teeth. Lower jaw slightly longer. Gill membrane with approximately ten rays.

BODY long, somewhat cylindrical, silvery, covered with thin scales. Back somewhat thicker, distinguished by two longitudinal lines on the sides.

Dorsal FIN positioned towards the tail, with approximately fourteen unarmed rays. Pectoral fins lanceolate, with twelve rays. Ventral fins round, short, with six rays, located in the middle of the length. Anal fin with approximately fifteen rays, opposite the dorsal fin. Caudal fin bifurcated, with fourteen rays.

Characteristic of the genus includes a gill membrane with approximately ten rays, a long cylindrical compressed body, and ventral fins closer to the anus than to the head.

56. *CYPRINUS pinna ani duplici, caude bifurca.* [Cyprinus, with a double anal fin and a bifurcated tail].[267] Act. stockholmens. 1740. p. 403. t. 1. f. 3. 4. 5. 6. 7. 8. Faun. Suecica 331. t. 2. *In Swedish, Chinesiska Gull-Fisken.*

57. *SYNGNATUS corpore quadrangulo, pinna caude carens.* [Syngnatus, with a quadrangular body, lacking a caudal fin].[268] Arted. gen. 1. syn. 1. sp. 3.

Hippocampus Raj. p. 45. Rondel. insect. c. 9. p. 114. Wilugb. p. 157. *Hippocampus aculeatus* Raj. p. 45.

Obs. *Dorsal fin with seventeen rays, pectoral fin with sixteen rays.*

Class V.

INSECTA.

58. *GRYLLUS apterus, thorace spinoso, femoribus punctatis.* [Gryllus, wingless, with a spiny thorax and dotted femurs].[269] *Locusta talpa lapensis, pedibus longis.* Peti. Gazoph. 21. t. 13. f. 7.

HEAD reflexed. Prominent eyes. Setaceous antennae, shorter than the body. Anterior lip of the mouth doubled, the lower part longer and somewhat rounded. Front maxillae inwardly toothed, thick; the posterior ones shorter, armed with a double clawed digit, and externally provided with a somewhat long antennula; the lower maxillae very simple, with club-shaped antennules.

THORAX flat above, compressed on the sides, elongated behind, rounded, with spinous angles all around in front [p. 44] and on the sides pectinately spiny, along with four spines on the disc. Below, between the front thighs, two spines.

Abdomen ovate, with twelve incisures, each segment provided with several spines towards the margin, and arranged in seven longitudinal rows along the back; below the abdomen, seven or eight transverse folds undulate at the sides. Anus trivalvular, equipped on each side with a sharp conical tube, below a two-valved sword, smooth, slightly curved, convex on the sides, and marked on each side with a groove, serrated at the apex on both sides along the length of the body.

LIMBS long, especially the hind ones. Femurs sprinkled with callous dots. Tibias spiny. No traces of wings or elytra; hence we believe it to be and are sure it is wingless. Size like that of a mole-cricket[270].

59. *ARANEUS hirsutus, thorace orbiculato convexo: centro transversim excavato.* [Araneus, hairy, with a rounded convex thorax: the center transversely excavated].[271]

Aranea maxima ceilanica insigne gerens Seb. thes. 1. p. 109. t. 69. f. 12. *Phalangium monstrosum* Laet. mer. 5.6. 9. t.

[264]*Exocœtus alatus* in SN 10

[265]*Membrana branchiostega* in the original.

[266]*Argentina* in SN 10. *Argentina sphyræna* was the only species in this genus, in that time monotypic.

[267]*Cyprinus aphya* in SN 10.

[268]*Syngnatus typhle* in SN 10.

[269]*Gryllus (Tettigonia) pupus* in SN 10.

[270]*Gryllo-talpæ* in the original.

[271]*Aranea avicularia* in SN 10.

570. *Aranea americ.* Seb. thes. 1. p. 110. t. 69. f. 3. Marian. in folio, p. 18. s. 18. *Phalangium americ.* Clas. posth. 47. t. 47[272]. Worm. mus. 244. t. 244. *Nahmdu guacu* Marcgr. bras. 248. Olear. mus. 29. t. 17. s. 3.

MOUTH with two jaws, pointed, curved, awl-shaped, thick at the base, black.

THORAX rounded, more convex anteriorly, bordered, with the disc almost in the middle or towards the posterior more excavated, with a deep transverse cleft, from which, hairs covering the thorax, are reflected as if from the center all around towards the margin. Eyes close together in the anterior and upper part of the thorax, forming a wart. [p. 45] Abdomen ovate, more convex above, hairy, terminated with two oblong appendages.

LIMBS eight, nearly equal, hairy (our authors mention two claws on each foot).

Size like that of a flying stag beetle, entirely covered with long, bristly hairs, rust-colored.

69. *SCORPIO pectinum denticulis XIII.* [Scorpion with a pecten of thirteen teeth].[273]

Scorpio Javanicus major pilosus e nigro coerulescente splendens. Peti. gazoph. 20. t. 13. f. 2. *Scorpio indicus niger magnitudine cancrum fluviatilem aequans, Herm.* Leyd. 144[274]. *Scorpio ceilonicus niger maximus,* Sibb. mus. 116. *Scorpio ceilonicus* Seb. thes. 1. p. 112. t. 70. f. 4.

ROSTRUM consists of two sessile chelae, toothed inwards. Two contiguous eyes in the center of the thorax. Then three distinct eyes on each side of the anterior margin of the thorax. Lower lip of the mouth four-cleft, connivent.

ABDOMEN with seven segments on the back and five in the abdomen, the first four of which are marked with an oblique lateral line. Pectines located where the thorax joins the abdomen, with two white ones underneath, each equipped with thirteen small teeth. Hairs scattered and erect throughout the body.

TAIL with six angular segments, rough with tubercles, the last of which is armed with a curved sharp spine, the penultimate segment being longer than the others.

LEGS, four on each side in addition to the chelae, each terminating in two acute claws and equipped with a callous tubercle at the bend of the joints. Chelae hispid with hairs, fingers sinuate on the inner side. [p. 46]

COLOR brown over the whole body.

SIZE of a river crab.

Obs. *Differs from the larger scorpion of the East Indies described by Swammerdam in that Swammerdam's scorpion has a tail with only four segments.*

61. *SCOLOPENDRA teres, pedibus utrinque XCVI.* [Scolopendra cylindrical, with ninety-six legs on each side].[275]

Millepeda orientalis omnium maxima. Seb. thes. 1. p. 131. t. 81. f. 5.

HEAD very small, obtuse, bordered at the tip. Antennae club-shaped. Eyes crescent-shaped, black, with minute rough dots.

BODY with forty-eight segments, excluding the tail and head, each segment pale with a yellow margin.

TAIL obtuse.

LEGS two on each side of each segment, short and ciliated underneath.

Length and thickness of a finger.

62. *SCOLOPENDRA plana, pedibus utrinque viginti.* [Scolopendra flattened, with twenty legs on each side].[276]

Solopendra capensis major lutescens, cauda longiore. Peti. gazoph. t. 20. f. 2. Milipeda african. Seb. Thes. 1. p. 130. t. 81. f. 3.

HEAD suborbicular, flat, underneath equipped with two very strong, very sharp, black-tipped jaws. Eyes located at the sides of the head behind the antennae, each consisting of three eyes. Antennae awl-shaped, with seventeen (or more, as they appear mutilated at the tips) subovate segments.

TRUNK consists of twenty-one segments, the second of which is the shortest and linear, the rest quadrangular; underneath the segments are similar to the dorsal ones but rounded at the back. The sides are rough, with a respiratory pore in every alternate segment on each side.

LEGS twenty on each side, short, with simple, sharp claws. Besides these legs, there are two longer ones on the tail, spiny at the base, or rather small horns.

COLOR yellowish.

Class VI.

VERMES.

63. *SEPIA*[277] Faun. Suec. 1281.

ROSTRUM almost like a bird's, surrounded by two black semiconical maxillae with an circular, lacerated mouth. Above the head, there are tentacles, six of which are connected at the base, awl-shaped, long, and equipped with

---

[272] *Phalangium americ.* description in pp. 46-47; figure in p. 46.

[273] *Scorpio afer* in SN 10.

[274] Probable reference to Hermann (1687), where there is no reference to any scorpion. However, in an anonymous publication, for which I could not locate an author or place of publication, with a probable date of 1700 (Anonymous, 1700), under number 144 (on an unnumbered page), "*Scorpius Indicus niger magnitudine cancrum fluviatilem æquans*" is cited.

[275] *Julus teres* in SN 10?

[276] *Scolopendra morsitans* in SN 10?

[277] *Sepia officinalis* in SN 10.

numerous orbiculate, umbilicated, petiolate cotyledons on the lower side. In addition, there are two tentacles very similar to the aforementioned ones, but longer, also equipped with cotyledons on the lower side.

SHELL soft, oval, and margined, within which the posterior part of the body is hidden. Within the substance of this shell, a cuttlebone is hidden from the back. The anus is a prominent cylinder below the head.

ARMS or filiform tentacles, solitary on each side, arising between the upper and lower tentacles near the head, very long, soft, dilated at the apex, and equipped with cotyledons very similar to but larger than those on the lower side.

64. *APHRODITA nitens*. [Aphrodita radiant].[278] Faun. Suec. 1284.

Eruca seu scolopendra marina corallina multipes. Seb. thes. 1. p. 141. t. 90. f. 1. 2. [p. 48]

MOUTH equipped with several fleshy bristles resembling antennae.

BODY ovate, convex above, covered with scattered, shining, purplish, awl-shaped spines and yellow-green, shining hairs of the same length; in the middle of the back, however, it is hairy without spines, and tomentose. In the center, there is a small opening. Under the skin of the back, there are folded membranes resembling the gills of crabs.

LEGS about forty on each side, or as many parallel cones inserted along the sides and margin of the body, ending in a bundle of spines at the tip and equipped with a soft, bristle-like appendage underneath.

We omit reviewing animals previously well-known and sufficiently described, such as crocodiles, flying dragons, etc., as well as incomplete or mutilated animals that have not yet undergone metamorphosis, such as butterfly larvae and Indian moths[279], among which the following are notable:

Larva ovate, with sixteen legs, and covered with erect wool.

Larva oblong, with sixteen true legs and four additional scattered ones, with a smooth body, spotted behind.

Larva oblong, with sixteen legs, and a body armed with distinct spines.

[s. n.]

[278] *Aphrodita aculeata* in SN 10.

[279] *Phalæna indicarum* in the original, name under which various nocturnal butterflies (moths) were grouped.

To the Most Esteemed Mr. Author,
A fellow citizen and a most faithful friend in all things

It is, I know not how, the nature of friendship that it cannot but affect those with whom it is shared with a deep joy at their reciprocal success in life. Hence you will easily understand why, since it has been my great fortune to have you as my most faithful friend from our earliest days, as they say, I should not let pass this excellent opportunity to demonstrate so beautifully the example of time well spent. The joy with which I am filled on this account is as great as it can be, and it lacks nothing but to see you as soon as possible showered with the most deserved rewards. So I pray and will always pray,

Yours entirely,
CHRISTOPH. DEPIEN FIL.
[s. n.]

To the Young Man of Elegant Genius,
distinguished in manners and letters,
D.n. LAURENTIUS BALK,
His most beloved kinsman.

Since it is almost universally said that all things are common among friends, it is not surprising that my pleasure aligns with the joy of others, with whom I am connected by kinship, familiarity, or some other bond, dear Brother. For, from the time we began to serve under the banner of this Academy together, I have enjoyed not only the relationship of kinship but also the sweetest companionship with you, and I should excel in my enthusiasm and affection towards you far more than others, as much as shared scholarly pursuits and mutual tests of trust usually strengthen friendships.

Moreover, there is a richer cause for rejoicing. Although we always see many others striving in honorable pursuits, we find ourselves in a rare situation where such exalted titles, as those proudly borne by your dissertation, are bestowed upon us. Indeed, your work first presents to us the Museum ADOLPHO FRIDERICIANUM, *reveals the magnificent benefaction of the most illustrious Prince to the whole literary world, and celebrates it. Furthermore, as it is now apparent how eagerly your Minerva desires to visit such lofty seats, and how the brilliant mind is engaged in grand, not ordinary labors, I foresee that as you advance through this august stage, fortune will one day bestow upon you a greater reward. In the meantime, I sincerely congratulate you on your well-begun efforts, and I continually wish for the same to progress equally successfully. For the sooner I see you reaching the final goal, the sooner I will begin to rejoice eternally. Farewell.*

Adrianus Ziegler

Tab. I.

Tab. I

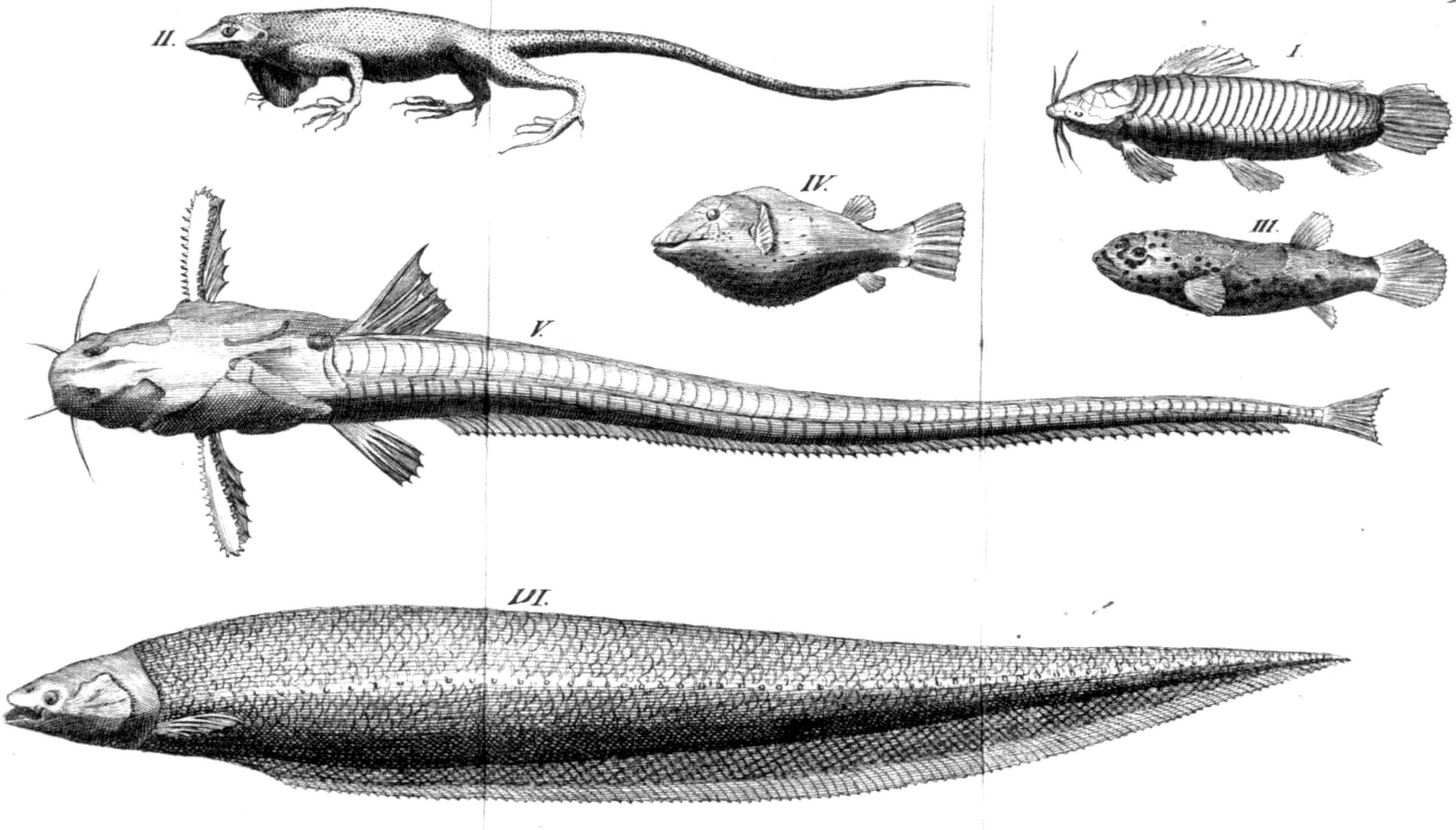

B. C. D.

SURINAMENSIA GRILLIANA

CUM CONSENSU AMPLISS. FACULT. MED. IN REGIA ACADEMIA UPSALIENSI,

PRÆSIDE

VIRO CELEBERRIMO ET EXPERIENTISSIMO,

D:N. DOCT. CAROLO LINNÆO,

S:Æ R:Æ M:TIS ARCHIATRO,

MED. ET BOTAN. PROFESS. REG. ET ORD. ACAD. JMPERIAL. REGG. MONSPEL. BEROLIN. STOCKHOLM. ET UPSAL. SOCIO,

SPECIMINIS ACADEMICI LOCO

CURIOSIS EXAMINANDA SISTIT

ALUMNUS OXENSTIERNIANUS

PETRUS SUNDIUS NIC. FIL.

STOCKHOLMIENSIS.

IN AUDIT. CAROL. MAJ. D. XVIII. JUNII. ANNI MDCCXLIIX.

H. A. M. C.

HOLMIÆ,

Typis LAURENTII SALVII.

Sundius, P. 1748. Surinamensia Grilliana. Holmiae. Laurentii Salvii. [5] + 1-34 + 1 plate.

In the dissertation on the donation of Claes Grill, Petrus Sundius (1748) named one *Caecilia*, one *Amphisbaena*, two *Anguis*, seven *Coluber*, one *Cenchris*, one *Crotalophorus*, and two *Lacerta*, and included references to three unnamed specimens (numbers 5; 7 and 8). Again, Linnaeus re-edited this dissertation in the first volume of the *Amoenitates Academicae* (Sundius, 1749) and, as in the previous case, was frequently cited from this source in the tenth edition of the *Systema Naturae* (Linnaeus, 1758). In turn, a century later, Lönnberg (1896) noted that the individuals 8; 9; 10; 11; 13; 14; 16; 17 and probably 18 were lost, while Wallin (2001) found only *Caecilia tentaculata, Anguis scytale, Coluber aesculapii, Coluber ahaetulla, Coluber cobella*, and *Lacerta chamaeleon* in the collections of the Uppsala Museum. *Boa constrictor, Caecilia tentaculata, Coluber Ammodytes* and the Egyptian *Gryllus cristatus* are all figured in a plate annexed to this dissertation.

W. G. h.[280] Surinamensia Grilliana. With the consent of the most eminent Faculty of Medicine in the Royal Academy of Uppsala, *under the presidency of the* celebrated and experienced, Mr. Doctor Carl Linnaeus, Physician-in-Chief of His Majesty's Royal College, Professor of Medicine and Botany, member of the Royal Academies of Monspel, Berlin, Stockholm, and Uppsala, the Oxenstiernian student Petrus Sundius son of Nicholas, From Stockholm presents, for curious examination, an academic specimen. In the auditorium of Charles the Great, on June 18th, 1748 C. H. B. N.[281] *Stockholm*, by the Press of Laurentius Salvius.

[s. n.]

To the renowned and most distinguished Lord Claudio Grill[282], citizen of the metropolitan city. known for his integrity in business dealings and his outstanding reputation, the very prudent possessor of the ironworks in Iggesund, etc. and a most worthy member of the Royal Academy of Sciences.

I cannot but greatly congratulate myself on the very auspicious and convenient occasion that has been offered to me to openly declare how much I owe to you, most kind Sir, both in my own name and on behalf of my father. For as long as our parents lived in Stockholm, you bound them to yourself by so many acts of kindness that I freely confess my inability to [s. n.] describe them. But lest you should think that this is the only reason why I have dedicated this dissertation to your illustrious name, behold another, greater and more public reason: our own Camoenae[283], our own Muses, acknowledge themselves to be most closely bound to you because of the rare and precious gift of Surinam, by which you have not long ago made their treasures more abundant and richer. Therefore, what has been done for us by your authority returns to you with the highest right. Indeed, we suspect that your generosity, which could only arise from the distinguished love you have for literature and its followers, is all the greater, [s. n.] the higher your virtue and erudition have placed you above the common ranks of powerful merchants.

As for myself, who judged that I could be of service to you in no better matter than in examining and illustrating this work of yours, even though I know that it does not fully respond to so many and such great services of yours, nevertheless, if you are not displeased with my humble labor, I

[280]The original *B. c. D., Bonus cum Deo,* can be translated as "With God's favor" or "With God's help." Thus, W. G. h.

[281]*H. A. M. C., Horis, ante meridiem, consuetis* in the original; translated as At the Customary Hours Before Noon, thus C. H. B. N.

[282]Clæs Grill (1705–1767). Swedish merchant, collector and patron of sciences and arts; member of the Royal Swedish Academy of Sciences.

[283]In Roman mythology, originally goddesses of childbirth, wells and fountains; also, goddesses of pœtry, often associated with the Greek Muses. See note 111.

will be glad, and I will never cease to offer pious prayers for your and the most illustrious family of the Grillians' perpetual happiness, remaining until my last days.

Of Your Illustrious Name
The most devoted cultivator,
Peter N. Sundius.

[s. n.]

TO THE MOST POLISHED IN CHARACTER AND LEARNING
AUTHOR,
A BELOVED NEPHEW FROM HIS SISTER.

What new wonders, what marvels do you relate,
Dreadful monsters found on exotic soil's gate?
What does your land of Suriname now bring?
Why, kind Peter, deal with beasts so menacing?
Let the sloth be slow on the ground, swift in the tree,
And let its discovery be pleasing music, free.
Long live the serpent king, the rattlesnake queen,
Named by its tail's rattle, ever keen.
Whether its deadly breath stops swift deer,
Or swallows whole goats, jaws open, no fear;
Let rattles trouble the serpent-born crowd,
As long as they hold no power, not allowed.
Let the Gymnotus be known to the Indian land,
Or the fiery snake, both grand;
These rare dishes do not our meals grace,
Let Caecilian be smooth-skinned in its place.
Let the Gryllus wear a hard shell's face,
Your serpents have scales in thousands' embrace.
Whether snake is scaly or spider is hairy,
To us these distinctions are necessary.
Indeed, we're delighted by all, and the whole
World benefits from knowing God's role.
To know which animals harm, which ones heal,
Sudden wounds' remedies, swift hand's appeal.
This book, to be studied both night and day,
Filled with nature's blessings, in every way.
But alas! known to few, revealed to less,
Though always present, open to bless.
How fortunate, Sundius! young man who sees,
Secrets and rites of the Great Goddess with ease.
Pallas and the Muses applaud your quest,
And behold! Apollo himself approves your test[284].

O. A. BURMAN.

To the author[285].

When you arrived here, you shared with me your intention of presenting an academic lecture, which naturally awakened a sense of joy in me, as it is usual among those who are so closely related by blood. I immediately felt obliged to express my happiness about the honor you are soon to receive. Therefore, please allow me to examine your beautiful work, and give me some specific reason to ponder upon it. As I opened your pages, I found that you had entered the kingdom of nature to admire its wonderful creations, which must be understood before their value, usefulness, and harm can be determined. I consider you fortunate, my brother, to engage in such an admirable activity, especially at a time when, under the guidance of an extraordinary leader, you have produced an excellent and informative collection of animals that are produced in another world, and that a widely acclaimed Swedish patriot has brought to us as a valuable contribution to our Nordic wisdom. May heaven bless your endeavors and allow you to enjoy the honor and the fruits of your labor soon! I hope that my sincere wishes will come true.

Stockholm, 14th June
1748.

NILS SUNDIUS Nilsson[286].

[p. 1]

PREFACE.

It is well-known that no region, whether easily accessible or not, lacks an abundance of its own species of stones, plants, and animals. This is why we have witnessed so many journeys undertaken to the most remote parts of the world, both on land and sea. Indeed, if we turn our attention and focus to the amazing works of the highest Creator of all things, we will find that even in our native soil, although there may be a vast number of species, there are only a few that we know of. However, the further we penetrate into remote areas, the more we are struck with admiration, as we often come across new things that are vastly different from those known to us in our homeland. Who, born under the Arctic or coming from our North, would not find new and foreign things when approaching the lands subject to the Equator? And what about the inhabitant of the middle of the world, would he not experience the same thing if he ever had the chance to feel our snow?

Indeed, there is no shortage of travelers who have gone to many regions, both by land and sea, and observed many things worthy of seeing, and have written them down in an orderly manner. However, most of these accounts contain only descriptions of places, the customs of the inhabitants, their trade, and similar topics, with very little attention given to what could shed light on natural history. It is not surprising that they have left these things untouched, as the vast book of nature, though open to all, can only be read by very few, and only in those parts that lead us to a knowledge of animals, which, even in the most well-known cases, we still lack accurate knowledge of.

And we will not be deceived if we say that this was the reason why the just evaluators of natural things have invested so much effort in collecting rare animals from here and

[284]Free translation in English verses.

[285]In Swedish in the original.

[286]Nils Sundius Nilsson (1687-1761). Vicar in the *Storkyrkoförsamlingen*, a congregation in the Diocese of Stockholm.

there, especially aided by travel, satisfying their desires and requests, and bringing home a considerable amount of foreign animals preserved in wine spirit[287], which will fascinate spectators of a certain nature or move them by some hidden impulse, even those for whom there is no or very little understanding of what they see, the rarity of the things itself attracting admiration.

Although we can rightly admire the most well-equipped museums of our age, such as Vincent's Museum in Harlem, Sloane's in London, Seba's Amsterdam Museum, and the St. Petersburg Academy, it is regrettable that many of their exhibits are either not accurately or not clearly explained, or still remain hidden in their own obscurity.

As far as we are concerned, such rare collections were not available to us until the donations of our most gracious heir, the Prince[288] of the Holy Roman Church, and formerly that of the Chancellor, Senator, and Count Carolus Gyllenborg[289], which enriched the Botanical Garden of Uppsala with extremely rare and exquisite collections, which we published in print some time ago, for the benefit and use of the public. (*)[290] [p. 4]

With such examples, and being himself inflamed with a great love and inclination towards good literature, Claudius Grill, a citizen of Stockholm, member of the Royal Academy of Sciences, and very distinguished by his family, fortune, and ancient customs of his municipality, in the previous year of 1747, donated to the most celebrated President a collection of Surinamese animals, not at all common and of great value, which added a considerable contribution to the Upsala Museum.

The author of this collection, made in the Surinam region of America, was Father Gerret[291], born in Suriname and founder and owner of the most prominent seminaries there. He was particularly celebrated for being among the first to successfully establish coffee plantations in America. He, along with his son of the same name, collected this treasure trove, and given his long-standing friendship and literary exchange with our own Claudius, a highly esteemed and beloved patron of the arts in our city, they sent him a distant gift; Grillius gave it to us, [p. 5] to be kept not in a better place than in the Uppsala Garden.

From this, it is clear that this entire collection of Gerret only includes those animals that were produced by the Surinamese land (*)[292], which is exhausted under the hottest climate. There is no doubt that this land also supplies more and rarer animals, most of which would certainly have been carefully observed and discovered if either the famous President had decided to undertake the task of the Medical Officer of the West Indies, which was offered to him by the most accomplished Boerhaave himself a decade ago; or if Johann Bartsch, to whom this task was later transferred, and who was educated in this field of science by the distinguished instructor, had been allowed to experience a gentler sky and better fortunes[293]. [p. 6]

Although I did not have the fortune to fulfill my wish of observing Grilli's animals in their natural habitat in Suriname, as an interpreter of his collection, I am still grateful for the opportunity to put to the test the powers of a young mind on such a noble subject. I am confident that my efforts will not be in vain and that my contribution will not be trivial or unhelpful to the learned community, as I will not merely rehash old and commonplace information, but rather present new insights and discoveries worthy of the attention of the curious and capable. Furthermore, it would be a disgrace to let such a precious and rare treasure, brought to us from a distant and inaccessible corner of the world, lie hidden and neglected in its box, either in the Tower of Pallas[294] or elsewhere. But let it be known that I have achieved my goal, and

---

[287]Ethanol.

[288]Refers to the first donation of the collection of the then prince (later king) Adolf Fredrik of Holstein-Gottorp (1710-1771), originally housed in his palace at Ulriksdal.

[289]Carl Gyllenborg (1679-1746). Swedish Count, statesman, collector and patron of sciences and art. The donation of his collection and library was in two installments, in 1744 and 1745.

[290]Footnote p. 3: (*) Museum Adolpho-Fridericianum. L. Balk 1746. Amphibia Gyllenborgiana. B. R. Hast. 1745.

[291]Pater Gerret. Not identified.

[292]Footnote p. 5: (*) Suriname, also known as "Belgium of the Americas" is a colony located on the Caribbean coast of the South American continent. The name is derived from a river that flows into the Northern Sea (commonly known as the North Sea), which washes the shores of the Americas. Suriname is situated at the sixth degree of latitude north of the equator and at 78°40' longitude west of the Uppsala Observatory or 43.5° west of the prime meridian that passes through the island of Ferro. This location makes it very close to the equator and gives it approximately equal amounts of daylight and nighttime throughout the year. Nevertheless, Suriname enjoys a milder climate due to the cooling effect of the nearby sea, which prevents the excessive heat that can scorch other regions subject to the same level of heat. NOTE WITHIN A NOTE: The Caribbean sea was considered part of the Northern Sea by diverse cartographers up to the middle of the XVIII century; see, among others, Gaztambide-Géigel (2004).

[293]When Linnæus was in Leiden, Herman Bœrhaave (1668-1738) offered him a trip to Suriname in his capacity as a physician, which he declined. This position was taken by Johann Bartsch (1709-1738), a German physician and botanist, who traveled to Suriname, dying six months after his arrival.

[294]In this context, *Pallas* refers to one of the epithets of Athena, Παλλὰς Ἀθηνᾶ, Greek godess of wisdom, strategy, and war. The "Tower of Pallas" symbolizes a stronghold or bastion of wisdom, knowledge, and scholarly pursuit. Using this metaphor can imply a place or institution where intellectual endeavors and learning are highly valued and protected.

that if my meager talents have managed to light a torch for others to carry on this great quest, then I have done my duty in this vast arena. [p. 7]

CLASS I.

QUADRUPEDIA.

1. BRADYPUS *manibus tridactylis, cauda brevi.* [Sloth with three-toed hands, short tail]. *Syst. nat.* 4 p. 34.

Ai seu ignavus. *Marcgr. bras.* 221. *t.* 221. *Pis. nat.* 321. t. 321.

Ai s. Jgnavus gracilis americanus. *Seb. thes.* I. *p.* 53. *t.* 33. *f.* 2.

Ai s. Haut Clusii. *Jonst. belg. t.* 74.

Jgnavus s. per antiphrasin Agilis. *Clus. exot.* 110. *t.* 373.

Jgnavus americanus major hirsutus, pilis longis & griseis. *Vincent. catal.* 2. *n.* 18.

Jgnavus s. Haut. *Nieremb. nat.* 163. *t.* 163.

Pigritia s. Haut. *Nieremb. nat.* 164. *t.* 164. *f.* 1 2 , *Olear. mus.* 7. *t.* 6. *f.* 2.

Jgnavus. *Olear. mus.* 7. *t.* 6. *f.* 2.

Hay. *Laet. amer.* 554. *t.* 554.

Unau. *Laet. amer.* 618. *t.* 618.

HEAD very small, nearly round, covered with hair all over; face covered with yellow hair, except for the nostrils. The two front *teeth* spaced apart [p. 8] (as if two intermediate teeth were missing), very blunt and erect. Several *molars* spaced apart from the front teeth. *Eyes* small, marked behind by black hair. *Throat* yellow. Nostrils small and blunt, with tiny holes in the front, and a smooth passage between them. *Ears* not visible, but the ear canals wind their way into the concha in a double channel-like passage, insinuating themselves inside.

NECK long.

ARMS longer than the thighs. *Hands* small, covered in fur all around, split into no digits, with three parallel, compressed, sickle-shaped claws, grooved underneath, equal in size, large.

THIGHS thicker, but far shorter than the arms.

FEET covered in fur all around, *heel* prominent like a bear, ending (but without toes) in three *claws,* equal, compressed, curved, grooved underneath and sharp, slightly shorter than those of hands.

TAIL very short, almost oval, covered with hair all around, with a slightly naked and pointed tip.

BREAST two, pectoral.

SIZE of a domestic cat.

Inhabits trees in the warmer parts of America.

Whole body, like that of a bear, covered with coarse and rough hair, with a color varying between brown, grey and white. [p. 9] Just as the supreme Creator wished this little animal to inhabit trees, so He equipped it with longer arms, spread-out thighs, and longer claws than other animals, so that it could more easily climb tree trunks. And lest it be harmed by the violence of the winds, He covered it with a shaggy and dense coat.

Feeds on the more succulent leaves of trees, therefore it is said that it rarely or never feels thirst.

The *gait* of the animal is very slow. For when it moves from one tree to another, as if abandoning its former home, if it falls to the ground, it can barely travel a distance of fifty paces during daylight hours. Hence, it has earned the name of "Idle", "Slothful", "Slow-footed", and "Sloth"[295], among travelers, and not undeservedly, because of its extreme sluggishness. Its longer thighs and arms, which are stiff, and its feet, which are not articulated with digits, cause it to move slowly.

The *sound* of the sloth is almost like that of a young cat, with a musical sequence descending from La, Sol, Fa, Mi, Re, Ut, in equal intervals. Therefore, Clusius prophesies that our Bradypus was the inventor of music. [p. 10]

CLASS II.

A V E S.

2. EMBRIZA (sic) *remegibus rectricibusque nigris, pectore viridi-caerulescente.* [Bunting with black wing and tail feathers, and a greenish-blue chest].

Passer caeruleo-fuscus. *Sloan. Jam. 2. p.* 311. *t.* 257. *f.* 3. *Raj. syn.* 187.

The BIRTHPLACE of the little bird is Brazil, from where it was obtained from the finest of Grill's partners, and it was nourished for a long time in an enclosed cage.

SIZE of a greenfinch[296]. *Faun. Sv.* 203.

BEAK white, splitting downward at the base, with an almost unnotched tip on each side.

WINGS, with all the remiges black.

TAIL also blackish, with little bifurcation.

CHEST and *throat* blue-green.

HEAD grayish-brown above, as is the back, which has green feathers interspersed.

FEET whitish, with three toes[297] like in related species.

ABDOMEN whitish. [p. 11]

---

[295]*Ignavi, Pigri, Tardigradi,* & *Bradypodis* in the original.

[296]*Ligurini* in the original.

[297]*tridactyli* in the original, corrected, by hand, to *tetradactyli* in the available copy at the library of the Hunt Institute for botanical documentation; also, printed as *tetradactyli* in Sundius (1749).

CLASS III.

AMPHIBIA.

3. CAECILIA *rugis CXXXV*. vid. Fig. I. [Caecilia with 135 wrinkles]. see Fig. 1.[298]

It is almost one foot long, a thumb's width in thickness, and maintains a cylindrical shape from head to tail. Its skin is completely covered by clouded and smooth scales, with tiny raised dots scattered on the back, and marked with one hundred and thirty-five wrinkles on each side, shallow and inclined. The head is smooth and sub-rounded, with the upper jaw slightly protruding and blunt. On either side of the nostrils, there is a tentacle or cirrus, so short that it almost escapes sight. Nostril openings very small, like pin-pricks, and far apart from each other. Eyes are also tiny, like dots, covered by skin but still translucent. Teeth like those of harmless snakes, small and fewer in the upper jaw. Tail blunt and almost nonexistent, densely packed with wrinkles, like in earthworms. Anus closed by two small dots, [located] near the extremity.

The following characteristics are peculiar to this animal and are not worthy of being observed in other species: [p. 12]

α — At first sight, it so much resembles an eel that one would easily believe it belongs to that family of fish. However, since it lacks both fins - which are essential characteristics of all fish - and gills, instead breathing through lungs and nostrils to exhale and inhale air in turn, it is evident that it cannot belong to that group.

β — It is distinguished from all other members of the serpent family by the fact that it has no tail, only a small opening for the anus near the end of its body, which is something never before observed in snakes.

γ — Furthermore, it is entirely devoid of scales, abdominal scutes, and complete and perfect rings like those of the Amphisbaena, with its entire body appearing completely naked.

δ — Also unique to it is the fact that the valve or upper lip of its mouth protrudes greatly and is blunt, resembling that of cartilaginous fish.

ε — Finally, two tentacles, such as those we know snails have, adhere near the apex of the mouth, and are frequently unseen.

We conclude, therefore, that this is a new genus of serpents, to be defined as:

Caecilia is a genus of snakes *with a naked body and lateral wrinkles. The upper lip of the mouth is prominent and has two tentacles. The snake has no tail.*

And we have not read about our serpent being described by anyone before, nor even named. [p. 13]

4. AMPHISBAENA *annulis abdominalibus* CC, *caudalibus* XXX. [Amphisbaena with 200 abdominal and 30 caudal rings][299].

Indeed, in terms of form, this Snake is quite similar to the one held in the *Prince's Museum*[300], although it differs in color. For its *head* is yellow, and its *belly* is also distinct everywhere with yellowish spots. So much so that anyone who wished to recognize color as the genuine characteristic would easily make this Amphisbaena a new species. However, since both the number of ringed segments and the eight anal points, and finally the crossed lateral lines, correspond exactly on both, it must be altogether concluded that they belong to one and the same species, as they actually are.

5. We have another of the same species in the Grillian Museum, with a *white head and white spots on a black body*, especially scattered on the abdomen, which is not much different from the one described above. However, since all other characteristics and even the segments themselves are congruent, it appears to be a sister to them.

The *head* of this snake is always whiter than the rest of the body, and the eyes are always clouded by a thin skin, and so small that they can hardly be seen.

6. ANGUIS *squamis abdominalibus* CCXXX, *caudalibus* XIII. Mus. Princ. p. 19. n. 23.[301] [Snake with 230 abdominal and 13 caudal scales].[302]

The shape of this *Scytale* corresponds to the one depicted in the Prince's Museum, with the same length but doubled thickness [p. 14] and as many annular bands, not rust-colored but conspicuously black. However, where the scales are white, they are scattered with small blackish dots, especially on the back.

It is clear that it is the same species as the one just mentioned. For although there are only two hundred and twenty-five scales on the head-to-anus part of our snake, while there are about two hundred and forty in the Prince's Museum snake, there are still the same number of subcaudal scales, namely thirteen, not one less or one more.

7. In ANOTHER (creature), indeed smaller but equally dark, there were found two hundred and thirty abdominal scales and thirteen tail scales.

8. In ANOTHER (creature) as well, there were in total two hundred and thirty of these (scales), but thirteen of these (scales) existed.

Like the Amphisbaena, this creature also has very small

---

[298]*Cæcilia tentaculata* in SN 10. See Fig. 1 in the plate at the end of this section.

[299]*Amphisbæna fuliginosa* in SN 10.

[300]Balk, 1746: 17, sp. 22. *ANGUIS annulis abdominalibus CC, annulis caudalibus XXX.*

[301]Species # 23 in p. 19 of Balk, 1746, correponds to *Anguis squamis abdominalibus CCXL, caudalibus XIII*, named *ANGUIS (Scytale) squamis abdominalibus CCXXX, caudalibus XIII* in Balk (1749) and *Anguis scytale* in SN 10, diagnosed as having 240 abdominal sclæs and 13 in the tail.

[302]Not mentioned in SN 10.

eyes that are barely visible.

The *tail* is barely the width of a finger and blunt, while its entire body measures a foot and a half. There are no marked colored bands surrounding the entire body, except one, the first, which surrounds the head and is interrupted ventrally.

This snake is easily distinguished from others because of its ashy coloration.

9. ANGUIS *scutis abdominalibus* CCL, *sqvamis caudalibus* XXXV. [Snake with 250 abdominal scutes and 35 caudal scales].[303] *Amph. Gyllenb. p.* 11. *n.* 6.

Magnitude, color, everything as in that of Gyllenborg's [collection][304], and around two hundred abdominal scutes and forty-eight caudal scales. [p. 15]

10. COLUBER *scutis abdominalibus CXIX, squamis caudalibus* CX. [Coluber with 119 abdominal scutes and 110 caudal scales][305].

Angviculus Surinamensis cyaneus. Seb. thes. 2. *p.* 44. *t.* 43. f. 2.

The serpent is two feet long, namely, one and a half feet long in the trunk; the tail is half a foot. The entire little animal is above a saturated blue and unspotted, but underneath a pale green. The sides are slightly compressed, and the abdomen is very flat underneath. From here towards the belly it is angular on both sides.

HEAD small, ovate, covered above with rather large, blunt scales. Nostril openings small. Eyes reddish. Teeth very small. A line is traced from the eyes towards the snout under the nostrils.

Dorsal SCALES ovate, flat, smooth, and more rounded on the sides, but greenish.

TAIL very thin, covered with obtuse and almost hexagonal scales.

ABDOMINAL SCUTES, one hundred and nineteen and *caudal scales*, one hundred and ten pairs.

This snake is externally similar to the *Ahaetullae Ceylonensium, Amphib. Gyll.* p. 7. n. 2., to the extent that it is easy to believe they are the same. Indeed, as regards its size, angles, tail and head, it maintains exactly the same form. [p. 16]

The color of this snake is not very different, although it is bluer and more spotless than our own. Although there is infinite variation in color within the family of snakes, this particular species cannot be determined by color alone, but rather by the arrangement of its abdominal scutes and subcaudal scales, especially since in this very Grillian Museum we have two snakes of the same species, one of which is barely half the size of the other. Nevertheless, when we counted the number of abdominal and caudal scales, not even the slightest difference was found between the larger and smaller snake, but both were distinct from the *ahaetulla* in terms of the number of scutes and scales.

11. COLUBER scutis abdominalibus CLV, sqvamis caudalibus XCIV. [Coluber with 155 abdominal scutes and 94 caudal scales][306].

Serpens americana lemniscata. Seb. 2. p. 12. t. 11. f. 4. *Can it be that?*

HEAD nearly round, much thicker in the neck than in the rest of the body, and shorter than in other [species], covered on the occiput by two large plates; between the eyes, three plates, then two, and finally two smaller ones. EYES pale. NOSTRILS very small. There are several blackish spots on the jaws. TEETH very small. There is a duplicated black line behind the eyes.

BODY one foot long, as thick as a swan feather. Dorsal region is slightly angular on both sides.

The TAIL is [the length of] a palm[307], round, narrow, and ends in an almost imperceptible point; hence, the scales [p. 17] under the tail are difficult to count and can only be seen with the eyes of a lynx.

Pale-grey COLOR, with slightly darker grey spots scattered around; on the back there are two linear, sparse dots here and there, and two interrupted, blackish, duplicated lines, which also distinctly distinguish the abdomen from the body, even on the tail. Moreover, the entire body is covered with numerous, barely visible brownish dots. The body is covered with smooth, pointed, oval-shaped scales.

Thus, this snake is distinct with a *back* that is keeled on the sides, a *tail* that is exquisitely pointed, and a head that is thicker than in most other species.

12. COLUBER *scutis abdominalibus CLXIV, sqvamis caudalibus CL.* [Coluber with 164 abdominal scutes and 150 caudal scales][308]. *Amphib. Gyllenb. p.* 7. *n.* 2.

This Ahaetulla is always consistent with itself and similar in appearance, both in its blue-golden color and the angle-shaped *scales* on both sides of its *abdomen* that form a white ventral area. It has a slender *head*, a black *line behind its eyes*, and a *tail* that tapers like a thread. However, it varies somewhat in only the scales on its tail. There are 162 to 168 of the former type, while the latter type may be somewhat more or less with 150 counted as the average.

Since several specimens of this kind of snake are found in Grillian collection alone, it is evident that the Coluber, which is so common in Eastern India, also abounds in the

[303]*Coluber lemniscatus* in SN 10.

[304]Refers to *ANGUIS scutis abdomimlibus CCL, caudalibus XXXV* in Hast, 1745.

[305]*Coluber cyaneus* in SN 10.

[306]*Coluber pallidus* in SN 10.

[307]*Spithama* in the original; see note 143.

[308]*Coluber ahætulla* in SN 10.

West, especially in Suriname. [p. 18]

13. COLUBER *scutis abdominalibus CCVIII, squamis caudalibus XC. Mus. Princ. p.* 28. *n.* 36. *Amphib. Gyllenb. p.* 13. *n.* 8. *Petola vulgo.* [Coluber with 208 abdominal scutes and 90 caudal scales (...) commonly Petola][309].

Others described this snake before us. The number of ventral scutes and scales varies slightly, as also noted in the collection of the Prince's Museum. In our specimen, there are *two hundred and eight* ventral scutes, while the caudal scales, which are not easily counted due to their extreme thinness, are approximately one hundred.

This snake is easily recognizable by its leaden-colored dorsum, transverse bands of reddish-brick, and immaculate white lateral and lower abdomen.

14. COLUBER *scutis abdominalibus CL, squamis caudalibus L.* [Coluber with 150 abdominal scutes and 50 caudal scales][310]. Mus. Princ. p. 24. n. 28. Amphib. Gyllenb. p. 9. n. 4.

Four specimens of this species of *cobella*, with different tongue colors, are preserved in this museum, but they can be easily distinguished from others by its brown back marked by twin white and curved lines on both sides.

And since many snakes of this species are found in our museums, it is rightly concluded that they are very common and not particularly harmful, since they lack large teeth.

One hundred and forty-eight to one hundred and fifty-three abdominal scutes, and between forty-eight and fifty-three tail scales are counted. [p. 19]

The *tail* reaches about one-fifth of the length and is very pointed.

The *throat* is covered by three pairs of scales.

We possess one [individual] of this species which varies considerably. The back is not brown, but cerulean, and the white marks are hardly noticeable. It has a distinguishable bright[311], transverse spot behind the eyes. The belly is white, and the lower part of the tail is bright, with uninterrupted, scattered bands, making it easy to assume that this is a new species, especially since it has 138 abdominal scutes and 62 tail scales. However, the trio of double scales under the throat, the paired sutures of the head, the dorsum moderately keeled, the interrupted abdominal bands, and the presence of some white marks on the back which vary between them, makes us not consider it entirely different, especially when it was found that the missing abdominal scutes correspond to the surplus of caudal scales. For example, if we add the one hundred and thirty-eight abdominal scutes and sixty-two caudal scales, we obtain the same number as if we add hundred and fifty scutes and fifty scales of our more common type, that is, in both cases, a total of two hundred.

15. COLUBER *scutis abdominalibus* CXC, *squamis caudalibus* XLII. [Coluber with 190 abdominal scutes and forty-two caudal scales][312].

Serpentis Aesculapii species, *Seb. thes.* 2. *p.* 19. *t.* 18. *f.* 4.

HEAD shorter than in the others, depressed, [p. 20] and somewhat round. *Nostrils* small; black band between eyes. *Mouth* lacks deadly teeth. *Throat* covered by five pairs of scales.

BODY covered with ovate, smooth, and shining scales. The *color* on the back is pale, adorned with more than twenty broad black bands, which often encircle the entire body. The *scales* on the back are arranged in fifteen rows. There are one hundred and ninety *abdominal scutes.*

TAIL reaches one-sixth of the body and is covered underneath by forty-two very distinct pairs of scales.

LENGTH about one and a half feet in total. *Thickness* of one digit.

16. CENCHRIS *scutis abdominalibus* CCXL, *scutis caudalibus* LXIV. [Cenchris with 240 abdominal and 64 caudal scutes]. see. Fig. III.[313]

Boiguacu. *Pis. bras.* 1648. p. 41. *Marcgr. bras.* 239. *Pis. natural.* 1658. p. 277. *Raj, quadrup.* 325.

Constrictor. *Kaempf, amoen.* 573.

Serpens indicus bubalinus, anacandia ceylonensibus. *Raj. qvadrup.* 332.

Giboya sive Jaboya. *Laet. occid.* 554.

Anguis stupidus, *Nieremb. nat.* 274. [p. 21]

Serpens peregrinus. Clus. *exot.* 113.

Serpens ceylonica maxima, pimberah dicta, *Seb. thes. 2. p.* 97. *t.* 91. *f.* 1. *An?*

Serpens oculea mexicana foemina. *Seb. thes. 2. p.* 104. *t.* 98. f. 1. *An?*

Serpens excellens ac speciosa brasiliensis e regione gvaira, regina serpentum habita. *Seb. thes. 2. p. 105. t. 99. f. 2.*

Serpens americana arborea, singulari artificio picta, magni aestimata. *Seb. thes. I. p.* 85. t. 53. *f.* 1.

BODY is thick, gray, with twenty-four large pale spots on the back; the anterior five on each side are notched, the posterior are transverse and have bifurcated edges but are whiter. The *tail* is also more reddish; the sides are beautifully variegated with gray and pale. Additionally, the entire body is scattered with small brown spots.

HEAD covered with very small scales, without larger plates between the eyes like other snakes. Region around nostrils significantly protruded, with a black spot underneath and a blunt tip. Black band behind the eyes. Upper plate with scales almost serrated.

Larger canine *teeth* none. Tongue fleshy, slightly bifurcated, acuminate-setaceous. Above the eyes, the head is

[309] *Coluber petola* in SN 10.

[310] *Coluber cobella* in SN 10.

[311] *Candidissima* in the original, adjective that can also be translated as translucent.

[312] *Coluber æsculapii* in SN 10.

[313] *Boa constrictor* in SN 10. See Fig. 3 in the plate at the end of this section.

raised on each side into a tubercle. [p. 22]

SCALES very small over the entire body, smooth, ovate-rounded.

TAIL barely reaches an eighth part of the total body.

This beast grows to such an immense size that it often surpasses four or more orgyias[314]. It is entirely smooth and beautifully adorned with large and variegated spots. For this reason, its skin is used as clothing by the Indians, and it is admired in curiosity cabinets, as well as in our own academic libraries, gardens, and museums. The natives of India and the Negroes from Africa consume its flesh. It is said that it does not harm travelers, which agrees with our description, in which we affirmed that it has no larger teeth that can inflict lethal wounds.

Piso[315]: *When hungry, it leaps out of the thicket, rearing up hideously by bracing itself on its tail, and fiercely wrestles with beasts and men alike, hissing when provoked, or leaps from a tree onto a traveler, and embraces with the strongest of grips, even to the point of killing him.*

Marcgravius[316] *saw a whole goat swallowed. By rolling around, it breaks (the prey) up, crushes the bones, covers (it) all in saliva, and swallows (it).*

Kaempf. amoenit. 573[317]. *They are called 'constrictors' because, I know not by what stratagem, they surround the beasts and, while they die, are said to constrict and finally swallow them suffocated. As for whether they also do this to bulls, as Clearyus related in his ephemeridibus germanicis*[318], *I have never learned, or heard from credible sources. They are the trifles of a chattering crowd.*

Nieremb. nat. 274[319]. It lives among rocks and caves, and feeds by attracting animals to itself with its breath. It lies in wait for deer, [p. 23] hidden among the foliage near the paths that the deer usually take. When a deer comes near, its spirit is drawn towards the snake like a magnet attracts iron, and the deer cannot escape the snake's hungry mouth (this does not happen to humans). The snake stops the deer's breath by constricting it, and then licks it and covers it with its saliva in preparation for easier swallowing, as otherwise, the roughness of the deer's fur would be an impediment. The snake gradually sucks the deer in from its feet up to its head, as it cannot swallow the antlers. The deer rots in the snake's mouth until it can be fully consumed. Nature has granted this kind of hunting ability to these foolish snakes, which would otherwise be entirely idle. The entire economy of this snake depends on the capture of its prey.

17. CROTALOPHORUS *scutis abdominalibus CLXXII, scutis caudalibus XXI, paribusque sqvamarum III.* [Rattlesnake with 172 abdominal scutes and 21 caudal scutes, plus three pairs of scales][320].

Domina serpentum. *Nieremb. nat. 268. t. 268. Jonst. serp. 26. t. 7. Olear. Mus. 18. t. II. f. 2.*

Serpens venenosus maximus virginianus cum ratulis. *Vincent. catal. 4. n. 38.*

Vipera americana caudisona, longitudine IV. pedes exaequans, coloris leucocephalei, hinc inde in dorso fusco colore adspersa, tintinabulum seu crepitaculum in cauda extrema a multis squamis compositum habens. *Mus. petropol. p. 454. n. 36.* [Vipera americana caudisona, with a length equal to 4 feet; color whitish, though on the back has brown splashes. At the end of the tail has a rattle formed by many scales.]. [p. 24, missnumbered as 42]

Although rattlesnakes are found in both Indias, it has so far been believed that each one is of the same species. As previously described in the dissertation on *Museum Principis* p. 20. n. 24, there are many other specimens that have never been examined. Now, we have a timely opportunity to present this information and I am certain that these specimens do not constitute a single species, but rather an entire genus. Therefore, the generic character of this snake will be the tail, while the specific ones will be the abdominal and tail scutes. It should be observed in this (animal) of ours that not only are scutes present under the tail, but also several pairs of scales above it, which are lacking in the other

---

[314]The *orgyia* (ὀργυιά) is a Greek measure of length, originally derived from the human body, and was the distance from extremity to extremity of the outstretched arms; equals 1.85 m.

[315]Piso (1648: 41-42): "...*Famelicus hic anguis vel ex dumetis prosilit, caudæque suæ nitens, horrende se erigit & strennue cum feris & hominubus luctatur, & sibulat irritatus...*".

[316]Marcgravius (1648: 239): "...*Serpens hic vocatur a Brasiliensibus* Boi guacu, *id est, magnus; nam vidi qui integram capream deglutiverat...*".

[317]Kæmpfer (1712: 573): "...*Pigrum hoc & inconditum serpentis genus (glabrum erat, albidum, magnis nigricantibus maculis variagatum) parum veneni, plus roboris habere dicitur: nam ex eorum classe sunt, quos sua lingua* Constringentes *nuncupat: ita dictos; quia nescio quo astu occupata quadrupedia circumdare & dum moriantur, constringere, suffocata denum deglutire dicuntur: an uros etiam, quod Cl.* Cleyerus in Ephem. Germ. *retulit, id nunquam comperi, vel fide dignis audivi. Nugæ sunt fabulantis vulgi; quemadmodum ea, quæ narratur* de Ruk, *tanti roboris & magnitudinis ave in Insula Madagascar, ut elephantum rapiat; quæ tamen in ipsa patria nec nomine nota est...*".

[318]Cleyer (1684).

[319]Nieremberg (1635: 274): "...*Vocant Hispanis hos angues* culebras bobas. *Cervis insidiantur, occultati frondibus circa vias qua cervi solent transire. Cum prope accedit cervus, vi spiritus sui attrahit ad se ut magnes ferru, nequeunte cervo evadere avidum os coubri (carent hoc imperio in homines). Detentum halitu cervum complectuntur, circumcompressione extinguunt, deinde adlambunt, & saliva sua linentes præparant & condiunt ut facilus glutiant, alioqui pilorum asperitas impedimento esset. Paulatim exsugunt illum integru a pedibus usque ad caput; quod quidem propter cornua deglutiere nequeunt, sed donec putrescat in ore deferunt. Accidit interdum, ut formicæ patulum os eius colubri subeant, eumque interimant, ulciscentes hoc modo cervicidam. Hoc genus venationis his stultis anguibus indulsit natura, alioqui nulla providentia esset tantæ ignaviæ. Tota œconomia & cesus huius anguis est spiritus...*".

[320]*Crotalus dursissus* in SN 10.

(animals). There is no doubt that if a chance were given to investigate the others as well, more species of this kind could be discovered.

Our (animal) is four feet long and the thickness of an arm, while another (animal) depicted in the *Museum Principis* is only two feet long and slightly thicker than a human finger.

COLOR varied, from white and yellow with black spots in diamond shapes on a white disk. In contrast, the other (animal) is completely white, with all ovate, keeled scales, especially on the back and towards the head.

AREA between the eyes black, with a broad white transverse line, which is not present in the other (animal).

HEAD very blunt, with a structure similar to the preceding. The teeth in the upper jaw are canine, hidden within a sac, and are pointed, awl-shaped, bent, and a finger-length long on each side, with which lethal wounds can be inflicted. [p. 25]

ABDOMEN before the anus covered with one hundred and seventy-two transverse scutes, but between the anus and the rattle there are twenty-one scutes, and just before these, three pairs of scales terminate the tail before the rattle.

TAIL tight, narrow, very short, and terminates in a rattle, consisting not of five, but nine articulations, and is flattened as in the other.

Therefore, many of the names and synonyms attributed to the rattlesnake in the *Museum Principis* will also be common to this species.

18. LACERTA *cauda tereti, corpore duplo longiore, pedibus pentadactylis, crista nulla, hypochondriis plicatis.* [Lizard with tapering tail, twice longer than the body, without crests and folded sides][321]. *Amphib. Gyllenb. p.* 22. *n.* 14.

This [lizard] agrees entirely with [the one in] Gyllenborg's collection.

19. LACERTA *cauda tereti brevi, pedibus pentadactylis, digitis duobus tribusque coadunatis.* [Lizard with short, tapering tail, pentadactyl limbs and digits joined in two and three][322]. Mus. Princ. p. 13. n. 16.

This chameleon from the West Indies is otherwise similar to the eastern one, except that it is of a darker color, has slenderer thighs, and a more pointed occiput, so that its back is more elevated above the thorax.

CLASSIS IV.

PISCES.

20. GYMNOTUS[323]. *Mus. Princ.* 51. *t.* 2. *f.* 6.

Also kept among the Surinamese collections is a fish of extraordinary size, measuring a full three feet in length and a palm's breadth in width, weighing ten pounds, with a very rounded and fat back, entirely covered in mucus, like a whole eel. It is called Poutaol by the Indians.

CLASSIS. V.

INSECTA.

21. *GRYLLUS crista thoracis quadrifida.* [Cricket with a thoracic crest divided into four parts], see *Fig. IV.*[324]

Gryllus ex Aegypto. *Olear. mus.21.t.* 12.f. 5.

Gryllus orientalis[325], *Frisch. insect. 9. p. 3 f. 1.*

Animal of a transverse-palm in length, testaceous.

HEAD blunt, bent, with the upper lip notched and the lower lip divided into two. There are two tentacles on each side; anterior pair with five articulated sections; the posterior with three. Lateral *maxilla* strong, toothed; gray eyes. *Antennae* three times shorter than the body, filiform, with twenty-seven segments, of which those closest to the head are shorter.

THORAX compressed, covered with a hard, rough crust, widened posteriorly into a quadrangular, obtuse, compressed ridge. Lower part *chest* perforated by two openings between each pair of legs, and is somewhat flat, and between the first pair of legs, it is furnished with a conical protuberance.

TIBIAE of the first four legs are round, subcylindrical, and somewhat spiny on the underside towards the apex. [p. 28] The *sole* of each foot merges into a four-fold tarsus, which forms a double-digit with a sharp, curved, black-tipped claw, and a rounded, convex cotyledon in between.

FEMORA of the hind pair are large, angular, smooth, and speckled with white on both sides, with slightly thickened knees. *Tibia* long and slightly cylindrical, armed behind with alternating sharp spines, black-tipped at the apex, and terminated by four spines. *Sole* with quadruple rounded tarsus; the previous two integers, while the other two are forked. *Digits* alike, projecting forward.

ABDOMEN subcylindrical, divided by seven or eight incisions, each with a foramen on each side. TAIL four-valved, pointed, with the upper valves shorter and resem-

[321] *Lacerta teguixin* in SN 10.

[322] *Lacerta chamæleon* in SN 10.

[323] *Gymnotus carapo* in SN 10.

[324] *Gryllus cristatus* in SN 10. See Fig. 4 in the plate at the end of this section.

[325] Orientalischen Heuschrecken in Frisch (1730).

bling a goat's hoof, and the lower ones slightly longer and sharper. Four *wings*, longer than the abdomen, of which the upper are narrower, paler, veined and reticulated, with a single stronger vein, and are therefore more like elytra; the lower wings, on the other hand, are eight times wider, more veined and reticulated, resembling a complex umbrella, blackening towards the posterior margin.

This species is the same one that, like clouds, migrates from Egypt, the Holy Land of Syria, and other Eastern regions to Europe, especially Poland, in some years, taking away all hope from farmers in one day or even one hour; [p. 29] thus, this species is common to Africa and America.

22. *GRYLLUS thorace lineari alarum longitudine; margine denticulis ciliato.*[326] [Cricket with a linear thorax the length of the wings; margin fringed with denticles].

Mantes. *Jonst. ins. t. 11. Aldrov. f. 1. 2. & t. 13. f. 21. Act. angl. n. 301. p. 2051. t. 20. f. 3.*

Gaayara. *Marcgr. bras. 246.*

This is definitely a species of cricket, although its structure differs somewhat from those of our [local species].

BODY yellow, length of a finger.

HEAD triangular, obtuse; forehead yellow. *Antennae* two, narrow, bristle-like. Between the antennae there are three *pores* located in a triangle. *Mouth* with an unbroken lip. *Maxillae* lateral, small. Brown tentacles beneath the mouth.

THORAX very long, linear, slightly convex, and the margin is enlarged with very small ciliated denticles. At the front, there is an enlarged and somewhat round apophysis. The femora of the first pair are somewhat triangular, with anterior teeth, and their length is scarcely greater than that of the thorax. They are attached to the anterior part of the thorax by an apophysis. The ulnae are almost as long as the femora but wider and compressed, with the posterior part bifurcated and spiny; digits solitary, mucronate and posteriorly serrato-spinous and, near the apex, provided with an articulated antenna-like structure.

ABDOMEN oblong, wide, depressed, with very short femurs of the second and third pairs near the base and without spines. *Fibulae* somewhat cylindrical and narrow, devoid of spines, and are joined to the tibiae. At the apex, there is a four-phalanged metatarsus with a bifid claw. Four *wings*, of which the upper ones are oval, wider, reticulate, translucent, yellowish, with a fulvous spot in the middle and a longitudinal rib on the outer margin; at the same time, the aforementioned margin is more yellow and elevated. Lower wings oval-oblong, membranous, striated, and tessellated with yellow spots.

We have another similar to the previous one, but smaller and more brownish, with cloudy wings, shorter than the abdomen, a more elongated tail, the upper lip ovate on both sides, toothed and the lower lip bicuspid below, with converging sides and a blunt tip. We recognize it as a pupa in progress, with the same structure of thorax, head, antennae and legs.

23. ARANEUS *hirsutus thorace orbiculato, convexo, centro transversim excavato.* Mus. Princ. p. 44. n. 59.[327] [Hairy spider with a rounded, convex thorax, transversely grooved in the center].

Nhandaguacu. Pis. bras. 44.

We keep two spiders of this species, of which one has the abdomen smaller than the thorax, and the other twice as large, from which we suspect that this one is female and that one is male.

A twin cylindrical appendage, consisting of three joints, adheres to each side of the posterior end [of the spider's body]. [p. 31]

Each leg has six joints. The jaws of the mouth are extremely strong, resembling the shape of a feline's claw, covered in hairs that are red at the point of bending, and the entire body, including the legs, is covered in black and dark hairs.

We have diligently sought after the descriptions of the *claws* of the feet from various authors. In truth, the tips of the feet are completely blunt and covered in dense fur, with a small, curved and simple claw hidden in the hairs, armed even in the front legs, but so small that it can barely be seen unless sought out with an iron stylus. Therefore, those who attribute and attach two, rather large, claws to each foot are mistaken.

Piso *relates that when it is angry, it inflicts a venomous sting into the skin, from which a livid swelling arises, accompanied by much pain, often incurable even by the Indians.*

24. *SCOLOPENDRA plana, pedibus utrimque viginti*[328]. Mus. *Princ. p. 46. n. 22.* [Centipede with a flat body, having twenty legs on each side].

Scolopendra magna peregrina. *Frisch. ins. 11. t. 2. f. 7. Loevvenhaek. epist. p. 102*[329]*. Olear, mus. 21. t. 12. f. 5. Aldrov. ins. l. 7. cap. 6*[330].

Yellow in color and entirely similar to that which is held in the *Museo Principis*, except that it is smaller. Antenna: awl-shaped, with twenty segments. [p. 32]

---

[326]*Gryllus gongylodes* in SN 10.

[327]Not identfied. *Mus. Princ. n. 60. 61* in Sundius (1749).

[328]*Scolopendra morsitans* in SN 10.

[329]Epistle 124, addressed to Hendrik van Bleyswyky, written on May 20, 1700. Analyze *De millepeda indica*. Sundius reference was not located; we refer instead to Leeuwenhœk, 1719: 173.

[330]The reference to *Liber Septimus de Insectis*, Caput VI *De Scolopendra*, is coincident with the Frenkfurt edition, Aldrovandi, 1623: 285.

APPENDIX

For the sake of illustrating natural history, especially that of snakes, and of a certain species in particular that is called *Ammodita*, I will include a snake captured by the distinguished and most noble Councilor of the Royal Chancellor, Sir Edward Carleson[331], when he was serving as Extraordinary Ambassador to the *Portam Ottomanicam*[332].

25. COLUBER *scutis abdominalibus CXLII. Squamis caudalibus XXXII.* vid. Fig. II. [Coluber with 142 abdominal scutes and 32 caudal scales][333].

Ammodytes *Raj. quadrup. 287. Jonst. serp. 14. t. 1. f. 3.*

It exceeds half a foot in length, is blue-green in *color* and almost sandy, with a wider blackened line, alternately toothed and furnished with several tips on each side, dividing the middle of the back. In short, whoever has seen the Viper described in *Faun. Svec.* 160. knows its color as well. However, ours can be distinguished from it in that it has 142 abdominal scutes, and thirty-two tail scales.

BODY covered with oblong, flat, and blunt scales arranged in nineteen rows.

HEAD almost triangular, covered with very many *small scales* located between the eyes, nostrils, and snout; absolutely no large scales, as is almost always the case with other species of snakes. The edges of the lips are variegated with white and black. There are two rather large, sharp and venomous *fangs* on each side of the upper jaw, each with its own venom sac, and none on the lower, but rather several very small and harmless ones.

SNOUT erect like a cone, two lines high, horn-shaped, but not made of horn, but rather of fleshy material, covered with very small scales, and bends towards the head. Between this and the eyes on both sides of the head, a certain raised *tubercle* can be seen. The *nostril* openings are located on either side at the base of this snout. Both *eyes* are covered with a single, wider scale in place of eyelids, and the back of the head has two distinct scales, slightly larger than the rest.

TAIL very short, only about the length of a transverse thumb in width, and tapered.

This snake was captured while preying on a lizard, as described in Fn. Sv. 1352[334]. The lizard had been swallowed up to its front legs, although it was large enough to be an equal match for its predator.

It is found in southern Europe, including Italy, Illyria[335], etc., as well as in Libya.

Regarding the excessive hardness of this snake's tail, reported by some authors, it is of no weight whatsoever: although it is somewhat harder than other parts of the body, I know that it does not become excessively hard.

Matthiolus in Dioscoride p. 950[336]. *Among snakes, the so-called Ammodites can be considered a monstrous wild beast. It looks like all snakes, except that its head is bigger, the jaws wider, and the tip of the snout has a certain projection, resembling a sharp wart. In the viper genus, there is certainly a monstrous wild beast that can be counted among them, which is called Ammodites. It is entirely similar to the viper, except that its head is wider, its jaws are broader, and at the tip of its upper snout, it has a certain prominence that resembles an acute wart. So, the street vendors, who offer remedies for snakes and their venom, commonly call this beast Aspide del corno, that is, horned asp. And they did not give it that name without reason, for it kills with no less speed than the Aspis. For we know that some who were bitten by it unexpectedly died within three hours. So Aëtius wrote accurately about it in book XIII, chapter 25. The Ammodities (sic) is indeed a cubit[337]-long in size (for it is not depicted or written as larger), and has a sandy color with black spots throughout the body. It has a stiff tail that is split at the top. Some Cenchrias, having the tail hard as a column[338], were called miliaris. The jaw is larger than in vipera; although similar in many other [characters], they can be differentiated by color: vipera is, in*

[331]Edvard Carleson (1704-1767), Swedish writer and economist. See letter L0728, from Bäck to Linnæus, to trace the probable origin of this specimen.

[332]*Portam Ottomanicam*, or Sublime Porte, synecdoche for the central government of the Ottoman Empire in Istanbul.

[333]*Coluber ammodytes* in SN 10. See Fig. 2 in the plate at the end of this section.

[334]Faun. svec. 1352. *Lacerta cauda tereti verticillata longitudine corporis, pedibus pentadactylis unguiculatis. Lacerta agilis* in SN 10.

[335]The former Illyria included Slovenia, the current Austrian state of Carinthia, other territories in Northwestern Croatia and the provinces of Trieste and Gorizia in Northeastern Italy.

[336]Matthioli (1570): "... *In viperarum quoque genere ex nimirum immanis fera recenseri potest, quam* Ammoditen *vocat (...) Est enim fera hæc viperæ omnino similis, præter quam quod capite sit ampliore, & maxillis latioribus, quodque ultimo rostro superiori parte eminetiam quandam habeat, acutæ verrucæ similem. Unde circumforanei, qui serpentes eorumque veneni remedia profitentur: hanc feram vulgo apellant Aspide del corno, hoc est, Aspidem cornutum. Neque sane ineptum illi nomen indiderunt: nanque non minore cum velocitate interficit, quam aspis. Scimus enim quosdam ab his inopinate demorsos, tribus tantum horis morte occubuisse. Proinde scite libro XIII. cap. XXV. de eo ita scriptum reliquit Ætius. Ammodites magnitudine quifrm cubitalis est (maior enim nom pigintur, neque scribitur) colore vero arenosus, per corpus autem maculis nigris insignitus: caudam habet præduram, superne ndiscissam. A quibusdam vero Cenchrias, hoc est miliaris, ob cauda instar milii duritem, apellatus est. Latiores autem maxillas, quam vipera habet, qumque in multis, alias ei similis sit, colore magis discerni potest: vipera enim subflava est. Cæterum ab Ammodite percussos festina ut plurimum mors sequitur., qui vero tam brevi no pereunt, his sanguis e vulnere manat, & vulnus intumescit: sed paulo post sanies effertur, sequiturque capitis gravitas, & animi deliquiuum. Qui vero optime dispositi sunt, & robusti, in triduo intereunt: quanquam repertii sunt, qui in septimum usque diem durarint...*".

[337]The cubit is the distance from elbow to middle finger.

[338]*Milii* in the original, was the marker that in roads indicated the distance to Rome, measured in miles, equivalent to a thousand steps.

fact, yellow. Moreover, swift death usually follows those who have been bitten by Ammodita. But in those who do not die so quickly, blood flows from the wound and the wound swells, but shortly afterwards pus is produced, and heaviness of the head and fainting of the mind follow. Those who are best prepared and robust, die within three days, although some have been found to last until the seventh day.

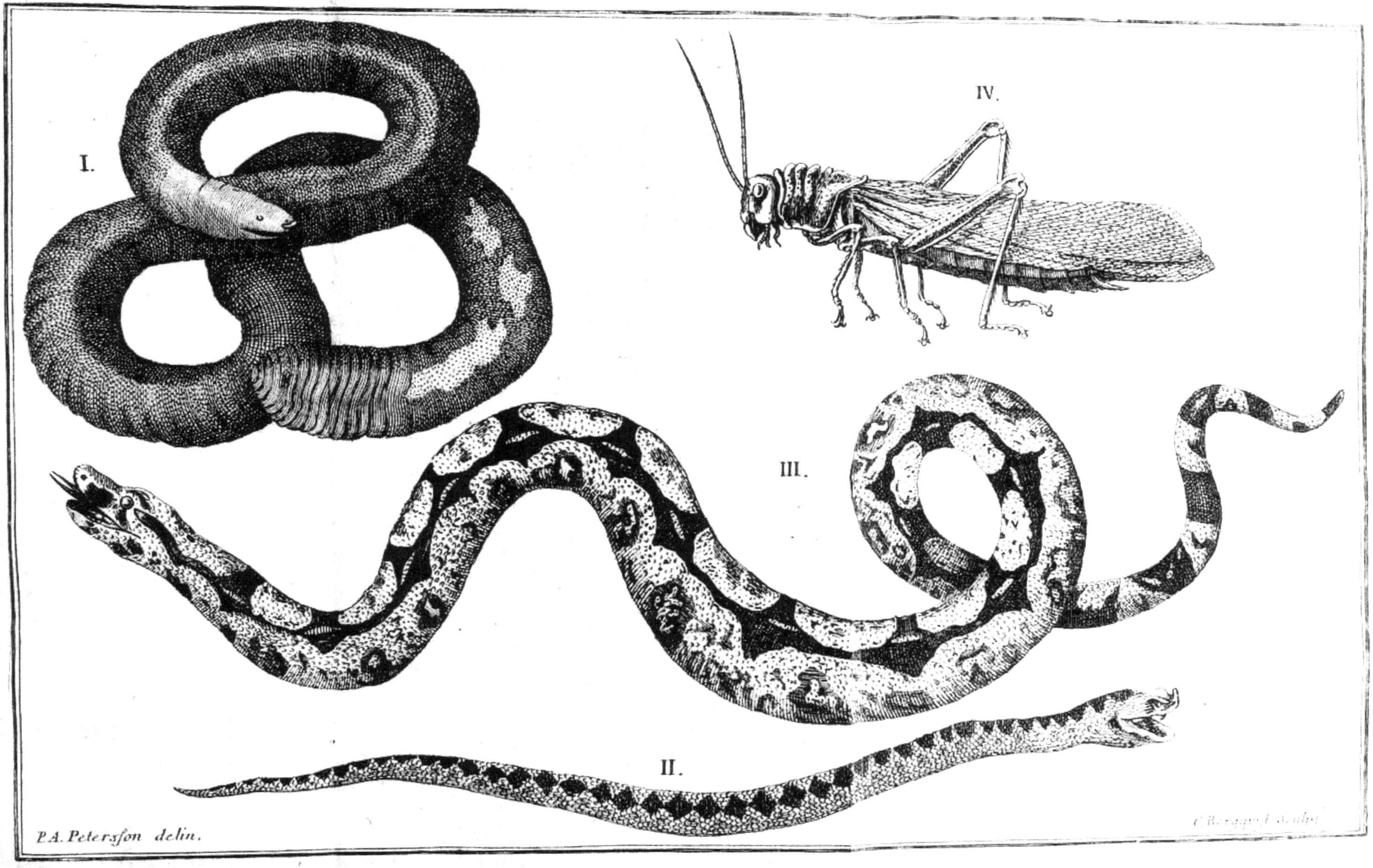
I.
IV.
III.
II.
P.A. Peterſson delin.

D. A.

LIGNUM COLUBRINUM

LEVITER DELINEATUM,

EX CONSENSU AMPLISS. FACULT. MED. IN REG. ACADEMIA UPSALIENSI,

PRÆSIDE

VIRO CELEBERRIMO ET EXPERIENTISSIMO

DN. DOCT. CAROLO LINNÆO,

S:Æ R:Æ M:TIS ARCHIATRO,

MED. ET BOTAN. PROFESS. REG. ET ORD.

ACAD. IMPER. MONSPEL. BEROLIN.

STOCKHOLM. ET UPSAL. SOCIO,

PRO GRADU DOCTORIS RITE OBTINENDO,

PUBLICO EXAMINI SUBMITTIT

STIPENDIARIUS REGIUS

JOHAN. ANDR. DARELIUS,

W-GOTHUS.

IN AUDIT. CAROL. MAJ. AD DIEM XI. MART.

ANNI MDCCXLIX.

H. A. & P. M. S.

UPSALIÆ.

Darelius, J.A. 1749. Lignum Colubrinum leviter delineatum. Upsaliae. S. n. [2] + 3-22.

This dissertation, defended on March 11, 1749, and subsequently reissued by Linnaeus in the second volume of *Amoenitates Academicae* (Darelius, 1751), addresses the identification and verification of the plant historically referred to as *Lignum colubrinum*. The primary aim was to clarify the plant's identity to prevent confusion, avoid adverse effects, and detect potential adulterations.

The original *Lignum colubrinum*, sourced from Ceylon (modern-day Sri Lanka), was recognized as an effective antidote for cobra bites. Initially described by Garcia de Orta in 1563, it was further detailed by Kaempfer in 1712. Darelius begins with a contextual analysis of poisons and remedies, followed by a detailed examination of the snake species known as *Naja* to the Singhalese, and its natural predator, the mongoose. The dissertation also reviews relevant literature, including works by García de Orta, Juan Fragoso, Kaempfer, Grim, Hermanni, Cristóbal Acosta, Camellus, Rumpf, Linnaeus, and Bauhin.

Among the various plants identified as *Lignum colubrinum*, Darelius concludes that the true plant is the one described by Garcia de Orta, later identified as *Radix Mungo* by Kaempfer and *Ophiorrhiza mungos* by Linnaeus. Sections V and VI of the dissertation hold herpetological significance, addressing the use of the plant as an antidote not only to snake venom but also to mad dog bites and putrid fevers.

[p. 1]

Academic Dissertation A brief sketch of Lignum Colubrinum, presented for public examination, with the consent of the esteemed Faculty of Medicine of the Royal Academy of Uppsala, under the presidency of the most distinguished and experienced Dr. Carolus Linnaeus, First Physician to His Royal Majesty, Professor of Medicine and Botany at the Royal and Ordinary Academies of Montpellier, Berlin, Stockholm, and Uppsala, and a Fellow of the Imperial Academy, Submitted by JOHAN ANDR. DARELIUS, of West Gothland, Regal Stipendiary, in order to obtain the degree of doctor, in the Great Carolinian Auditorium, on the 11th of March, in the year of our Lord 1749. H. A. af P. M. S. Upsala.

[p. 2]

INTRODUCTION

I.

Medicine, strictly speaking, exclusively concerns itself with repelling diseases, which must be combated with medications; thus, it is evident that it rests on a dual foundation, the knowledge of diseases and of remedies. Therefore, if a physician's theory is not built upon these pillars, not only will all his practice be uncertain, but often it will also be dangerous and harmful. In brief, he will stagger like an empiricist throughout the day.

II. Remedies against diseases are extracted from the three kingdoms of Nature, and are either alimentary or toxic; the latter, by modifying the weakened body, restore lost health to its former state; the former, by preserving the body, protect it against any attack of diseases. Many believe that medications and poisons are distinguished in such a way that poisons, even when administered in small doses, destroy the body, but not so medications. We consider this distinction a harmful heresy, which must be eradicated from the domain of physicians. For since all poisons are like very powerful weapons, with which, according to the daily experience of many physicians, even the most severe diseases are combated, who would prudently deny a physician well-versed and experienced in his art the right to prescribe them according to circumstances? However, we must not forget the sadly famous and very true saying: the most excellent medicines, in the hands of the inexperienced, are like a sword in the hand of a madman. Therefore, poisons administered with due proportion and dosage are considered by us as heroic medicines[339]. The opinion of some about poisons has caused more harm to Materia Medica than any other common error, as is sufficiently demonstrated by the use of opium, quinine, arnica, mercurials, etc., which the medical state has greatly abhorred at certain times, not without considerable loss of health and life for many people.

III. To avoid seeming to stray from the topic, as they say, we must confine ourselves to the boundaries of current medicine, considering those remedies that are available today in pharmacies and that can and should be of practical use for the Physician. Some of these remedies are provided by Europe, others are imported from the Arabs, some from the far reaches of the Indies, and others from various other places. Wherever they come from, it is undoubtedly the duty of a good physician not to use any of them unless he has carefully investigated their properties and nature. Therefore, the greatest concern and dedication of the physician will be to properly know the medicines in pharmacies and to choose the most suitable ones for each known disease. That learned physician who combines this selection with a thorough study of the history of diseases will be justly hailed as a wise physician and a successful practitioner.

IV. To achieve complete knowledge of *Materia Medica*, what we call the history of each medicine is especially required. [p. 4] We do not deny the value of the most learned physicians of all times in this genre of studies, and we sincerely acknowledge that we owe much to the zeal and experience of these illustrious individuals in the knowledge of medicines. However, I do not know how it has happened that some things have become so deeply rooted that we completely ignore their origin or use. It is undeniable that the ancients knew excellent remedies, whose memory has now been erased by oblivion; I believe that no one who has reviewed the works of DIOSCORIDES, where the properties and uses of medicines are described, can deny this. Since in those times little attention and care was given to the description of simples[340], and the ancients did not master the art of explaining and delineating in the same manner, it is not surprising that posterity is unaware of many of these things or can hardly discover them with great difficulty. Bearing this in mind, today's learned physicians have dedicated great effort to transmit precise and clear knowledge of the Simples of the Pharmacies, so that future generations do not fall into the ignorance and negligence we lament regarding previous generations. For this reason, we can form clear concepts of all simples, excepting some that the peoples of regions very distant from ours have transmitted to us. However, obtaining some knowledge about the exotics is more difficult, as rarely do others besides sailors and uneducated people have the opportunity to see them. For this reason, today, in terms of medical knowledge, we are unaware of remedies such as Myrrh, Bdellium, Sagapenum, Elemi, Carana, Ammoniac, Star Anise, Ebony, Aspalathus, Lignum Aloe, Frankincense, Indian Myrobalan, Citric Myrobalan, Belleric Myrobalan, Chebula, and other remedies[341] that were highly praised and common in other times, just as around the birth of Christ. The most sensible Physicians have dedicated non-superficial efforts to discover these unknown [p. 5] remedies. An example is the Cinchona Bark[342], which was discovered, examined, and designated for specific uses a few years ago (*)[343], as well as *Cascarilla*, which has recently been confirmed as a species of *Clutia*, not to mention *Contrayerva*, *Senega*, and

[339]*Medicamenta heroica* in the original. A group of medicines with powerful and often drastic effects, which were used to treat serious or life-threatening diseases. These medicines were considered the last hope for patients and were administered with caution due to their strong side effects, like mercury, arsenic, opium, bloodletting, etc.

[340]Simples, in the original; refers to unprocessed or minimally processed mineral, plant or animal materials used for medicinal purposes. They might be used directly or combined to create more complex remedies (e.g., syrups, infusions, ointments).

[341]Myrrh: Aromatic resin from the *Commiphora* genus, used as an antiseptic, cicatrizant, analgesic, and expectorant. Bedellium: Reddish-brown resin similar to myrrh from the *Boswellia* genus, used as an anti-inflammatory, antispasmodic, and expectorant. Sagapenum: Yellow-green resin from the *Ferula* genus, used as an analgesic, anti-inflammatory, and expectorant. Elemi: White or pale yellow resin from the *Canarium* genus, used as an antiseptic, cicatrizant, and expectorant. Carana: Dark brown resin from the *Bursera* genus, used as an anti-inflammatory, antispasmodic, and expectorant. Ammonia: Colorless gas from nitrogenous organic matter, used as a respiratory stimulant, expectorant, and rubefacient. Star Anise: Star-shaped dried fruit from the *Illicium* genus, used as a digestive, carminative, and expectorant. Ebony: Hard, black wood from various *Diospyros* trees, used as a tonic, astringent, and antidiarrheal. Aspalathus: Reddish-brown wood from the *Pterocarpus* genus, used as a diuretic, expectorant, and antiasthmatic. Alœs Wood: Yellow-green wood from the *Alœ* genus, used as a laxative, purgative, and tonic. Olibanum: White or pale yellow resin from the *Boswellia* genus, used as an expectorant, anti-inflammatory, and cicatrizant. Indian Myrobalan: Black dried fruit from the *Terminalia* genus, used as a laxative, purgative, and astringent. Citrine Myrobalan: Yellow dried fruit from the *Terminalia* genus, used as a laxative, purgative, and choleretic. *Belleric Myrobalan*: Brown dried fruit from the *Terminalia* genus, used as a laxative, purgative, and tonic. Chebulic Myrobalan: Black dried fruit from the *Terminalia* genus, used as a laxative, purgative, and astringent.

[342] *Cortex Chinæ* in the original, referring to the bark of diverse species of *Cinchona*, such as *C. calisaya, C. officinalis,* or *C. pubescens*, t.

[343]Footnote on p. 5: (*) *Conf. Experientiss. Prof. Lundens. D. EB. ROSENII Diss. de China.*

others[344]. Considering this, we have proposed to the learned world a not unpleasant task, if we could bring to light a single remedy that has so far lain buried in obscurity due to neglect and abandonment. In particular, we have set out to examine and describe the LIGNUM COLUBRINUM. Pharmacists, by confusing it with others, have sometimes dispensed it indiscriminately, which has led Physicians, not obtaining the desired effect, to be forced to abandon its use, although it is a very valuable medicine. It is not surprising, since all adulterated remedies suffer the same fate. Thus, it is no wonder that when pharmacists frequently substitute Hypochaeris for Arnica, Centaury for Scabiosa, and Lathraea for Dentaria, these three excellent remedies fell into disuse[345]. Similarly, Lignum Colubrinum, having been imperfectly described by travelers, was confused with two other different plants, assuming that it would produce the same effect as ours. But to form a clear concept about this Lignum, it is necessary to start from the beginning.

V. Among the admirable phenomena of nature, the poisons with which the most provident Creator has endowed certain animals, such as scorpions, centipedes, ants, bees, and snakes, are not of inferior order. The manner in which the snake bites has been a subject [p. 6] that has notably troubled the great learned minds. Rhede[346] seems to have hit the mark by teaching that venomous snakes bite with the large teeth of the upper jaw, those that are movable, hollow, slightly perforated at the tip, and that contain at their base a vesicle filled with a liquid which, injected into the wound during the bite, causes the symptoms that follow snake bites. Moreover, this liquid is so benign that it can be ingested without danger by a healthy person. However, when instilled into the wound and mixed with the blood as if through a siphon, it creates symptoms specific to each species of snake: thus, the bite of the asp induces sleep, that of the cerastes causes tetanus, that of the viper causes jaundice, that of the seps causes gangrene, that of the dipsas causes intense thirst, and that of the prester causes swelling (*)[347].

NAJA

VI. Among snakes, one of the most venomous is the *Naja*, commonly known as such among the Sinhalese.

Coluber scutis abdominalibus 193, squamis caudalibus. [Snake with 193 abdominal scales, 98 caudal scales]. *Syst. Nat.* 6. n. 89. § 7.

Serpens Indicus coronatus Diademate seu conspicillo insignitus. [Indian snake crowned with a diadem or spectacle mark]. *Raj. Quadr.* 330.

Vipera Cobras de Cabelos, Naja dicta. [Viper called Cobras de Cabelos, known as Naja. *Kaempf. Amoen.* 565. tab. 567].

Anguis conspicillo insignita. [Snake with spectacle mark]. *Amoen. Acad.* 112.

Serpens Indicus coronatus diademate seu conspicillo insignitus, Lusitanis Cobras de Capello dictus. [Indian snake crowned with a diadem or spectacle mark, called Cobra de Capello by the Portuguese]. *Seb. thes. 2.* t. 90. f. 1.

Serpens malabarica diademate coronata sive conspicillo insignita, Cobra de Capello dicta. [Malabar snake crowned with a diadem or spectacle mark, called Cobra de Capello]. *Seb. thes. 2.* t. 85. f. 1.

Serpens ex regno peru conspicillo insignita. [Snake from the kingdom of Peru with spectacle mark]. *Seb. thes. 2.* t. 85. f. 1. [p. 7]

Serpens Naja siamensis cum conspicillo seu Cobra de Capello, vel Cabelo dictus. [Siamese snake with spectacle mark, called Cobra de Capello or Cabelo]. *Seb. thes. 2.* t. 89. f. 1.

Serpens cum conspicillo minor. [Smaller snake with spectacle mark]. *Seb. thes. 2.* t. 89. f. 2.

Serpens indicus cum conspicillo lepide circulatus. [Indian snake with beautifully ringed spectacle mark]. *Seb. thes. 2.* t. 89. f. 3.

Serpens brasiliensis cum conspicillo cordis oculati formam habente. [Brazilian snake with a spectacle mark in the shape of an eye heart]. *Seb. thes. 2.* t. 89. f. 4.

Serpens indicus Naja seu Lusitanis Cobra de capello dictus maximus conspicillo notatus, mas. [Indian snake Naja, called Cobra de Capello by the Portuguese, with very marked spectacle, male]. *Seb. thes. 2.* t. 97. f. 1.

Serpens indicus seu Cobra de Capello femina sine perspicillo. [Indian snake or Cobra de Capello female without spectacle mark]. *Seb. thes.* 2. t. 97. f. 2.

Serpens Ceilanica conspicillo notata seu Cobra de Capello. [Ceylon snake with spectacle mark, called Cobra de Capello]. *Seb. thes. 2.* t. 97. f. 3.

Serpens conspicillo ornata ex Nova Hispania. [Snake with spectacle mark from New Spain]. *Seb. thes. 2.* t. 97. f. 4.

Cobra de Capello femina sine perspicillo. [Cobra de Capello female without spectacle mark]. *Seb. thes. 2.* t. 90. f. 2.

Known as Naja among the Sinhalese.

And as Cobras de Cabelo to the Portuguese residing in India.

It inhabits East India, especially Ceylon.

The specific characteristics of this snake are: the expan-

[344]Cascarilla: likely refers to *Croton cascarilloides* and *C. eluteria*; acts as a bitter tonic to stimulate digestion, an astringent for treating diarrhea and dysentery, a febrifuge to reduce fever, and an expectorant to expel phlegm. Contrayerva: Is *Dorstenia contrayerva*, used as stimulant for treating fatigue and weakness, diuretic, expectorant, and antispasmodic. Senega: Is *Polygala senega*, commonly employed in the treatment of bronchitis and pneumonia.

[345]Hypochæris: Likely *Hypochœris radicata*, used as a diuretic and wound healer. Arnica: Refers to *Arnica montana*, applied topically to reduce inflammation and swelling from bruises and sprains. Centaury: Likely *Centaurium erythræa*, used as a digestive tonic, febrifuge, and wound healer. Scabiosa: Refers to *Scabiosa columbaria*, used as an expectorant and wound healer. Lathræa: Parasitic plants in the broomrape family (Orobanchaceæ), lacking chlorophyll and obtaining nutrients from other plants' roots; unlikely widely used medicinally due to their parasitic nature. Dentaria: Various species used as a diuretic, expectorant, and historically as an antisyphilitic.

[346]Redi (1664).

[347]Footnote on p. 6 (*) *Amœn. Acad.* p. 112. See Hast (1749).

sion of the skin on both sides of the neck when irritated, not unlike the [patagium] of the flying squirrel when it is about to leap; the skin extends about a spithama[348] on each side; the mark on the back of the head resembles a pair of spectacles, which is constant, although its color and sometimes that of the whole body may vary. The abdominal scales were counted by the illustrious Müller and by the illustrious Knight of Sagramoso in the Petropolitan Museum[349], as well as by Prof. Kalm in the Sloane Museum of London, who, as I have seen in letters sent to the Nob. D. Praes.[350], all agree on 193 abdominal scutes and 60 pairs on the tail[351]. [p. 8]

KAEMPFER, who personally saw this snake in India, reports the following: "It is a fierce beast and extremely dangerous to humans, as its bite injects a lethal poison into the wound, causing immediate anxiety and fainting, and if an antidote is not used promptly, the victim dies in convulsions. Even when the antidote is administered late, it is rarely possible to prevent gangrene in the affected part, which is difficult to heal. Therefore, it seems that this type of snake is the only one that the barefoot natives fear and avoid. This viper reaches a length of 3 or 4 feet, with moderate thickness, scaly skin beautifully striped, tense, rough, and of a blackish-brown color, with a whitish abdomen. It has the peculiarity of inflating the skin on both sides of its neck when provoked, extending it in a flat or compressed circle, similar to a wing or band. That band on the back shows an exact image of a pair of spectacles delineated in white, with the circles represented on both sides of the extended band and the arch connecting them on the same head. p. 565[352].

RAY *quadr.* 318[353]. The venom of this snake is notable for an observation worthy of mention. An Indian brought a large snake[354], claiming it was harmless, and allowed the previously beaten snake to bite his bare arm. He then placed the blood that flowed out on his thigh; after that, he took another smaller snake, a highly venomous cobra de capello, and by squeezing the snake's neck, extracted approximately one gram of liquid from its maxillary vesicles and placed it on the clotted blood on his thigh. As soon as it mixed with the blood, it immediately caused a surprising effervescence and bubbling, acting like a ferment and changing its color to yellow. In summary, this snake in India is so deadly that any person, though not much concerned about other species of snakes, is horrified at the sight of this one, which is the most virulent of all. Hence, when a young man is bitten by a Naja, [p. 9] and another tries to suck out the venom, he himself dies within an hour, according to the same author. This is why snake charmers prefer to enchant these specimens over others, as Kaempfer reports on p. 566, when they want to show something astonishing to the spectators.

MUNGOS

VII. We also deem it worthy of mention to say a few things about a mongoose called Mungo.

Mustela glauca *Syst. Nat.* 6. n. 9. §. 9.

Viverra indica, Qwil & Qwirpele dicta. [Viverra indica, called Qwil and Qwirpele]. *Raj. quadr.* 197.

Viverra indica ex griseo rufescens. [Viverra indica, of reddish-gray color]. *Raj. quadr.* 198.

Mustela seu Viverra indis Mungutia, Lusitanis Mungo, Batavis Muncus. [Viverra or Viverra called Mungutia in India, Mungo by the Portuguese, and Muncus by the Dutch]. *Kaempf. Amoen.* 574.

Qwil or Qwirpele, *Garz. Arom.* 163.[355]

HABITAT in the warm regions of Asia.

It has the figure of a squirrel, though a bit larger and with slower movements, and is adorned with bluish-gray (glaucous) hair, longer on the tail and decorated with black spots. Kaempf. 574. This animal pursues the Naja in the same way a cat chases a mouse. As soon as it sees a snake, it begins to fight with it; if at any moment it is bitten by the snake, it runs to find a specific plant, and after eating it, immediately returns to renew the fight. The Indians believe that the mongoose eats a root they call Mungos and thus frees itself from the poison. The illustrious traveler Kaempfer, who had a domesticated viverra as a companion, tablemate, and

---

[348]See note 143.

[349]*musæo Petrop.* in the original. Refers to *Serpens Indicus ex fusco et albo variegatus, ad duos pedes et dimidum longus, rostro oblongo et capite plano, diademate seu conspicillo ignitus. Lusitanis dicitur* Cobras de Capello Germanis Brill-Schlange. (Anonymous, 1742: 454).

[350]Refers to Carl Linnæus.

[351]Letter identified as L0923 in linnæus.c18.net, dated in London on June 14, 1748. There are differences in the number of scales. While Darell noted 193-60, Kalm indicates 183 abdominals and 60-61 caudals ("...*Scuta abdominalia på denna äro 183; men om man wil lägga de til, som under hakan ligga imbricatim, som dock ei tyckas kunna kallas scuta, så blir det 2. mera; squamæ caudal äro 60. à 61. Ty den yttersta war så otydel., at iag ei kunde se, om den bestod af en el'. 2. squamis...*").

[352]"...*Ferox & hominibus infensissima est bestia, quæ mordendo exitiale virus instillat vulneri. Nam ab eo, qui mordentur, subito anxietates cordis & lipothymias patiuntur, neglectoque alexipharmaci maturu usu, convulsu moriuntur: vel sumpto serius antidoto, partis læsæ gangrænam raro effugiunt, difficulter sanabilem: inde nullum aliud serpentum, quam hoc, genus nudipedes Indos timere & cavere videris. Obtinet vipera hæc (nam ad hanc classem referunt) trium vel quatuor pedum longitudinem, crassitiem mediocrem; cutem squamatim & decore striatim, tensam, asperam, coloris ex fusco nigricantis, in abdomine albidi. Habet hoc lacessita singulare, quod utroque cervicis latere cutem inflare, & in orbem planum, sive compressum, ad instar alæ vel vittæ extendere soleat. Vitta illa in dorso exactissimam imaginem exhibet conspicilli albido colore delineati, cujus orbiculi in expansa utrinque vitta, &, quæ hos connectit arcus, in ipso capite expressi sunt...*". See Cobras de cabelo descriptio, in Observatio IX. Tripudia Serpentum in Indiâ Orientali, in Fasciculus III of *Kæmpf. Amœn*-Kæmpfer, 1712.

[353]See *Raj. Quadr.* - Ray, 1693.

[354]What follows is part of a story narrated to Ray by a friend, "...*homine intelligente & multiscio...*".

[355]Pagination coincides with the fourth edition of *Aromatum...* See *Garz. Arom.* – Horto (1593).

escort on all his excursions, claims to have witnessed this fight, although he could not determine with certainty which root the viverra sought. Whatever the truth may be, whether this mongoose has discovered an antidote to the venom of the cobra or not, it is certain that there is an Indian root that acts as a sacred anchor against the bite of the Naja, and we have set out to investigate it. [p. 10]

VIII. Regarding which root in India is as specific against the bite of the Naja as the Senega root is against the bite of rattlesnakes in America, there are almost as many opinions as there are authors who have mentioned this subject. And we should not be surprised: European doctors, due to the great geographical distances and the fact that exhaustive botanical expeditions have not yet been carried out in these regions, have not been able to accurately compare the description of this root. Therefore, García calls it Lignum Colubrinum, Kaempfer calls it Mungum, Grimmius calls it Ekawerya, and others give it different names.

GARCÍA

IX. García de Orta, in his history of the aromatic medicines of Indian plants, was the first to recount the story of *Lignum Colubrinum* in Book I, Chapter 44. Not only is it effective against bites and stings from venomous animals, but also against worms, eruptions, rashes, impetigo, cholera, fevers, etc. On the island of Ceylon, there is a very dangerous species of snake called the cobra de capello. There is also a species of animal about the size of a viverra that is a mortal enemy of this snake, and which the locals call Quil and Quirpele. Whenever this animal is about to confront the snake, it first chews the exposed root of a plant, then wets its front paws with saliva and rubs its body. It then attacks the snake and does not let go until it kills it. Instructed by this spectacle, the Cingalese have learned to use this root to counteract the venom. Thus, it is clearer than sunlight that García and Kaempfer were talking about the same snake, the same viverra, and the same root. [p. 11] However, confusion about the identity of the true Lignum Colubrinum arose because García described three different species under the same name, all of which are found on the island of Ceylon. We will select the main characteristics of these species to be able to conclude which is the true Lignum Colubrinum.

α) The first and most praised Lignum Colubrinum is the one from which the viverra seeks help, and the locals call Rametul. It reaches a height of two or three spans, with few twigs, usually 4 or 5 and slender. The root, which is the most used, spreads with multiple heads or knots, so that there is always some part of it protruding from the ground. This root is grayish white, very solid, and has a bitter taste. The leaves are similar to those of the peach, but greener. The flowers are very separate from the leaves, in clusters of a beautiful red color. The fruits are similar to those of the elder, but red and hard, grouped in clusters, like in the *Periclymenum* genus (strawberry tree).

β) The second type, similar to the first, is praised for its effectiveness against venoms. It is a tree similar to the pomegranate, with short and strong thorns that make it rough, with white, dense, cracked, solid, and bitter bark, although not as thick as that of the first. Its leaves are yellow and of very beautiful appearance. It is said that this tree, if it grows near another, extends over its highest branches and embraces it like a gourd. Its wood, along with its root and bark, is usually used, but the root is preferred. Personally, I never had the opportunity to see this tree.

γ) The third type, native to Ceylon, has thin, hard, black, and fragrant roots, with few thin branches [p. 12] about 4 or 5 cubits long, which cannot stand by themselves unless tied, and spread along the ground. The leaves are scarce and elongated, resembling those of the lentisk (myrtle), not green but spotted or sprinkled with white spots on a black background.

FRAGOSO

X. Fragoso, in his *historia aromatum ex utraque india*, Lib. 3. c. 37.[356], speaks of Lignum Colubrinum, with which the Cingalese have learned to resist venoms. On the island of Ceylon, there is a species of highly venomous crowned snake. A small animal the size of a viverra, before confronting this snake, chews this wood or its root, which protrudes slightly from the ground, and then rubs its body with its front paws moistened in its saliva. Three types of this Lignum are found in Ceylon: α) The first and most praised is a shrub 2 or 3 spans high, with 4 or 5 slender branches, leaves similar to those of the peach, flowers in clusters of brown color, and fruits like those of the elder, but red and hard. The root is similar to the small roots of our vines, knotted, dense, ashy-colored, and bitter. β) The second type is similar to a pomegranate tree, with short and strong thorns, yellow flowers of rare beauty. γ) The third type is a small tree with few branches that, if not tied up high, cannot support themselves and spread along the ground. It has few leaves, similar in shape to those of the lentisk[357], elongated, not green, but mottled with whitish spots on a black background, while the roots are thin, black, and fragrant. It is evident that Fragoso in everything followed García, adding only that he saw a fragment of the first species with an outer bark similar to that of the white poplar, and an interior [p. 13] of a color intermediate between green and yellow. This confuses the matter, as the species he saw was the second, not the first.

KAEMPFER

XI. Kaempfer, in his work *Amoenitates Exoticae*, p. 577, describes the root called Mungos as follows. The plant, to which the Indians have given the name due to its bitterness, and which the Portuguese call viverra, is native to Java, Ceylon, and Sumatra. The root is simple, about a span long, the thickness of a finger, very twisted, with a spongy, red, and

[356]See Chapter XXXVIII, *De Ligno Colubrino*, in p. 85v. of Fragoso (1601), first published in Spanish in 1572: 181, *Discurso Primero del Palo Colubrino*.

[357]Presumably *Pistacia lentiscus* Linnæus, 1753a.

rough bark; its interior is woody, hard, without fibers, fragile, whitish, and extremely bitter, similar to gentian but with a subtler and less unpleasant taste. The stem is straight, slender, uniform, and a foot tall, with alternate twigs. The leaves are opposite, sparse, narrow, an inch long, and lanceolate. The peduncles arise from the axils of the shoots, are two or three transverse fingers long, thin, and deciduous. There are about 50 very small flowers, arranged in several clusters and forming a small umbel the size of a nail, bright red, naked, tubular, and with 5 or 6 lobes. The fruit rests on a pedicel, is didymous (twin), with divergent peaks, round, slightly compressed, and spindle-shaped, bony, the size of a coriander seed, with a fleshy skin of black or pale color, and a white, fatty, and shiny core.

GRIM

XII. Nic. Grim, a Swedish physician who lived a long time in India, in his "Laboratorium ... Ceylon" p. 116[358], says the following: The root of the snakes, called Ekawerya, is a very beautiful plant, growing tall, with slender branches and long, pointed leaves. The root, with its knots underground, is extremely woody, white and chestnut-colored, with ashy bark and a bitter taste. This is the first species described by García de Orta among the Lignum Colubrinum. [p. 14] Its properties are no less than those of the great snake tree, and the natives highly praise it against all kinds of venoms. From this, it can be concluded that the species described by Grimm as Lignum Colubrinum is the first (IX. x.), similar to gentian, and that his great snake tree is the second species, as will be shown below.

HERMANN

XIII. Paul Hermann, in his Cynosura Mat. Med., p. 234[359], asserts that the Lignum Colubrinum found in pharmacies is more of a root than a trunk, originating from a tree in Timor. The root is woody, the thickness of an arm, solid, heavy, extremely bitter, odorless, with a brown ferruginous bark sometimes stained with ashy spots. In his Musaeo Zeilanico, p. 55[360], Hermann also describes the Naghawalli colubrina Zeilanica, a plant similar to Periclymenum, with spotted leaves, brought from Mount Adam. It is named after the Nagha snake, whose bites are wonderfully and specifically cured by the leaves of this plant. In the same work, on p. 37, he describes the Ekawerya, a herbaceous Periclymenum from Ceylon, with variegated leaves adorned with spots of various colors. In the Prodromus paradisi Batavici, page 363, it is deduced that this plant, according to Hermann, is the Ekawerya of Grimm (XII). Plukenet illustrated it in his Phytographia, plate 112, figure 6[361].

ACOSTA

XIV. Thus, there is evidence upon which we can investigate this obscure topic of Lignum Colubrinum, to finally determine with certainty which plant presents the true one; also, what the other two species of Lignum Colubrinum according to García mean, and what the Mungos of Kaempfer, the Ekawerya of Grim, and the Naghawalli of Hermann represent. In addition to these, Acosta, Camellus, and Rumpf have described some more species of Lignum Colubrinum. [p. 15]

Acosta describes two species from Malabar, whose figures are shown by Dalechamp in his Appendice historiae pl., Lugdun[362].

1. LIGNUM Colubrinum primum Acosta, p. 276. Dalechamp. histor. 1911, f. 1.

Clematitis malabariensis, with vine-like leaves, dracunculus-colored. According to Bauh. pin. 301, it seems to be a species of Aristolochia or Dracunculus.

2. LIGNUM Colubrinum secundum Acosta, p. 76[363]. Dalechamp. hist. 1911, f. 2.

Clematitis malabariensis, with a sinuous root. According to Bauh. pin. 301, described imperfectly.

OTHERS

XV. Camel[364] also mentions two species of Lignum Colubrinum or Munungae from the Philippine Islands, proposed by Ray in his hist. plant., vol. 3, Dendrol. 119[365]. However, both are described so imperfectly that they do not clarify this issue at all.

Rumpf[366] also offers a species of Lignum Colubrinum, which he calls Cajuular, very different from the others. We do not intend to reconcile these various botanical opinions at the present time, contenting ourselves simply with having shown what is the genuine Lignum Colubrinum Oficinalis.

---

[358]See Grim (1677). Reference to *Ekawerya, Slangen Wortel*, is on p. 76 (the book has only 108 numbered pages).

[359]Pagination coincident with Hermann (1726).

[360]Pagination coincides with the first (Hermann, 1717) and second (Hermann, 1726) editions.

[361]Pluckenet (1691).

[362]See p. 1911, Lignvm Colvbrinvm, Cap. CXL, in Dalechamps (1587).

[363]Acosta (1578).

[364]Georgius Josephus Camellus (1661-1706) was a botanist, Jesuit missionary, pharmacist, and naturalist known for his contributions to the knowledge of the flora and fauna of the Philippines. His writings were published in the Philosophical Transactions and in the third volume of John Ray's Historia Plantarum (Camel, 1704; Ray, 1704).

[365]Ray (1704).

[366]Rumph (1741)

LIGNUM COLUBRINUM 1. GENUINE[367].

XVI. García was the first to discover the Lignum Colubrinum, indicating that it grows in Ceylon and is consumed by a certain mongoose to counteract the bite of the *Naja* (as mentioned in § IX). Kaempfer also mentioned the same (§ IX), having seen the Naja, the Mongoose, and the Mungos root in India. Kaempfer's description perfectly coincides with García's (9) in all respects, leaving no doubt that Kaempfer's Mungos (§ VII) and García's highly praised Lignum Colubrinum are one and the same plant. Hermann, who lived in Ceylon, mentions on p. 55 of his book that the Naghawalli or Colubrina Zeilanica has the specific and remarkable ability to cure the bites of Colubri, Nagha, or Najas. In Hermann's personal herbarium, we have seen the Naghawalli collected by himself in Ceylon, along with its name and properties handwritten, confirming that there is no difference between this plant and the Ophiorhiza described in Fl. Zeil. p. 402. Comparing it with the plants of García and Kaempfer, it is clear that they are the same plant. The bitter root protrudes above the ground; the stem reaches up to one or two spithamas in height; the leaves are lanceolate like those of the peach; the flowers are red and clustered in clusters. This last detail is especially notable and has been meticulously observed by García, Kaempfer, and Grimm in this exceptional plant. Kaempfer also notes that the flowers are grouped in a smaller umbel the size of a fingernail, which agrees with the flowering of Ophiorhiza, where the clusters grow until they begin to ripen fruits, which are twin (twin) and similar in size to the cilantro seed. In contrast, the Ophyoxylum described in Fl. Zeil. p. 398 has no red flowers or so tiny as to be visible under a microscope, nor does it carry leaves opposite in pairs, but four opposite leaves decussate and vertical. Therefore, it is understood that Hermann made a mistake in considering his Ekawerya as the true Lignum Colubrinum. Therefore, we conclude that this should be called the more genuine and most optimal Lignum Colubrinum.

XVII. It is regrettable that this exceptional root, due to confusion in its identification, has become so rare and obsolete that it is not found in any European pharmacy, replaced by fragments of wood from a completely different tree. To enable future doctors to recognize this root, here we provide its distinctive characteristics: it is a simple root, no longer than a spithama nor thicker than a finger, curved, wavy, flexuous, woody, hard, fragile, white in color, extremely bitter; its bark is rough, spongy-looking, and red in color. Any root that does not possess all these characteristics cannot be considered as the Lignum Colubrinum Laudatissimum for life-saving treatments.

LIGNUM COLUBRINUM 2, COMMON IN PHARMACIES[368]

XVIII. According to García, the Lignum Colubrinum 2 is a tree with short thorns that, if growing near another tree, embraces its branches. García asserts that its leaves are yellow, though in reality they are green. It is important to note that the leaves of all Strychnos species turn yellow only through desiccation, and García himself admits to never having seen the tree in person. This tree corresponds to the species of Nux Vomica, as securely affirmed by Hermann in Mat. Med., Dalechamps[369], and others.

STRYCHNOS foliis ovatis, cirris simplicibus. Linn, Mat. Med.[370] [Strychnos with ovate leaves, simple tendrils].

Solanum arborescens indicum, foliis napecae majoribus magis mucronatis. Breyn. prod. 2. p. 301[371]. [Solanum arborescens indicum, with larger and more mucronated leaves].

Clematitis indica spinosa, foliis luteis. Bauh. pin. 301[372]. [Thorny Clematitis indica, with yellow leaves].

Modira-Caniram. Reed Malab. 7. p. 10. F. 5[373].

Lignum Colubrinum. Dal. Pharm. 358[374].

Grows in Timor, Ceylon & Malabar.

This is the Lignum Colubrinum found in our European pharmacies. Its true characteristics are: [p. 18] woody root, solid, the thickness of an arm, solid, heavy, bitter, extremely bitter, odorless, with a rusty bark marked with ash-colored spots. The Lignum Colubrinum in European pharmacies is composed of fibers or rather white, numerous, and compact sheets, arranged transversely from the center to the periphery, as well as other distant circular white sheets, among which there are square, hollow cells, interwoven here and there with transverse sheets. It also contains some very narrow, tiny, perpendicular cylindrical tubes that reside in each cell. However, since many cells hardly have one or two tubes, they are often seen as empty, making the lignum substance very spongy and porous, yet at the same time hard.

L. COLUBRINUM EXOTIC[375]

XIX. This is the third type of Lignum Colubrinum ac-

[367] *Verum* in the original. Refers to *Ophiorhiza mungos* L. (Rubiaceæ).

[368] *Officinalis* in the original.

[369] Cited as *Dalæus* in the original.

[370] Linnæus (1749: 27).

[371] Brein (1739). The work has only 108 pages (plus 54 pages + 34 plates of the *Icones rariorum*...fascicle), and *Solanum arborescens indicum*... is mentioned on page 98.

[372] Bauhin (1671).

[373] Rhede (1686).

[374] Pagination coincides with the fourth edition, Dale (1739).

[375] *Peregrinum* in the original.

cording to García. It has slender branches, approximately 5 cubits in length, drooping, and its leaves are mottled or speckled with blackish or whitish spots. This plant, faithfully depicted by Plukenet, is described as follows: Periclymenum Zeilanicum herbaceum, with variegated leaves adorned with spots of various colors. Herm. prodr. 363. Pluk. Almag. 287. t. 212. F. 6.

Clematitis indica foliis mastichi, maculatis albis. [Clematitis indica with mastic leaves, speckled with white spots]. Bauh. pin. 301.

Ekawerya. Herm. Zeyl. p, 37.

Grows in Ceylon. Not found in our pharmacies. The roots are thin, hard, black, and fragrant.

VIRTUES OF GENUINE LIGNUM

XX. From the above, we see that the three species of Lignum Colubrinum differ from each other in genus, species, and virtue. To prevent the virtues of one species from being attributed to others, we have compiled the medicinal qualities of each of these three species in detail. [p. 19]

Radix serpentum or the most praised *L. Colubr.* (§. XVI.) This root, which is lacking in our pharmacies, enjoys the following virtues according to the most trustworthy authors:

García, p. 164: The root is administered with wine or cordial water[376] to persons bitten by a Naja snake. It is also rubbed and spread on wounds, and more in §. IX.

Kaempfer, p. 577: The root is taken both for preservation and cure, against poisons inflicted or introduced, in the amount of a drachm[377] or half a drachm, with a gulp of fresh water or any other liquid, whether ground into powder, by raw friction, or in a more refined manner. A paste made from root powder with saliva or water can also be applied. In putrid fevers and malignant diseases, the root has been used with good results, and with many infallible successes against the bite of a rabid dog in both animals and humans.

Grimm, Lib. 116: Reports that the Sinhalese praise it as a remedy against all poisons, venomous diseases, poisonous snake bites, rabies, continuous fevers, etc. The dosage varies from half to one drachm. This celebrated physician, who lived for a long time in Colombo, Ceylon, used Lignum Colubrinum many times, as can be seen in his Comp. Med. Chym. published in Vienna in 1684[378], where he says the following.

Preparations of Lignum Colubrinum:

In alcohol (spiritus), in sudorific mixtures, 1 to 3 scruples[379]. Used in scurvy, paralysis, colic, cirrhosis, arthritis, etc. (pag. 99, 101, 178, 218, 354).

In sudorific mixtures, 20-30 drops. For convulsions, jaundice, dropsy, intermittent fever, arthritis, syphilis, etc. (pag. 78, 125, 238, 327, 328, 329, 355, 366).

In stable salts[380] obtainable in India, p. 464. [p. 20]

Compositions in which Lignum Colubrinum is used from Pharmac. Indica:

Bezoardic Tincture. [Tinctura Bezoardica] p. 423[381].

Bezoardic Vinegar. [Acetum Bezoardicum] p. 426.

Antidotal Extract. [Extractum Alexipharmacum] pag. 437.

Diaphoretic Liquor. [Liquor diaphoreticus] p. 417.

Lignum Colubrinum is commonly combined with Lignum Ebenum[382] due to their affinity in properties.

LOCHNER (*)[383]: praises the Mungos root especially in malignant purpura. The dosage of the powder is 2-3 scruples, in tea or any other suitable liquor. pag. 84.

A compound of this root is the Stone of Goa[384], according to the author on pag. 87, whose formula, as it is uncommon, is provided here.

Stone of Goa:

Recipe Mungos Root. [Rec. Rad. Mungos]:

African Angelica. [Angelic. afric.]

Virginia Snake[385]. [Serpent. Virg.]

Contrajerva; 31 ana[386]. [Contrajerv. ana 31].

Zwelfer's Deer Skull Powder[387]. [Pulv. e typhis Cerv. Zwels.], 4 scruples.

69 marine chelicerae. [from crab]. [Chel. 69. marin.]

Goat Bezoar Horn. [Corn. Caprae bezoard].

Prepared elk. [Alcis philos. ppt.] ana 3 scruples.

---

[376]*Aqua cordiali* in the original. A preparation of aromatic herbs and spices macerated in the distilled water for a period of time, allowing their essential oils and other compounds to infuse into the liquid.

[377]One drachma is equivalent to 3 scruples, or 60 grains, or 1/8 of an ounce, or 3.888 grams.

[378]Grim (1684).

[379]1 scrupulus = 20 grains. Approximately 1.28 grams.

[380]*Salia fixa* in the original,opposed to *sal volatil.*

[381]In the analyzed edition, the recipe for *Tinctura Bezoardica* begins on page 422.

[382]Probably *Diospyros ebenum* J. Kœnig ex Retz., 1776. The use of ebony wood before its taxonomic (nomenclatorial) definition can be traced to Dioscorides.

[383]Footnote on p. 20 of the original: (*) *Lochnerus M. FR. Mungos Animalculum & Radix. Norib. 1715 in E.N.C. cent. 3 append.* See Lochner (1715).

[384]*Lapis de Goa* in the original, a man-made bezoar.

[385]*Polygala senega* root?

[386]*ana* (sometimes abbreviated as "aa"), from Greek ἀνά, each or per each. In Latin *ana* refers to *partes æquales*, "equal parts". It instructs that this and the preceding ingredients should be taken in equal quantities.

[387]Probably a recipe attributed to Johann Zwölfern. See Zwölfern (1692).

Viper spine. [Spin. Viperin.[388]] 1 scruple.

Ambergris. [Ambr. griseae].

Quality oriental musk as needed. [Mosch. orient. ana qv. insig.].

Mix to make very fine powder, ground on porphyry and combined with some gummy binder to form a subglobular mass covered in gold. [m. f. pulv. subtiliss. sup. porphyr. levigand. & mediante gummi aliquo combinand. in formam subglobos. auro obducend]. [p. 21]

BONT (*)[389]: praises the root of Lignum Colubrinum, more bitter than aloe, against tapeworms, roundworms, ascarids, venomous bites, intermittent fevers, and against Timor fever, which is a continuous fever caused by sandalwood. [p. 30]

VIRTUES OF L. COMMON IN PHARMACIES

XXI: The medicinal virtues of Lignum Colubrinum vulgaris or Lignum Colubrinum 2 according to García, stored in our pharmacies (§. XVIII) were exposed by Hermann in Mat. Med. 235. The use of this Lignum is not recommended until it is at least one year old; fresh, in the first year, it causes madness, colic, vomiting, and finally convulsions. It is recommended for intermittent fevers, quartan fevers, and for eliminating worms; its use should be prudent due to its potential harm.

RAY, hist. 2. p. 1807. Recommends this type of Lignum Colubrinum against poisons and uses it in the same way as Lignum Colubrinum Laudatissimum.

HEIDE centur. obs. 7.[390] A lathe operator took a powder in the afternoon to treat daily fever. That night he felt quite well, but when he tried to get up in the morning and move his body, he experienced tremors in his limbs; called, I administered a hypnotic mixture and within a few hours the patient was relieved of tremors. Administering 3 scruples of this Lignum Colubrinum to a cachectic woman, in addition to tremors, caused stupor, so much so that the patient had no awareness of the world or being [p. 22] alive; she was treated with a hypnotic mixture. An eminent lady reported that her servant had been in a state of satiety after taking this wood.

Therefore, this tree is used against intermittent fevers, snake bites, and worms. The dose should be one scruple in decoctions, or 3 scruples in essences, as it is rarely administered in its pure form.

Although Lignum Colubrinum causes tremors and stupor according to the mentioned authors, we do not consider it should be excluded from pharmacopeia, just as wine is not eliminated from food, which in excess has similar effects. A prudent physician must correctly determine the doses of remedies and thereby provide healthy medications even from poisons, as evidenced with opium, senega, and other heroic medicines.

VIRTUES OF L. EXOTIC

XXII: The third type of Lignum Colubrinum according to García (§. XIX). It is said to be suitable for countering poisons and dulling poisoned arrows, but since it has not gained as much fame as the first types of this wood and has not been correctly recognized or examined by botanists, we do not consider it appropriate to continue with this, which has already been left behind, out of respect for the truth.

XXIII: The Ophioxylon, with leaves in quartets, as described in Fl. Zeil. 398 and Linn. Mat. Med. 474 according to Hermann and Burmann, is often used against snake bites, poisons, and venomous diseases; however, as it is more unfamiliar to physicians, we do not consider it ours. The dose is one 𝔊[391] in infusion.

NOTE: The amendment (*Emendanda*) is not included as the corrections were incorporated into the text during translation.

[388] *Spina Viperæ*, the spiny part or bone of a viper.

[389] Footnote on p. 21 of the original. (*) BONTIUS Jacob. Hist. nat. & Med. Ind Orient. Aust. 1658, only has few data on Lignum Colubrinum. p. 46. Refers to the Cobra de Capello as a small snake, the thickness of a finger, black above and yellow below. He observed a Moor who was bitten by it on the thumb, and although he allowed his arm to be tightly bound due to the pain, he fervently begged that the dressing be loosened, and when it was, he immediately died. See Bontius (1658).

[390] Edition not found. In those consulted (Heide, 1684; 1686), the story of the turner's fever is on page 121.

[391] The symbol 𝔊 (a letter G in fraktur) is an 18th-century abbreviation for *drachma* (ʒ), a unit of measurement in apothecary systems, equivalent to about 3.9 grams in the metric system.

D. D.

RADIX SENEGA,

QUAM,

CONSENS. AMPLISSIM. FACULT. MEDIC.

IN REGIO AD SALAM LYCEO,

PRÆSIDE

VIRO CELEBERRIMO & EXPERIENTISSIMO

DN. DOCT. CAROLO LINNÆO,

S:Æ R:Æ M:TIS ARCHIATRO,
MED. ET BOTAN. PROFESS. REG. ET ORD.
ACAD. IMPER. MONSP. BEROLIN.
STOCKHOLM. UPS. SOCIO.

PRO SUMMIS IN MEDICINA HONORIBUS
OBTINENDIS,

PUBLICÆ CENSURÆ SUBMITTIT

JONAS KIERNANDER, AND. FIL.

O-GOTHUS.

IN AUDIT. CAROL. MAJOR. AD DIEM. VIII. APRIL.
ANNI MDCCXLIX.

H. A. & P. M. S.

HOLMIÆ,
TYPIS LAURENTII SALVII.

Kiernander, J. 1749. Radix Senega. Holmiae. Laurentii Salvii. [3] + 4-32 + 1 pl.

Kiernander's dissertation (also included it in the second volume of *Amoenitates Academicae*, Kiernander, 1751) explores the botanical and medical properties of *Polygala Senega*, commonly known as Senega or snakeroot, particularly focusing on its efficacy as an antidote to snake venom, specifically that of the Rattlesnake in America. Divided into eleven sections (I. *Introductio*, II. *Crotalophorus*, III. *Polygalae*, IV. *Senega*, V. *Qualita*, VI. *Vires*, VII. *Usus*, VIII. *Preparata*, IX *Therapia*, misnumbered as IV, X. *Succedanea* and XI. *Epilogus*), the dissertation provides a comprehensive study that begins with historical references and botanical descriptions drawn from early accounts such as Catesby's. It details the medicinal applications of Senega root, emphasizing its use as a potent sialagogue, diuretic, and expectorant. The dissertation also discusses various preparations of the root and their therapeutic uses in treating conditions like dropsy, gout, rheumatism, and *Marasmus Virginicus*, an endemic disease in Virginia. Notably, Kiernander's work highlights Senega root as a specific remedy against snake venom, a discovery attributed to Dr. Tennent's identification of *Polygala Senega* as the true Rattlesnake root. Comparisons with *Polygala vulgaris* underscore the unique potency of Senega root in medical applications, despite the latter's similar but less pronounced properties.

G. h. g.[392] The Root of Senega which, by the consent of the most noble Faculty of Medicine in the Royal Salamis Lyceum, under the Presidency of the most celebrated and experienced man, Dr. Carl Linnaeus, Archiatrus of the most serene Kingdom of Sweden, Professor of Medicine and Botany at the Royal academy, member of the Imperial Academies of Montpellier, Berlin, Stockholm, and Upsala, for the highest honors in Medicine to be obtained, is publicly submitted to review by Jonas Kiernander, son of Andreas, from O-Gothus. in the Major Carolina Auditorium on the eighth day of April, in the year 1749. H.A. & P.M.S[393]. Holmiae. Printed by Laurentius Salvius.

[p. 2]

To the President and Chief Physician of the Royal College of Physicians, to the Royal Physicians and Assessors, to the most noble and experienced patrons and promoters, esteemed to the highest degree, this Medical Dissertation is submitted as a token of due respect and commendation, with a prayer for all-encompassing success. D.D.D.[394] The Author.

[p. 3]

INTRODUCTION

I.

Nature seems to have provided serpents with less defense than other animals, as she cast them onto the earth without the same aids. While other animals were endowed with legs for running away, wings for flying, or fins for escaping the attacks of their enemies, serpents received none of these. Yet, upon closer inspection, we clearly see that provident nature has equipped serpents with such dreadful weapons that often superior animals shudder at their sight. Even at the mere

[392]D. D. in the original; probably "Deus dedit," which translates to "God has given". Thus, G. h. g.

[393]Not clear, probably H.A. could stand for "Hof-Bokhandler" (Court Bookseller) or "Hof-Akademiens Bokhandlare" (Bookseller of the Royal Academy), and P.M.S. could represent "Privilegierad Majestats Boktryckare" (Privileged Royal Printer).

[394]Dedit, Dicavit, Dedicavit, which translates to *He has given, dedicated, and devoted.*

sight of serpents without venom, many animals are nearly paralyzed with fear[395]. These weapons, though not large, are extremely dangerous; indeed, [p. 4] nature herself has instilled fear of these creatures into other animals. And rightly so, since in the entire animal kingdom, the bite of a serpent is one of the most treacherous evils, as noted by Pliny[396].

2. The weapons of these animals are generally considered to be their teeth, which are positioned in the upper jaw, usually two on each side. These teeth, although somewhat larger, are similar to others but differ in that they cannot chew food with their assistance. Contrary to the properties of regular teeth, they are mobile and attached to a thin horizontal bone, which allows them to be raised or lowered. At the tip, they are perforated, functioning like a syringe when they inject their deadly juice into the veins of the wounded animal.

3. Although the teeth of serpents that inflict lethal punctures are similarly shaped throughout the family, the venom or the liquid distinguished as poison varies, mixed with the mass of the blood. This is supported by experience, which teaches us that the bites of different serpents are accompanied by different kinds of illnesses. Some cause jaundice, others gangrene, some coma, and others hemorrhages, etc.

4. To attempt to demonstrate the action of liquids on liquids with subtle reasoning is not our purpose, [p. 5] any more than it is to determine why milk received into the veins, according to the method of infusion once most common among physicians, kills just like poison.

5. The genus of serpents, due to their exceedingly dangerous armament, often brings great calamities even to the most blessed lands. Europe harbors the viper, Africa the asp, Asia the cobra, and America its rattlesnake. Europe is considered happier than the rest because its resident enemy is less dangerous. Although the bite of the viper sometimes leads to sad outcomes, it is not always deadly, while the others scarcely ever attack the inhabitants without causing sudden loss of life.

6. The rattlesnake, inhabiting American lands, is exceedingly cruel to the people there. Its penetrative power often proves fatal to the inhabitants, who have always been anxious to find a certain antidote for its venom. They believe that the Wise Creator has so arranged it that where poison is spread, there also is a remedy stored. Not wrongly, the simple and wild Pennsylvanians happily noticed a divine gift in a root growing among them. They guarded it as a secret and, for no reward, revealed it to the Christians inhabiting Virginia. Until Mr. Tennent (with a reward supposedly set by the supreme English Senate) brought it out of darkness for the salvation of many people. Having obtained the medicine, he rightly believed that such great power in purifying the blood mass was not devoid of all strength to combat other diseases as well.

7. Lest this excellent specific against the bite of the rattlesnake remain further hidden from our learned world, I have decided to outline it in this academic dissertation, driven by the hope and confidence that our work will benefit not only those who handle medicinal materials but also practical physicians who have neither seen the root nor experienced its virtues, nor read the writings recently published in the English language about this root[397]. [p. 6]

II. CROTALOPHORUS (The rattlesnake).

8. The genus *Crotalophorus* includes serpents with scales covering their belly and underside of the tail, as well as a rattle made of cup-like segments at the end of the tail. *System. N. 6. N.* 91[398].

9. Two species of this genus are known:

I. *Crotalophorus* with 172 abdominal scales, 81 tail scales, and 3 small scales. *Surinamens. Grill.* 17. [p. 7]

To the Virginians, the **Rattle Snake.**

To the Swedes in Pennsylvania, the **Skallerorm.**

II. *Crotalophorus* with 165 abdominal scales and 28 tail scales. *Mus. Princip.* 27.

Vipera caudisona americana minor. *Catesby carol. 2.* p. 142. t. 142. an ?.[399]

Both species inhabit parts of *America*, both South and North, particularly in Virginia, Pennsylvania, & Carolina, Brazil, Mexico, and Surinam.

10. I willingly omit a detailed description of these animals, as it is thoroughly provided in the cited dissertation[400], and their anatomy has been accurately presented by Tyson in *Act. Londin.* N 144, later recited by Ray in *De Quadrupedibus* p. 291. My goal is not to depict the serpent, but rather to convey the idea of a remedy against its venom.

11. Caudisona or *Crotalophorus*, commonly referred to as the **Rattle Snake / Skallerorm**, is so named because of a unique appendage attached to its tail, an addition granted

---

[395] Linnæus included a paragraph with similar content in the tenth edition of *Systema Naturæ* (Linnæus, 1758: 194).

[396] Pliny, Book XXV, Chap. 55. *The employment of these plants for injuries inflicted by serpents.* Referring to Aristolochia, Pliny wrote: "... But it will be as well now to mention the various uses made of these plants, and the effects produced by them, beginning with that most dangerous of all evils that can befall us, stings inflicted by serpents..."

[397] Tennent (1738).

[398] Linnæus (1748a).

[399] Kiernander's characterization refers to *Vipera caudisona americana*, in p. 41 of the second volume of first edition of Catesby's Natural History (1743). The description of *Vipera caudisona americana minor* is on p. 42. Page and plate 142 do not exist.

[400] *Diss. citatis* in the original; refers to Balk (1746).

by the Supreme Creator to no other animal known to this day. This appendage consists of almost concave, elastic bony segments, [p. 8] each of which is inserted into the next at its apex so they cannot fall off. Thus, when the snake bends this way and that, or even prepares to resist, it emits a sound like a rattle, which serves to warn humans to withdraw promptly.

12. Mr. Catesby, the most recent author who saw and illustrated the snake in Carolina[401], provides the following details which are relevant and should not be omitted here. The rattlesnake has a brown head, red eyes, and a somewhat brown back with irregular black stripes running down it. The rattle itself is also brown, composed of many connected cells made of a membrane almost as hard as horn. The cells are pyramidal and wavy, and their articulation is lubricated with abundant synovia, facilitating easy movement and enabling them to strike together and produce the distinctive sound when the snake shakes its tail. When this snake bites with its teeth, only the most dangerous symptoms can result, especially if the length of the teeth penetrates the largest and most vital vessels, which often happens. If a major vein is pierced, all medicine is futile unless the injured part is immediately amputated. [p. 9] Sad experience has shown that a person injured in this way can scarcely survive more than 5 or 10 minutes, sometimes barely more than two, before the medicine can exert its internal effect, and the person dies. However, if the snake's fangs penetrate fleshy and muscular parts without wounding any vital organ, a certain root, if immediately available, is a sacred anchor of hope. This horrid snake, the rattlesnake, is the most dreadful among vipers, being the largest and most cruel of them all. The largest I have seen was eight or nine feet long, weighing eight or nine pounds. This dreadful animal stealthily crept into the house of Prefect Blake in Carolina, where it would have undoubtedly hidden for a long time if all the domestic animals had not warned the inhabitants with unusual clamor about the presence of an enemy. Dogs, pigs, turkeys, and all domestic birds, inflamed with equal hatred, showed singular consternation by raising their fur and crests; and, to make their aversion more evident, they surrounded it, not approaching closer than necessary to remain safe from attack. In the meantime, the serpent ignored their threats and slowly retreated. Among all vipers, this rattlesnake uses the greatest slowness in its movements. It attacks no animals except those which serve as prey or those which provoke it in some way, and it warns of its retaliation by shaking its rattle. In that part of America, this serpent is believed to be the most venomous: and there is little to contradict this belief, as it is the largest, and its bite causes the largest wounds, into which it injects a large amount of venom. [p. 10] The remedy used by the natives to prevent the further spread of the venom, besides the amputation of the injured part, is sucking the wound, which sometimes yields the desired effect if the venom is minimal. Those once freed from this affliction feel periodic pain annually at the same time of the year they were bitten, similar to the effects of a tarantula bite. They apply various roots against the bite, such as Asarum (13), Prenantes (13), and a certain root (16: ii), which the natives of Virginia and Carolina bring from the interior of their region, carrying it with them in a pouch. They chew it, swallow the juice, and apply the chewed root to the wound. Throughout America, it is generally believed that this serpent possesses an enchanting, or attractive, power by which it draws animals to itself to devour them. Although I have never witnessed this myself, many have reported it, and their accounts agree amicably. They say that animals, especially birds and squirrels, which the serpent mostly eats, do not notice it at first, but then, jumping from branch to branch, flying towards the enemy without realizing the imminent danger, they approach closer and closer, their gestures and clamor indicating their disturbance, descending from the treetops to the serpent, which, with open jaws, seizes the arriving prey and instantly devours it. Those bitten by this rattlesnake experience intense pain, [p. 11] inflammation of the wound, difficulty breathing, coughing, bloody sputum, a hard and inflammatory pulse, sometimes weak and rapid. The wound becomes gangrenous, the injured area swells, and the flesh decays into pus. The bite is much more dangerous under Sirius than at other times of the year[402]. All this according to Mr. Catesby.

13. After the more sagacious Europeans had settled in America, they observed those wounded by this rattlesnake suffering from these dreadful calamities. At the same time, they learned that the indigenous Americans had kept an antidote specific to this for themselves alone, and that it was a root. Having tried various plants, they understood that there are several plants which could provide some remedy for this wound, such as:

1. ARISTOLOCHIA with weak, angularly flexuous stems, heart-shaped oblong leaves, and solitary recurved flowers. *Gron. Virg. 57.*

Serpentaria Virginiana *officinarum.*

The Snake Root of Virginia[403].

2. ACTAEA with very long racemes. *Gron. Virg. 37.*

Black or wild Snake Root for New Englanders.

3. ASARUM from the land of Maryland, with the leaves of the yellow violet or asarine.

Black Serpentaria: Pluck. Alm. 53. T. 15. F. 3. Raj. Hist. 2. p. 1857.

Black Snake Root in English. [p. 12]

---

[401]Catesby (1743).

[402]*sub Sirius* in the original, or "in times of Sirius". The heliacal rising of the binary star Sirius, or *alpha Canis Major* (α Cma), occurs at the beginning of the summer in the Northern Hemisphere.

[403]In Swedish in the original.

4. PRENANTHES with numerous flowers, hastate and angular leaves. *Linn. Hort. Cliff.* 383. (α)

Autumnal Prenanthes, with light purple flowers nodding downward, arranged in a spike along the stem, rough incised leaves, single stem. Gron. Virg. 89. Witts Snake Root.

5. VERATRUM with a very simple stem. Gron. Virg. 195. Reseda with lanceolate leaves, very simple stem. Gron. Virg. 59.

6. OSMUNDA with pinnately divided fronds, spicate fructifications. *Gron. Virg.* 196. Fern Rattle Snake-Root.

14. None of these (13) were truly genuine, as was quite clearly demonstrated by the distinguished Mr. Tennent of Scotland, who, unless I am mistaken, through his own efforts, finally obtained this knowledge from the American natives around the year 1736, so that they candidly taught him to recognize the true species. This Mr. Tennent later openly revealed in his writings that it is the same species of Polygala, which, however, was little known to botanists until then, and which is not easily distinguished from other species included under the vast genus of *Polygala*. Therefore, in order to clearly present the genuine appearance of this Polygala, I have decided to enumerate the truly discovered species up to this day, so that there may be no further error [p. 13 (missnumbered as 14)] in selecting the true one: and besides, I will add an illustration of it, as the one provided by Plukenet is very rarely seen in our libraries.

III. POLYGALA.

15. Since the character of the genus Polygala is fully defined in the *Gen. Plant.* of the celebrated Mr. President, No. 686[404], I refrain from describing it anew here: only emphasizing that, just as Polygala Tournefort and Penaea Plumier are included under the same genus, so also those that have a filamentous bundle under the keel, like the common *Polygalas*, and those that do not bear such a bundle, like the *Penaeas*, should be referred here. Hence arises a very clear distinction, of dividing the species into two groups, the former containing the crested ones, and the latter the unbearded ones.

16. The species of Polygala that have been duly known to botanists and to us up to this day are as follows:

1. POLYGALA with crested racemose flowers, simple herbaceous procumbent stems, and linear-lanceolate leaves.

Polygala with linear-lanceolate leaves and diffuse herbaceous stems. Virid. cliff. 70. Hort. *Cliff.* 352. *Roy. lugdb.* 393. *Fl. Svec.* 586.

(α) Polygala with all acute leaves. Hall. Helv. 606. Polygala major. Bauh. pin. 215. [p. 14]

(β) Polygala with lower leaves somewhat round, upper leaves narrow and acute. Hall. Helv. 606.

Polygala vulgaris Bauh. pin. 215.

(γ) Polygala minor, with slightly round leaves around the root. Vaill. paris. 161[405]. t. 32. f. 2.

(δ) Polygala, with more acute leaves, Monspeliaca. Bauh. pin. 215. Vaill. paris. 167. Mag. monsp. 207.

Onobrychis tertia purpurea. Dalech. hist. 491.

2. POLYGALA with crested racemose flowers; the keel shorter than the crests, with a suffrutescent stem, and linear-lanceolate leaves.

Polygala africana, with awl-like leaves and large flowers. Old. afr. 32.

Polygala frutescens, with narrow, branched leaves, and flowers appearing umbellate at the top. Raj. app. 640.

Polygala with oblong acute leaves and broader purple flowers. Burm. afric. 204. t. 73. f. 5.

Polygala africana glabra, with the helmet of the flowers adorned with two fringed crests from the involucrum, etc. Pluk. alm. 300. t. 53. f. 2.

3. POLYGALA with crested racemose flowers; the keel shorter than the crests, with a suffrutescent stem, and linear-awl-like leaves.

Polygala with linear-awl-like leaves. Hort. Cliff. 353.

Polygala frutescens with linear leaves and larger purple flowers. [p. 15] Burm. afr. 202. t. 73. f. 2.

Polygala africana frutescens with narrow larger leaves. Tournef. inst. 175.

4. POLYGALA with crested racemose flowers, with an erect very simple suffrutescent stem, and awl-like leaves.

Polygala aethiopica, with very strict glabrous leaves and scarlet flowers. Pluk. mant 153.

Polygala with linear acute leaves and smaller purple flowers. Burm. afr. 203. t. 73. f. 3.

Polygala capensis, with very narrow leaves and smaller flowers. Pet. sicc. 247.

5. POLYGALA with crested alternate flowers, with an erect branched suffrutescent stem, and linear obtuse rough leaves.

Polygala with linear obtuse leaves and very small whitish flowers. Burm. afric. 204. t. 73. f. 4.

Polygala aethiopica, with narrow hirsute leaves and dull purple flowers. Pluk. mant. 153. t. 404. f. 6.

6. POLYGALA with crested racemosely scattered flowers; with a crescent-shaped keel, fruticose stem, and lanceolate obtuse leaves.

Polygala with lanceolate obtuse leaves and frutescent stem. Hort. Cliff. 353.

Polygala frutescens, with oblong glabrous leaves and purple flowers. Burm. afric. 200. t. 13. f. I.

Polygala arborea myrtifolia, from the Cape of Good Hope, with myrtle-like leaves and white flowers with a purple interior. Comm. Hort. I. p. 87. t. 46. Raj. app. 640.

Polygala arborea, with subrotund leaves like myrtle and

[404] Numeration coincides with the second edition of the *Genera plantarum* (Linnæus, 1747b: 259).

[405] Pagination coincides with Vaillant, 1727.

large fruits like tordylium. Pluk. mant. 153. t. 437. f. 4. [p. 16]

β. Flowers with the keels lacking crests.

7. POLYGALA with beardless flowers, and whorled leaves.

Polygala with whorled leaves. Gron. Virg. 80.

Polygala quadrifolia var. cruciata, with flowers turning red from green, compact in a globe shape. Pluk. alm. 301.

8. POLYGALA with beardless flowers, ciliated capsules, and an erect herbaceous stem. Fl. Zeyl. 268.

9. POLYGALA with beardless flowers, multiflorous lateral peduncles, a diffuse very branched herbaceous stem, and acuminate leaves. Fl. Zeyl. 270.

10. POLYGALA with beardless spicate flowers, an erect herbaceous filiform branched stem, and linear leaves.

Polygala with filiform stems, alternate linear leaves, and spicate peduncles. Gron. Virg. 172.

Polygala quadrifolia minor from Maryland, with small whitish flower spikes. Raj. app. 639.

Polygala mariana quadrifolia minor, with a small whitish spike. Pluk. alm. 153. t. 438. f. 4.

11. POLYGALA with beardless spicate flowers, an erect very simple herbaceous stem, and broadly lanceolate leaves.

Polygala with a simple erect stem, ovate-lanceolate alternate entire leaves, and a terminal erect raceme. Gron. Virg. 80.

Polygala virginiana, with oblong leaves, white flowers in a thyrsus, and an alexipharmic root. Mill. lexic.[406] [p. 17].

A small plant from Maryland, with an unbranched stem and a singular spike composed of white florets at the top. Raj. app. 640. see Pluk. phyt. 439. & 453.

Seneka officinarum. Geoff. Mat. 2. p. 137.

Senega, Seneka, Seneca among the Indians of Pennsylvania.

Habitat in Virginia, Pennsylvania, Maryland.

12. POLYGALA with beardless globose-capitate flowers, an erect very simple herbaceous stem, and slightly obtuse lanceolate leaves.

Polygala with alternate lanceolate leaves, a very simple stem, and a terminal capitate corymb. Gron. Virg. 80.

Polygala rubra virginiana, with a small compact spike. Pluk. alm. 300.

Polygala mariana, with red spicate flowers. Pet. mus. 462.

13. POLYGALA with beardless oblong-capitate flowers, an erect herbaceous stem, and acute lanceolate leaves.

Polygala s. Flos ambervalis, with yellow flowers clustered in an oblong head. Pluk. alm. 301. Raj. app. 639.

Polygala virginiana, with yellow capitate flowers. Pet. hort. 5.8.

14. POLYGALA with beardless lateral solitary flowers, an arboreal stem, and obtuse petiolate leaves.

Penaea arborescens, with rough boxwood-like leaves. Plum. Spec. 22. [p. 18]

15. POLYGALA with beardless racemose flowers, an arboreal stem, older leaves oblong-ovate, younger leaves somewhat ovate.

Polygala with oblong-ovate leaves on older branches, somewhat ovate leaves on younger ones, and an arboreal stem. Hort. Cliff. 353.

16. POLYGALA with beardless scattered flowers; with a round-tipped keel, a fruticose stem, and lanceolate leaves.

Polygaloides Hall. Helv. 601.

Chamabuxus with colutea-like flowers. Bauh. pin. 471.

Chamabuxus with colutea-like yellowish flowers. Tournef. act. 1705. p. 312.

β. Polygala fruticosa with boxwood-like leaves and red flowers turning purple. Tourn. inst. 175.

17. POLYGALA with beardless lateral flowers, an arboreal spiny stem, and mucronate oval leaves.

A small shrub with angular branches ending in spines. Leaves small, oval, glabrous, entire, scattered, subsessile, slightly mucronate at the tip. Axillary, sessile, whitish flowers, solitary or in pairs. Habitat in Ethiopia.

18. POLYGALA with beardless flowers, an erect herbaceous stem, alternate linear leaves, and subtriflorous peduncles. Fl. Zeyl. 269. [p. 19]

IV. SENEGA.

17. SENEGA, or the plant that exhibits the true root called Rattle Snake Root, or the antidote of *Crotalophorus*, is the eleventh species of Polygala, of which we add a brief description here, while in the Plate, Fig. 1 shows the root, and Fig. 2 shows the herb nearly life-sized.[407]

ROOT: Perennial, fibrous, woody, flexuous, variously branched, with the apex or base expanding into various tubercles from which the stems emerge.

STEMS: Numerous, very simple, about a foot tall, cylindrical, scarcely noticeably pubescent, weak.

BRANCHES: None.

LEAVES: The upper ones are lanceolate-ovate, the mid-

[406]Unspecified edition/translation. Until 1749, there were three editions and several translations. Here we cite the second edition, *in folio* (Mill. lexic. a, Miller, 1740) and the Dutch translation (Mill. lexic. b, Miller et al., 1745).

[407]See Plate and figures at the end of this section.

dle ones narrower and more lanceolate, all ending in scarcely noticeable petioles, entire, slightly rough on the edges, glabrous and green on both sides but paler underneath, acuminate; the closer to the base of the stem, the smaller, so that the lowest scarcely exceed two lines in length, while the upper ones reach two transverse inches in length and a single inch in width. The longitudinal rib of the leaf is prominent underneath, cylindrical, sending out very faint vessels at right angles towards the margin. [p. 20]

STIPULES: None.

SPIKE: Pedunculated, terminating the stem, with a peduncle shorter than the leaves.

FLOWERS: Arranged in a slender spike, white, sessile, about forty, growing on a striated scape (*)[408]. Thus, the appearance of the plant is almost like that of *Persicaria* or the aquatic *Epilobium*.

V. PROPERTIES.

18. Only the ROOT has attracted the attention of physicians, as the remaining parts cannot match its virtue. Hence, I deemed it necessary to provide a full description of the root, so that it may be better recognized by physicians and pharmacists, to avoid substituting false roots for the genuine one.

α. The appearance of the root is woody, flexuous, somewhat span-long[409], narrower than a finger, divided into multiple branches, somewhat knotty, uneven, with a less evident keel [p. 21] running along its length and contracting towards the same side of the root. This root consists of a thick bark, almost saturated with resin, and a woody, solid, or non-porous substance, somewhat cylindrical.

β. The color of the bark substance is whitish-yellow, covered with a grayish cuticle, while the medullary part is snow-white.

γ. The taste of this root is at first farinaceous, then slightly acidic, and finally penetrating, stimulating, burning, almost constricting the throat with a slight cough.

This sharp, burning, and simultaneously acidic taste is so unique that a similar one can scarcely be found in the entire vegetable kingdom.

δ. No odor.

19. I intended to include an analysis or adequate chemical resolution of the root in this dissertation, but since I lacked a sufficient quantity, I could not carry out this task as desired. However, with the quantity provided by the esteemed President, I conducted some experiments to determine which parts constitute the plant, which contain the main virtue, and which form of medication would offer the best help to the sick. Both tinctures, spirituous and aqueous, turn red with the solution of litmus, [p. 22] and green with syrup of violets. Thus, the litmus solution indicates an acid, while the syrup of violets indicates an alkali, in a contrary manner; this is derived from the fact that the acid hidden in the root is detected more quickly by the litmus solution than by the syrup of violets; the litmus solution is combined with an alkali, while the syrup of violets is combined with an acid. Hence, it is evident that both an acid and an alkali are present. The taste further demonstrates the acid, while the precipitation of mercuric chloride, though slow, nonetheless occurs, appearing almost white and nebulous. From these observations and from the tinctures, both aqueous and spirituous, and their taste and mode of action on the interior parts of the mouth, I conclude that the root consists of earthy, resinous, gummy, and saline particles, and that the greatest virtue resides in the saline-gummy particles.

VI. VIRTUES[410].

20. The virtues of this plant are evacuative, resolving, and possibly vulnerary, which I will explain more clearly.

21. Sialagogue: Anyone who has tasted even a small portion of this root perceives how copiously saliva flows; it also induces very abundant expectoration in the throat.

22. It becomes an emetic when administered in a larger dose than appropriate; [p. 23] although I do not believe that a prudent practicing physician would prescribe it with this intention, as its burning taste and almost fiery acrimony contraindicate it.

23. It is observed to be a purgative, although not very specifically: sometimes it gently moves the bowels, sometimes it acts up to a cathartic level; sometimes it has no effect in this way at all, perhaps this diversity of effect is due to the varying temperament of the subjects.

24. The diuretic power is almost naturally attributed to the root, placing its singular efficacy beyond all doubt. By promoting urination, it eliminates purulent, putrid, or other substances harmful to the animal economy from the blood mass, thus securing its place among the most excellent diuretic medications.

25. The diaphoretic effect is not so conspicuous, as it acts almost exclusively on diuresis; when one excretion is

---

[408]Footnote on p. 20: (*) The parts of the flowers could not be determined from the dried specimens at hand, so I was unable to provide a detailed illustration of the genus. However, since the description seems to have been written by the distinguished B. Jussieu, we have little doubt about the genus after it has been reviewed by such a sharp-eyed botanist; moreover, the non-branching, spicate, flaccid habit of the plant with simple leaves confirms the same genus.

[409]Subspithamalis in the original, being a spithama an ancient unit of length equivalent to the distance from the tip of the middle finger to the tip of the thumb when the hand is outstretched. See note 143.

[410]*Vires* in the original; it can also be transalted as potency, strength, force, power.

increased, the other is always diminished.

26. It is an excellent resolvent in both hot and cold diseases; it marvelously breaks down the tenacious and gelatinous cohesion of humors and resolves their antiphlogistic diathesis.

27. It is an excellent expectorant, stimulating the lungs to expel secretions with a gentle stimulus. [p. 24]

28. Finally, Tennent attributes to it vulnerary and detergent properties, with evidence drawn from pleurisy cases; in such instances, it must act like a balsam, cleansing and expelling urine.

VII. USAGE.

29. In cases of *Crotalophorus* bite. This remedy is highly praised by the Americans based on long-standing experience and recent medical observations, provided it is administered at the right time, before the venom spreads throughout the entire humoral mass. Hence, we recommend it to all conscientious physicians as an experiment against the bites of our own vipers, as it is well known that many of our compatriots often succumb to their venom. To date, no certain antidote has been discovered to rely on in this immediate danger; not even the highly praised olive oil has met our expectations. See *It. Gothl.* p. 306.

30. In pleurisy, pneumonia, pleuropneumonia, and paraphrenitis. Observing that those affected by *Crotalophorus* bites were troubled by symptoms similar to those afflicted by these inflammatory fevers, Dr. Tennent dared to experiment with the root in these cases. The experiment was so successful that no other remedy is yet known [p. 25] to match its efficacy in curing these diseases. Thus, he did not hesitate to call this the only specific for these diseases. It would certainly be desirable for this root to be kept in our pharmacies as well, since these diseases prematurely kill many of our people, especially the common folk. After the author's often-mentioned observations were published in Virginia, it began to be used with such success in these cases that now rarely three or four out of a hundred die, whereas previously more than half succumbed to the same fate.

31. In dropsy and asthma, its efficacy is also proven by experiments due to its exceptionally resolving power.

32. In arthritis, gout, and rheumatism, the root has also been found significant, guided by experience.

33. In *Virginian marasmus*. This disease is endemic to Virginia and particularly fatal to Europeans who have arrived there. It has certainly shown great effectiveness in combating this. This was the reason why Dr. Tennent applied it in the early stages of phthisis, sometimes with favorable outcomes for the patients.

VIII. PREPARATIONS.

34. In raw or dried form, the barbarians apply it to the bite of the Caudisona snake. [p. 26]

35. Powder is prescribed in doses of 35 grains, but in liquid form, it more quickly helps those afflicted.

36. Spirituous tincture is of a purplish-yellow color, with a vinous odor, and a very hot taste, burning like fire. This heat and taste are so persistent in the esophagus that if they do not always induce vomiting, they always excite the urge to vomit, along with much spitting and mucus. Hence, Tennent never used this.

Aqueous tincture initially has a nearly vinous and pleasant taste and odor, but later it acquires a peculiar taste almost characteristic of the root itself, eventually causing a persistent cough that is difficult to control while it acts by nearly pricking and provoking saliva. This tincture is whitish in color and generally so mucilaginous that it hardly penetrates filter paper or linen while being filtered. The author, Dr. Tennent, never used this either.

Vinous tincture, prepared with unfermented wine, is of a slightly purplish color, with a pleasant odor and a taste almost like the root. The dose is three tablespoons. This was the most commonly used by Tennent. It is prepared as follows:

Take 4 ounces of powdered Senega root.
Add 1 pound of Canary wine.
Let it stand in a warm place for 3 days.
Decant, filter, and keep for use. [p. 27]

37. Spirituous extract has the color, taste, and odor of the spirituous tincture but milder.

Aqueous extract is of a slightly brownish color, with a pleasant vinous balsamic odor, and a taste that is initially sour, then sharp, pungent, and stimulating, almost the same as the root itself. Tennent never used these two in practice.

38. Decoction, which was used as often as the vinous tincture, is prepared as follows:

Take 3 ounces of cut Senega root.
Add 2 pounds of pure spring water.
Boil until a third part is consumed.
Strain through a cloth with pressing, and offer warm.
The dose is three tablespoons.

IV. THERAPY.

39. For Rattle Snake Bite (29): The Americans chew the root, swallow its juice, and then apply the chewed root to the wound. The same effect is achieved by a decoction of the root sweetened with milk, and even by the vinous tincture, as all the particles of the root act with combined strength in these preparations.

40. In Inflammatory fevers (30): These are treated by

Dr. Tennent using the following method: [p. 28] At the onset of the fever, first perform a venesection on the arm or leg, drawing 10 to 12 ounces of blood. After a quarter of an hour, administer a dose of the tincture (36) or decoction (38), repeating the same dose every three hours until the symptoms remit. The ordinary drink should be an infusion of tea or mallow roots, sweetened with honey or sugar. The food should be a porridge made of flour with water, a small bread-based broth, or baked apples. If the symptoms are too persistent to diminish after the third dose, another venesection should be performed, proceeding again if the symptoms seem to worsen anew; however, the vein should be punctured only once within 24 hours. It rarely happens that this method of treatment is necessary beyond the second or third time. When the symptoms are mitigated, the dose of the medicine should be repeated only every four to six hours; if they increase, it is necessary to administer it within a shorter interval. If diarrhea seems to precipitate the patient due to the use of the medicine, the dose should be less frequent or even stopped by administering the following drink prescribed here:

Take 2 ounces of cinnamon water.
40 grains of prepared crab's stone.
Half a scruple of salt of tartar.
Half an ounce of quince syrup.
Mix and administer as a single dose. [p. 29]

Care must be taken not to use this or any other binding agent if the diarrhea is critical, or if it persists without loss of strength. Otherwise, use:

Take 1.5 drachms of powdered Senega root.
15 grains of Diascordium.
1 ounce of cinnamon water.
Mix for a single dose.

Dr. Tennent adheres to this method until a reddish sediment appears in the urine. If the patient seems relieved with a white sediment, the author is less certain of the complete crisis of the disease, fearing relapses.

41. In cold diseases (31) and painful conditions (32) without fever, the root, with its preparations, is used for treatment; the necessity of phlebotomy is then judged based on the diversity of the subject.

X. SUBSTITUTES.

42. Polygala vulgaris (16. 1.) root has almost the same virtues as our Senega; however, it is far inferior in both taste and efficacy. Parisians have successfully used this in pleurisy cases [p. 30] with positive outcomes. It induces copious expectoration, and the blood appears less viscous and much brighter the next day after use, as reported by Godofr. Mat. Med. vol. 2, p. 141. I am quite surprised that Godofred[411] prescribes only 12 grains of this, as it can be safely used in double the quantity of Senega. It is highly desirable for our local doctors to experiment with its effectiveness in treating snake bites until we have an ample supply of true Senega.

43. Wild Valerian (or garden valerian, which has much superior virtues): Tennent combined this in tincture with Senega root at a ratio of one and a half ounces. He highly recommended this combination in painful diseases (32:41) and wasting diseases (33), where patients quickly recovered. Unlike true Senega, it not only promotes excretions but also has an anodyne effect, as evidenced by its odor.

44. Bittersweet[412]: The stems in infusion powerfully promote all excretions, such as laxative effects, inducing sweats, menses, lochia, and especially urine, making it closely resemble Polygala and even Valerian in these properties. [p. 31] Given that all excellent antidotes against snake bites are diuretics, it would be highly beneficial if our doctors could also test its effectiveness in neutralizing our viper venom and treating inflammatory fevers.

XI. EPILOGUE.

45. Thus, I have briefly and, as I believe, clearly presented the history of this species of *Polygala*, also known as *Senega*. We rightfully attribute the first recognition of this plant, as previously mentioned, to the distinguished English doctor John Tennent, who willingly shared it with the learned world to the great benefit of many patients in his treatise titled *Essais on the pleurisy*, published in Philadelphia in 1736 in 8vo. In his Physical Disquisitions, he provided a further account of its virtues, although I have seen only the first two printed in London in 1735 in English. The fourth, fifth, and sixth parts are either currently in the press or have long been published, as he promised in his first disquisition, though I cannot say for certain. [p. 32] Nonetheless, I do not consider my efforts wasted in making this excellent and almost divine remedy widely known for the significant benefit of my compatriots.

That's all.

[411]Godofredus is a Latinized version of the Germanic name Godfrey, and Geoffroy is its French variation.

[412]*Dulcamara* in the original; probably *Solanum dulcamara*, commonly known as bittersweet or bittersweet nightshade.

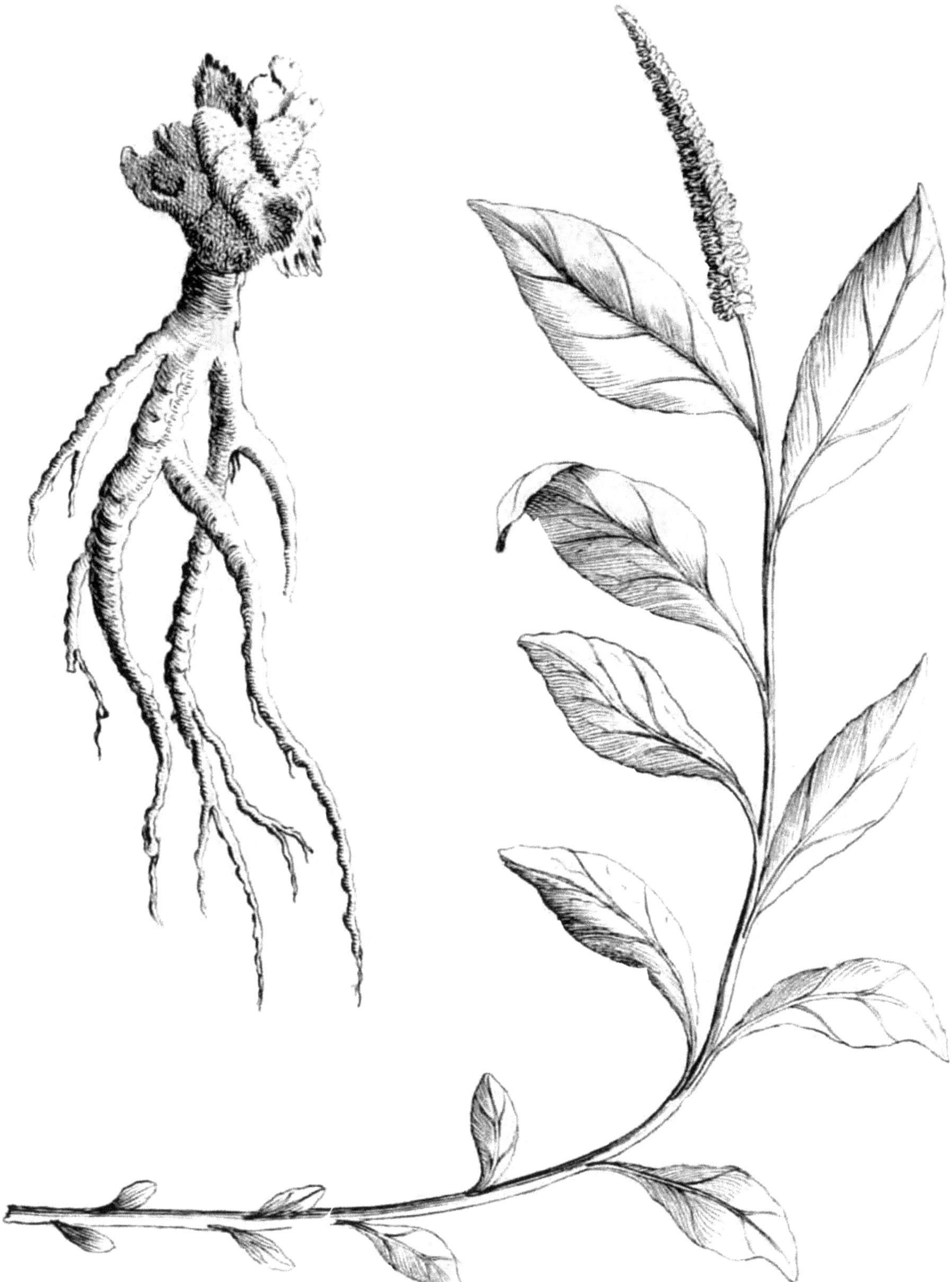

Polygala Senega dicta

Comments on 25 Dissertations with Miscellaneous References on Amphibians and Reptiles

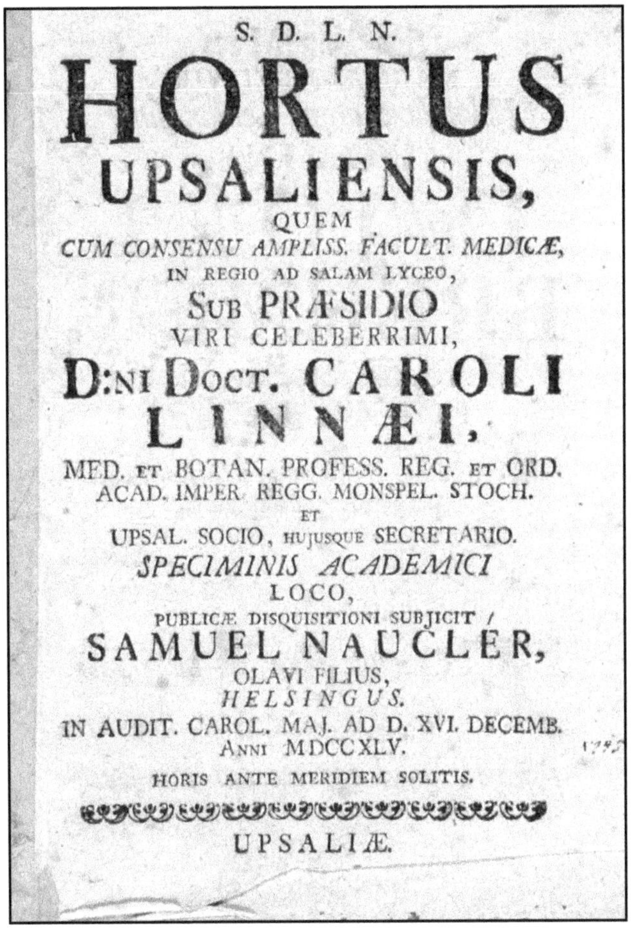
S. D. L. N.
HORTUS
UPSALIENSIS,
QUEM
CUM CONSENSU AMPLISS. FACULT. MEDICÆ,
IN REGIO AD SALAM LYCEO,
SUB PRÆSIDIO
VIRI CELEBERRIMI,
D:NI DOCT. CAROLI
LINNÆI,
MED. ET BOTAN. PROFESS. REG. ET ORD.
ACAD. IMPER. REGG. MONSPEL. STOCH.
ET
UPSAL. SOCIO, HUJUSQUE SECRETARIO.
SPECIMINIS ACADEMICI
LOCO,
PUBLICÆ DISQUISITIONI SUBJICIT
SAMUEL NAUCLER,
OLAVI FILIUS,
HELSINGUS.
IN AUDIT. CAROL. MAJ. AD D. XVI. DECEMB.
ANNI MDCCXLV.
HORIS ANTE MERIDIEM SOLITIS.
UPSALIÆ.

1. Nauclér, S. 1745. Hortus Upsaliensis. Upsaliae. S. n. [3] + 1 plate + 1 map + [2] + 3-45 + [3].

Naucler's dissertation, defended in 1745 and subsequently published in the first volume of *Amoenitates Academicae* (Nauclér, 1749), explores the medicinal potential of botanical specimens, focusing particularly on *Polygala* species.

••••

[p. 2 (part)].

(...) VEGETABLES, through their notable use and excellence in medicine, in economy, in manufactured works, in commerce, and finally in culinary matters, have greatly stimulated people of all ages to diligently cultivate the study of Botany.

Mortals did not consider it sufficient to make familiar to themselves the use of herbs known in medicine up to this time, since experiments on other herbs, of which the ancients knew little or nothing, are daily observed. Nevertheless, the use of these herbs has shown that they are in no way inferior to the former ones and have not had any lesser use. Botanists, distinguished by their use and experience, [p. 3] have prepared various systems and compendia to expand this science, thus laying certain foundations by which knowledge of botany could be attained and by which, through so-called affinities, better judgments about their analogies could be made.

For example, recently in Pennsylvania, Mr. Tennent tested the strength of a root, which the native tribes held as a specific cure against the bite of the Vipera caudisona, and furthermore observed the notable virtue of this remedy in inflammatory fevers. The Parisians immediately argued by analogy of affinity that as it was *POLYGALA caule simplici erecto, foliis ovato-lanceolatis alternis integerrimis, racemo terminatrice erecto* [*Polygala* with a simple erect stem, ovate-lanceolate alternate entire leaves, and a terminal erect raceme.] Gron. Virg. 80., so undoubtedly the common *Polygala* throughout Europe, or *POLYGALA foliis lineari-lanceolatis caulibus diffusis herbaceis* [*Polygala* with linear-lanceolate leaves, diffusely herbaceous stems] Virid. Cliff. 70., would indeed soon yield the same benefits in use. And being taught by experiments, they saw the same success happily achieved in inflammatory fevers. (...)

Nauclér further emphasizes the importance of collecting specimens towards the end (p. 43), detailing the location and contents of the Museum. Notably, he mentions donations from Adolph-Friderik and Carol Gyllenborg[413], which included diverse "Indian" snakes, lizards, frogs, and turtles.

[p. 42]

[413]Studied by Balk (1746) and Hast (1745), respectively; see above.

MUSEUMS

(...)The residence of the Prefect can display various items for the delight of viewers. Here, one can see the PORTRAITS of the most excellent Botanists, for example: Vailant, Tournefort, Rivin, Ray, Morison, Sloane, C. Bauhini Columnae, Till, Jungermann, Plukenet, Breyn, Boerhaave, Burmann, Kanig, Paui, Camerar, Subs, and both Rudbeck, which adorn the walls of the atrium, along with various illustrations of the most illustrious Academic Gardens in the European world.

In the Prefect's own Museum, you will find a very select library and collection of Botanical books, as well as the largest herbarium or collection of dried plants that the Kingdom of Sweden has ever possessed. Furthermore, the curious eyes are offered an immense multitude of fish, prepared in the same way as the herbs, by being mounted on paper; the majority of these were prepared by the illustrious Botanist, J. F. Gronovius. Here is also preserved a collection of stones exceeding the number of three thousand.

The Academic Museum, located in the southern part of the residence or Chamber (D), consists mostly of the donations from His Sacred Royal Highness and Prince the Most Serene Adolph Frederic, and from the Most Illustrious Chancellor of the University of Uppsala, the Most Excellent Count and Lord Carl Gyllenborg. A significant portion of these donations consists of indian serpents, such as: *Vipera Caudisona, Amphisbaena, Ahaetulla Zeilonensis, Cobella Americana, Petola Americana*, etc. Lizards, including *Crocodylus, Iguana, Draco volans, Ameira, Tejugvacu, Temapara, Lemniscata, Salamandra, Cordylus, Gecco*, plus various frogs & turtles. Many exotic fish, among which are *Cyprinus auratus* of the Chinese, *Torpedo*, Trigonal Fish, *Orbis echinatus*, etc. Numerous birds, depicted by the Chinese so accurately and vividly that one might easily believe these images to be breathing and alive; many kinds of fruits, in the crafting of which the aforementioned Chinese nation has given a remarkable display of their art and industry, so much so that even the sight of these fruits can arouse the desire to taste and eat them in the viewer. Additionally, there are preserved here various corals and petrified wood, as well as numerous and diverse fossils, noted for their excellent species and beauty.

Thus today, the immense variety of plants, stones, and animals, arriving here within a short interval of time, wonderfully nourishes the eyes of both naturalists and the curious in the Uppsala Garden. (...)

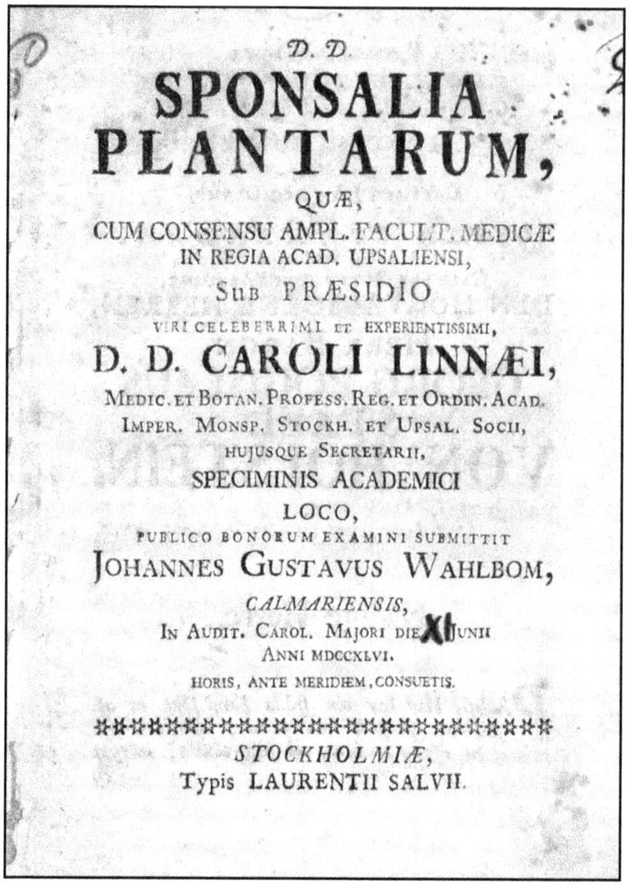
D. D.
SPONSALIA
PLANTARUM,
QUÆ,
CUM CONSENSU AMPL. FACULT. MEDICÆ
IN REGIA ACAD. UPSALIENSI,
SUB PRÆSIDIO
VIRI CELEBERRIMI ET EXPERIENTISSIMI,
D. D. CAROLI LINNÆI,
MEDIC. ET BOTAN. PROFESS. REG. ET ORDIN. ACAD.
IMPER. MONSP. STOCKH. ET UPSAL. SOCII,
HUJUSQUE SECRETARII,
SPECIMINIS ACADEMICI
LOCO,
PUBLICO BONORUM EXAMINI SUBMITTIT
JOHANNES GUSTAVUS WAHLBOM,
CALMARIENSIS,
IN AUDIT. CAROL. MAJORI DIE XI JUNII
ANNI MDCCXLVI.
HORIS, ANTE MERIDIEM, CONSUETIS.
STOCKHOLMIÆ,
Typis LAURENTII SALVII.

2. Wahlbon, J.G. 1746. Sponsalia plantarum. Stockholmiae. Laurentii Salvi. [12] + 8-60 + 1 plate + [2].

Dissertation defended on June 11, 1746, and reprinted in the first volume of *Amoenitates Academicae* (Wahlbon, 1749). It traces the historical evolution of the doctrine of plant sexes, highlighting the shift from ancient misconceptions, where notions of male and female plants often contradicted modern observations.

••••

[p. 23 (part)]

§138

§. XII. (...) Amphibians enjoy their own laws, for they mate like all animals, yet with this distinction: the male of snakes, like the crab, possesses two penises, and the *Vipera caudisona* boast four spiny ones.

Frog reproduction, however, remains unclear and awaits clarification from Reaumur's much-anticipated observations. The idea that penises develop on the thumbs in spring and fertilize the females by depositing sperm near their armpits is difficult to believe[414], especially when we compare this analogy of frogs with rays. Nevertheless, we have no doubt that birth in these animals occurs after mating. (...)

[414] This hypothesis was originally presented by Hast (1745), and later included in SN 10; see note 174.

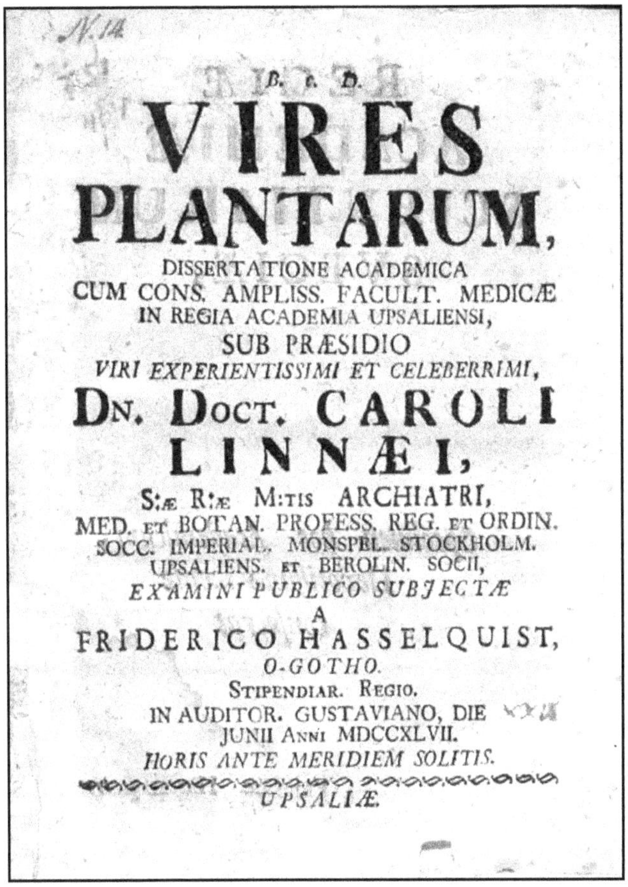
Q. B. D.
VIRES
PLANTARUM,
DISSERTATIONE ACADEMICA
CUM CONS. AMPLISS. FACULT. MEDICÆ
IN REGIA ACADEMIA UPSALIENSI,
SUB PRÆSIDIO
VIRI EXPERIENTISSIMI ET CELEBERRIMI,
DN. DOCT. CAROLI
LINNÆI,
S:Æ R:Æ M:TIS ARCHIATRI,
MED. ET BOTAN. PROFESS. REG. ET ORDIN.
SOCC. IMPERIAL. MONSPEL. STOCKHOLM.
UPSALIENS. ET BEROLIN. SOCII,
EXAMINI PUBLICO SUBJECTÆ
A
FRIDERICO HASSELQUIST,
O-GOTHO.
STIPENDIAR. REGIO.
IN AUDITOR. GUSTAVIANO, DIE
JUNII ANNI MDCCXLVII.
HORIS ANTE MERIDIEM SOLITIS.
UPSALIÆ.

3. Hasselquist, F. 1747. Vires plantarum. Upsaliae. S. n. [2] + 4-37 + [3]

Dissertation defended on June 20, 1747 and reprinted in the first volume of *Amoenitates Academicae* (Hasselquist, 1749). The dissertation explores the concept that plant virtues can be inferred from their botanical relationships, expanding on ideas discussed by Petiver and Hoffman. Hasselquist's analysis of Linnaeus's "Philosophia Botanica," particularly Chapter 12, further supports this deduction through natural and artificial botanical orders.

••••

[p. 15]

§ IX.

(...) We contend that the properties of plants can most safely be determined according to Natural Classes, provided we have previously become familiar with the properties of certain plants within the same class through experience. Without this knowledge, any aid expected from this method will be in vain.

This method, however, is commendable not only for its usefulness but also for its necessity. By investigating the nature of one plant, such as from Didynamia, Angiosperma, Diadelphia, or Umbellifers, I can reason about many other plants, even if there are a hundred species whose qualities I might otherwise have to ignore or investigate with uncertain experiments.

A notable example of this congruity is the celebrated root known as *Rattle-Snakeroot* by the English. This root was a secret of the wild inhabitants of Virginia, used against rattlesnake bites, pleurisy, pneumonia, and other phlogistic fevers. It remained hidden for a long time until it was discovered by Tennent to be a species of *Polygala*. Consequently, the Parisians applied the unusual and non-officinal root of our *Polygala* in such phlogistic fevers with the same happy result, though its potency was weaker.

Therefore, if the species of a genus is known for its properties, as a botanist, I can infer the properties of others within the same genus. Those unfamiliar with herbal science, especially those who do not know natural genera, are left without this aid. From this foundation, a botanist can select among congeners.

There is no other way to achieve a firm knowledge of the properties of plants except through this sole and unique Empiria. (...)

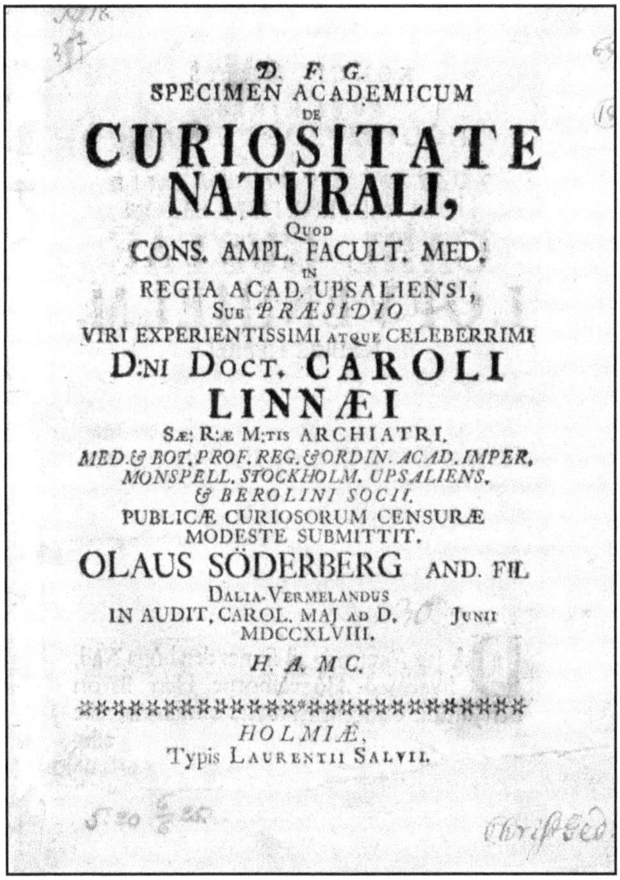
D. F. G.
SPECIMEN ACADEMICUM
DE
CURIOSITATE
NATURALI,
QUOD
CONS. AMPL. FACULT. MED.
IN
REGIA ACAD. UPSALIENSI,
SUB PRÆSIDIO
VIRI EXPERIENTISSIMI ATQUE CELEBERRIMI
D:NI DOCT. CAROLI
LINNÆI
S:Æ R:Æ M:TIS ARCHIATRI.
MED. & BOT. PROF. REG. & ORDIN. ACAD. IMPER.
MONSPELL. STOCKHOLM. UPSALIENS.
& BEROLINI SOCII.
PUBLICÆ CURIOSORUM CENSURÆ
MODESTE SUBMITTIT.
OLAUS SÖDERBERG AND. FIL.
DALIA-VERMELANDUS
IN AUDIT. CAROL. MAJ AD D. JUNII
MDCCXLVIII.
H. A. M. C.
HOLMIÆ,
Typis LAURENTII SALVII.

4. Söderberg, O. 1748. Specimen Naturale De Curiositate Naturali. Holmiae. Laurentii Salvii. [1] + 1-25 + [1].

Dissertation defended on June 30, 1748, and reprinted in the first volume of *Amoenitates Academicae* (Söderberg, 1749) under a slightly modified title, serves as a compelling call to engage in the study of natural history. It presents a series of interconnected arguments and observations, drawing from the profound manifestations of wisdom and goodness evident throughout nature. Emphasizing the inherent dignity and practical significance of natural history in benefitting humanity, the author asserts its rightful place as a subject of paramount importance in human inquiry.

••••

[p. 17]

(...) When we consider the works that we can accomplish, they cannot be compared in the least with those of the Creator, whose craftsmanship is most exquisite. We borrow our clothing from plants, insects, and animals, and adapt them to our pleasure; yet we strive in vain to match the splendor of the tiger, the peacock, the snake (i.e., *Aëtullae*), the Chinese carp, and the cantharides. From this, a firm and indisputable truth emerges: that Solomon, in all his majesty, cannot be compared to a single Jacob's lily or amaryllis. [p. 18] Therefore, a person who is not concerned with the sober contemplation of things seems truly unworthy of being created by God and impious in his actions. He is like a beast; he despises the entire and immense work of Creation, who was created to admire his Maker. (...)

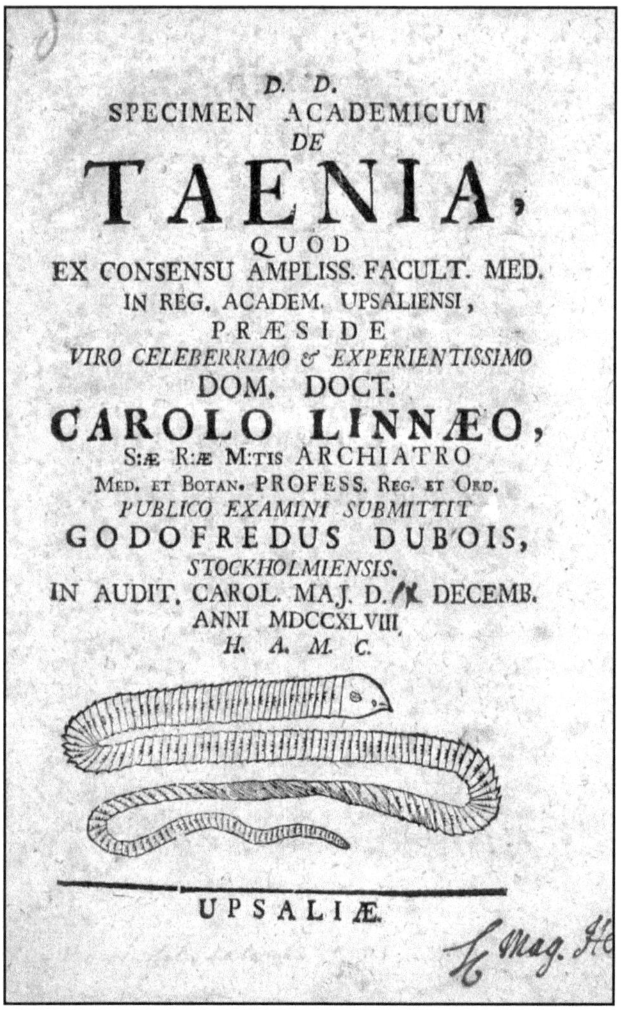

D. D.
SPECIMEN ACADEMICUM
DE
TAENIA,
QUOD
EX CONSENSU AMPLISS. FACULT. MED.
IN REG. ACADEM. UPSALIENSI,
PRÆSIDE
VIRO CELEBERRIMO & EXPERIENTISSIMO
DOM. DOCT.
CAROLO LINNÆO,
S:Æ R:Æ M:TIS ARCHIATRO
MED. ET BOTAN. PROFESS. REG. ET ORD.
PUBLICO EXAMINI SUBMITTIT
GODOFREDUS DUBOIS,
STOCKHOLMIENSIS.
IN AUDIT. CAROL. MAJ. D. IX DECEMB.
ANNI MDCCXLVIII.
H. A. M. C.

UPSALIÆ.

5. Dubois, G. 1748. Taenia. Upsaliae. S. n. [3] + 1-30 + 1 plate.

Dubois's dissertation, defended on December 9, 1748, and published in the second volume of *Amoenitates Academicae* (Dubois, 1751), focuses on tapeworm morphology and related biological phenomena.

••••

[p. 26]

(...) The *Taenia* also exhibits this characteristic[415]. Initially, it is a mere segment, but then it grows by adding more and more segments, often a great number of them, in a manner similar to the tail of a rattlesnake (*Crotalophori*). Thus, each segment of our worm becomes a distinct animal if it is separated from the whole. However, as long as the segments remain connected, they share a common sensation, in the same way that a composite polyp does.

Physiologists teach us that movement is determined by the brain. Thus, we conceive that each brain has its own determination, or, as the well-known proverb states, "As many heads, so many senses".[416] We have indeed seen a two-headed snake (*Coluber*) (b)[417], whose single body moved according to the determinations of two heads, and yet it was sustained, like the dog-headed snake (*Serpens cynocephalus*) (c)[418]. It is credible that there exists a longitudinal and common brain for all the segments of the tapeworm, which occupies the worm longitudinally, like the spinal cord. And that this is separated at the joints of the segments when the tapeworm's segments detach, since all the segments live and move when separated from the body. A snake with its head cut off often moves its tail for a whole day. A large sea turtle (*Testudo marina major*) can often swim around for entire eight days even after being decapitated. This undoubtedly occurs through the action of the nerves in its spinal cord. We commit the investigation of this very matter to future centuries, as there are still many things obscure to us, which perhaps will be brought to light (d)[419]. (...)

[415] It refers to the regeneration of polyps.

[416] *Quot capita, tot sensus* in the original.

[417] Footnote in the original: (b) A few years ago, a two-headed snake of this kind was observed in the collection of B. Roberg, Professor of Uppsala.

[418] Footnote in the original: (c) For here we do not mean a snake with a head at each end, like the ancient Amphisbæna, but a snake with a double head at one end.

[419] Footnote in the original: (d) Plin. Lib. II c. 3. When I contemplate nature, it persuades me to believe nothing incredible about it.

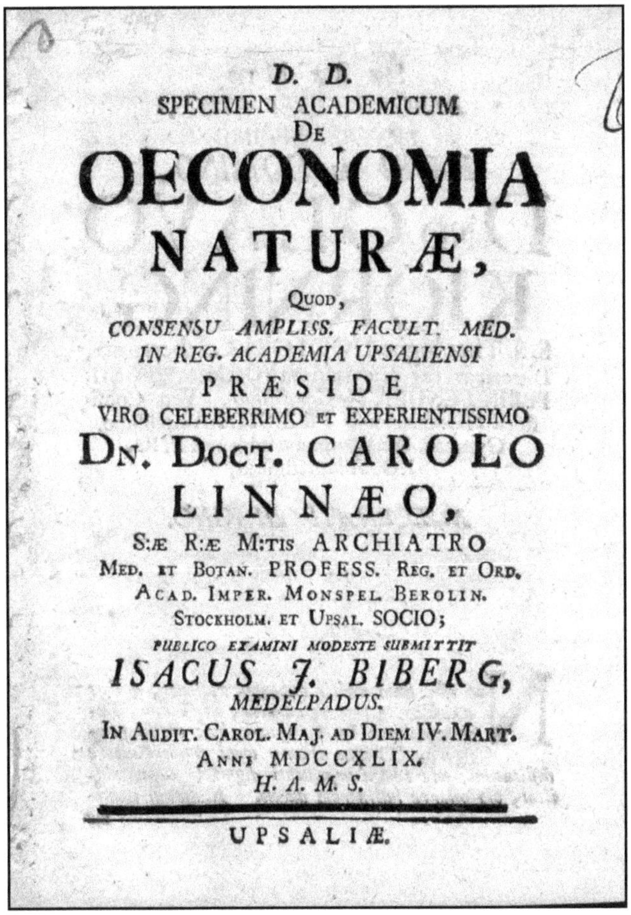

D. D.
SPECIMEN ACADEMICUM
DE
OECONOMIA
NATURÆ,
QUOD,
CONSENSU AMPLISS. FACULT. MED.
IN REG. ACADEMIA UPSALIENSI
PRÆSIDE
VIRO CELEBERRIMO ET EXPERIENTISSIMO
DN. DOCT. CAROLO
LINNÆO,
S:Æ R:Æ M:TIS ARCHIATRO
MED. ET BOTAN. PROFESS. REG. ET ORD.
ACAD. IMPER. MONSPEL. BEROLIN.
STOCKHOLM. ET UPSAL. SOCIO;
PUBLICO EXAMINI MODESTE SUBMITTIT
ISACUS J. BIBERG,
MEDELPADUS.
IN AUDIT. CAROL. MAJ. AD DIEM IV. MART.
ANNI MDCCXLIX.
H. A. M. S.
UPSALIÆ.

6. Biberg, I.J. 1749. Oeconomia naturae. Upsaliae. S. n. [7] + 1-48.

Biberg's dissertation, defended on March 4, 1749, and later reprinted in the second volume of *Amoenitates Academicae* (Biberg, 1751), and into English (1759) and in French (1789) translations, spans theological and expansive themes.

••••

[p. 28]

§ XII.

(...) Many amphibians give birth to live offspring, such as the viper, Fn. 260[420], and the toad known as *Rana Bufo*[421], etc. However, species that lay eggs deposit them in locations where the heat of the sun can sustain them in place of the mother. Thus, other frogs and lizards, Fn. 254, 256, 257[422], lay their eggs in warm waters. Natrices, Fn. 259, leave their eggs in manure piles and other warm places, entrusting them to nature as a provident mother to care for[423]. Crocodiles and marine turtles are forced to ascend from the waters to the shores to bury their eggs under the sand to be vivified by the sun's heat. (...)

[p. 33]

(...) Amphibians, fish, and insects, which cannot benefit from parental care, nevertheless owe it to their fortune that they are deposited in such locations where they can easily obtain nourishment, as we have seen in § XII. (...)

[p. 37-38]

(...) The Creator has shown no less care for some amphibians, particularly snakes and frogs, which, since they are neither equipped with wings to fly nor feet to swiftly and comfortably run, are given the means to capture prey in such a way that some animals almost willingly rush into their mouths. Thus, the Crotalophorus, an inhabitant of America, merely by gaping its jaws, looks at a bird, fly, or squirrel sitting in a tree; these creatures, completely stupefied and as if having no refuge left, run into its gullet[424]. Conversely, we revere the utmost kindness of the Creator towards humans when we consider the rattle attached to the tail of this serpent. By its means, humans can often protect themselves from this most venomous enemy, as, hearing the sound of its rattle, they flee; otherwise, a person struck by it, within the space of six hours or often within half an hour, would have their entire body corrupted by putrefaction. (...)

[p. 39]

§ XIII.

(...) Most amphibians occupy caves or the bottoms of lakes and marshes. (...)

[p. 41]

§ XVIII.

Destruction.

(...) In the preceding sections, we noted that not all animals feed on plants, but some also subsist on certain small animals as food and sustenance: Specifically, there are those that live solely by preying and daily tear apart many defenseless creatures; thus, animals are destroyed in such a way that the weaker are usually continually harassed by the

[420] Anguis scutis abdominalibus CXLIV, squamis caudæ XXXIX; Linnæus (1746). *Coluber berus* in SN10.

[421] The reference to viviparity in *Rana bufo* is intriguing, and it was reiterated by Linnæus on page 210 of the tenth edition of Systema Naturæ, where he notes: *Vivipara mare, ut fertur, obstetricante* (Linnæus, 1758).

[422] 254, *Lacerta pedibus inermibus, manibus tetradactylis, palmis pentadactylis, colore livido, linea dorsali fusca duplici* (*Lacerta vulgaris* in SN 10). 256, *Lacerta pedibus inermibus fissis, manibus tetradactylis, palmis pentadactylis, cauda ancipiti.* (*Lacerta palustris* in SN10). 257, *Lacerta pedibus inermibus, manibus tetradactylis, palmis pentadactylis, cauda teretiuscula* (*Lacerta aquatica* in SN10). All in th synonymy of Lissotriton vulgaris (Linnæus, 1758).

[423] This comment was included in the characterization of *Coluber natrix* in SN 10 (Linnaus, 1758).

[424] Comment included in the characterization of *Crotalus horridus* in SN 10 (Linnaus, 1758) and in the dissertations by Balk (1746), Kiernander (1749), Acrell (1762).

stronger. In this way, the aphid feeds on plants; aphids are preyed upon by the aphid-eating fly; this fly is ambushed by the asilid fly; dragonflies sustain themselves on asilid flies; these are caught by spiders, which in turn are preyed upon by birds, which finally are forced to yield to hawks. Similarly, the copepods (Monoculus) delights in putrid water, which becomes food for mosquitoes (Culicibus), these for frogs (Rana), frogs for pikes (Luciis), and pikes for seals (Phocis). (...).

[p. 44]

(...) Storks and falcons free Egypt from frogs, which cover the whole land after the inundation of the Nile, as well as Palestine from mice. Thus, Bellon[425] mentions that so many storks come to Egypt that the fields and meadows become white with them; nevertheless, the Egyptians love them, since frogs breed there in such abundance that if the storks did not devour them, nothing would be more common, and also because they capture and devour snakes. (...)

[p. 47].

Every species of any created thing is worthy of having its own dedicated investigator. If we were to count, according to some calculation, 20,000 plants, 3,000 worms, 12,000 insects, 200 amphibians, 2,600 fish, 2,000 birds, and 200 quadrupeds, the world would present 40,000 living species. (...)

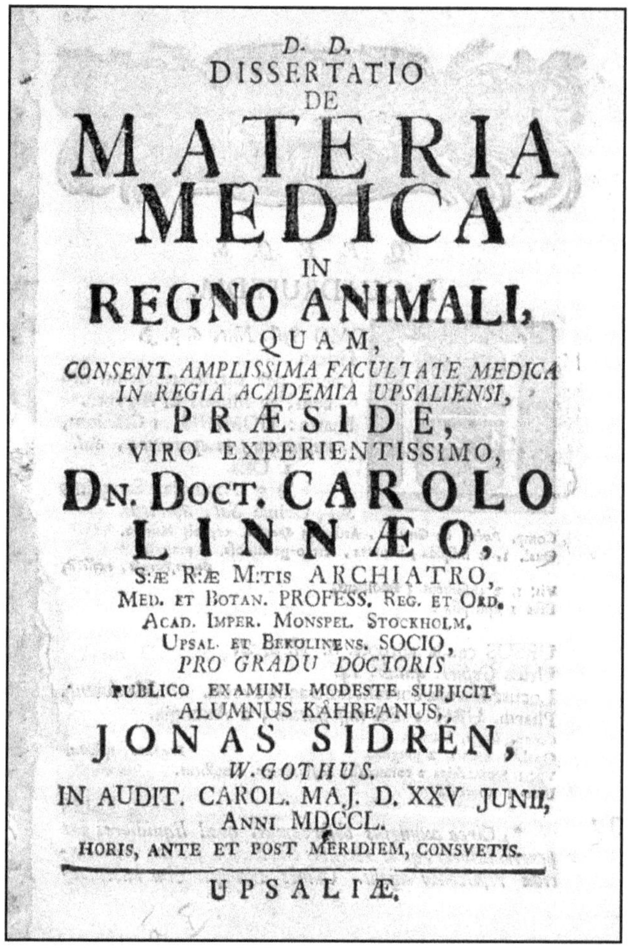

D. D.
DISSERTATIO
DE
MATERIA
MEDICA
IN
REGNO ANIMALI,
QUAM,
CONSENT. AMPLISSIMA FACULTATE MEDICA
IN REGIA ACADEMIA UPSALIENSI,
PRÆSIDE,
VIRO EXPERIENTISSIMO,
DN. DOCT. CAROLO
LINNÆO,
S:Æ R:Æ M:TIS ARCHIATRO,
MED. ET BOTAN. PROFESS. REG. ET ORD.
ACAD. IMPER. MONSPEL. STOCKHOLM.
UPSAL. ET BEROLINENS. SOCIO,
PRO GRADU DOCTORIS
PUBLICO EXAMINI MODESTE SUBJICIT
ALUMNUS KÅHREANUS,
JONAS SIDRÉN,
W-GOTHUS.
IN AUDIT. CAROL. MAJ. D. XXV JUNII,
ANNI MDCCL.
HORIS, ANTE ET POST MERIDIEM, CONSVETIS.
UPSALIÆ.

7. Sidrén, J. 1750. Materia medica in regno animali. Upsaliae. S. n. 1-20.

Dissertation defended on June 25, 1750, and reprinted in the second volume of *Amoenitates Academicae* (Sidrén, 1751). Deals with 67 subjects used in the manufacture of medicinal preparations. Amphibians and reptiles are considered on pp. 11-12.

••••

[p. 11-12]

III. AMPHIBIA

RANA manibus tetradactylis fissis, plantis hexadactylis palmatis, pollice longiore [Frog with hands having four split digits, feet with six webbed digits, and a longer thumb][426]. Fn. Sv. 250.

Rana aquatica s. innoxia. Gesn. ovip. 46

Habitat: Swamps throughout Europe. *Tame.*

Pharmacy: Dried frog sperm, water, oil cooked.

Compounds: Plaster of frog sperm; of frogs with mercury; Cataplasm.

Quality: *Common.*

Effect: cooling, repelling, cooked water cooling.

Usage: Inflammation, Erysipelas, Burns.

[425]Belon (1605: p. 102). It will be later repeated by Ekmarck (1757).

[426]*Rana temporaria* in SN 10.

RANA manibus tetradactylis fissis, plantis hexadactylis palmatis, pollice breviore [Frog with hands having four split digits, feet with six webbed digits, and a shorter thumb][427], Fn. Sv. 253.

Bufo. Rond. aquat. 2. p. 221.
Habitat: Shady places in Europe. *Tame.*
Pharmacy: Dried toads.
Compounds:
Quality: *Rarer.*
Effect:
Usage: Whitlow, Hemorrhage? Head congestions?

LACERTA cauda tereti, collo crassitie capitis, pedibus pentadactylis marginatis. [Lizard with a rounded tail, a neck as thick as the head, and feet with five bordered digits][428].
Stincus. Rond. pisc. 2. p. 231.
Habitat: Egypt. *Wild.*
Pharmacy: Marine stincus.
Compounds: Electuary of diasatyr; Mithridate.
Quality: *Common.*
Effect: aphrodisiac? diuretic?
Usage: Purgative?

Coluber scutis abdominalibus 176, squamis caudalibus 60 [Snake with 176 abdominal shields and 60 caudal scales][429]. S.N. 34.
Natrix torquata Raj. quadr. 334.
Habitat: Dung heaps in Scania. *Tame.*
Pharmacy: 1. Dorsal spine. 2. Exuvia. 3. Fat.
Compounds:
Quality: 3. *Frequent.*
Effect: 1. Absorbing. 2. Diuretic?
Usage: 1. 2. Dystocia? Dropsy? Wound?

Coluber scutis abdominalibus 145, squamis caudalibus 36 [Snake with 145 abdominal shields and 36 caudal scales][430]. S.N. 34.
Vipera Gesn. Serp. 124.
Habitat: Sweden. *Tame.*
Pharmacy: VIPERS 1. Dried, Powder, Troches, Salt, 2. Bones. 3. Fat.
Compounds: Ther. Androm., Ointment of Tutia, D. Sloane.
Quality: *Common.*
Effect: 1. Antidote, diuretic, restorative, 2. Absorbent.
Usage: 1. Syphilis, Leprosy, Scabies, Malignant ulcer, Poison? 3. Ophthalmia.

D. A. G.
INSTRUCTIO
MUSEI
RERUM NATURALIUM,
QUAM,
CONSENSU EXPERIENT. ET NOBILISS.
FACULT. MEDIC. UPSAL.
Sub PRÆSIDIO
VIRI NOBILISSIMI ATQUE EXPERIENTISSIMI
DN. DOCT. CAROLI
LINNÆI,
S:Æ R:Æ M:TIS ARCHIATRI, MED. ET BOTAN. PROF.
REG. ET ORD. ACAD. IMPER. MONSPEL. STOCKH.
UPSAL. BEROL. TOL. ET LONDIN. SOCII,
Nec non
EQUITIS AURATI DE STELLA POLARI.
PLACIDÆ BONORUM CENSURÆ SUBJICIT
ALUMNUS REGIUS,
DAVID HULTMAN,
UPLANDUS.
IN AUD. CAR. MAJ. D. XIV. NOV. AN. MDCCLIII.
H. A. M. S.
UPSALIÆ,
Exc. LAUR. MAGNUS HÖJER, Reg. Acad. Typogr.

8. Hultman, D. 1753. Instructio musei rerum naturalium.
Upsaliae. Laur. Magnus Höjer. [3] + 1-19 + [1].

Dissertation defended on November 14, 1753, and reprinted in the third volume of *Amoenitates Academicae* (Hultman, 1756). Contains guide lines for the formation of natural history collections, indicating the appropriate methods for the conservation of various plants and animals and highlights the most important repositories of the time in Sweden.

••••

[p. 8]
(...) Amphibians are preserved with difficulty unless in spirit of wine[431] so that the appearance and color remain, but not in a very strong one, for if it is very strong, there is a danger that they may be peeled. Therefore, it is mitigated by adding a small amount of Sugar. However, Tortoises can

[427] *Rana bufo* in SN 10.

[428] *Lacerta stincus* in SN 10.

[429] *Coluber natrix* in SN 10.

[430] Not identified. Sidren notes "*S. N.* 34", referring to the sixth edition of the *Systema Naturæ* (Linnæus, 1748a), where only 26 species of *Coluber* were described. The one with the scale count 145-36 bears number 20, and was identified as Vipera. *Fn.* 260. ⊠. In turn, the species *Fn.* 260 corresponds to *Anguis scutis abdominalibus CXLIV, squamis caudæ XXXIX. Amph. gyllenb.* (Linnæus, 1746), and later was included in the synonymy of *Coluber berus* in SN 10, but with a scale count of 146-20 (Linnæus, 1758).

[431] See note 287.

be excepted from this as they remain intact even when dried and are mostly free from insect damage. (...)

[pp. 12-13]

The Museum of the Uppsala Academy, which houses amphibians, fish, birds, worms, and is preserved in the buildings of the Academic Garden, is maintained by our Most Gracious King, by the noble Senator of the Kingdom and Chancellor Count Carl Gyllenborg (d)[432], by Director Claudius Grill (e)[433], and by Director Magnus Lagerström of Gothenburg, who collected items from the East Indies and transferred them to the Noble President. Through his own efforts and the assistance of friends, he enriches this museum daily with his hands.

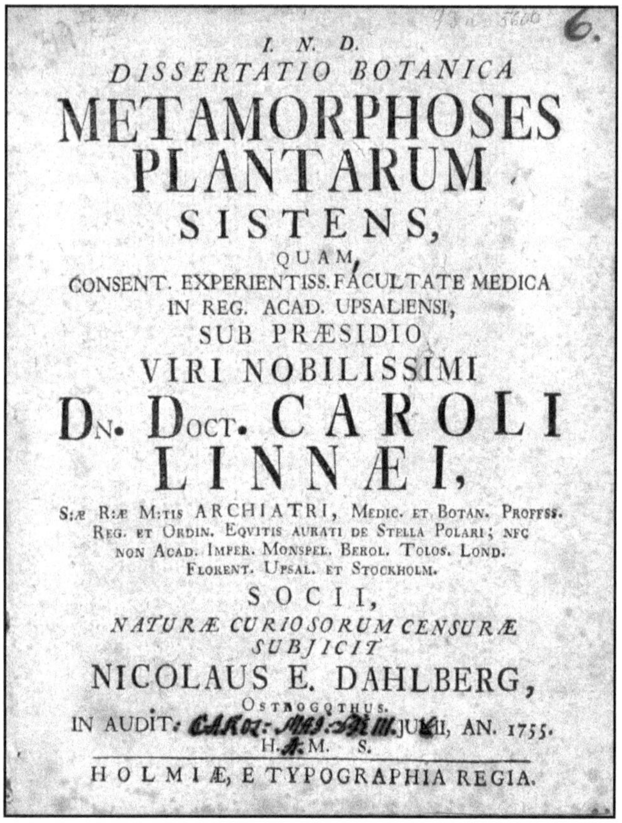

I. N. D.
DISSERTATIO BOTANICA
METAMORPHOSES PLANTARUM
SISTENS,
QUAM,
CONSENT. EXPERIENTISS. FACULTATE MEDICA
IN REG. ACAD. UPSALIENSI,
SUB PRÆSIDIO
VIRI NOBILISSIMI
DN. DOCT. CAROLI LINNÆI,
S:æ R:æ M:tis ARCHIATRI, MEDIC. ET BOTAN. PROFESS. REG. ET ORDIN. EQVITIS AURATI DE STELLA POLARI; NEC NON ACAD. IMPER. MONSPEL. BEROL. TOLOS. LOND. FLORENT. UPSAL. ET STOCKHOLM.
SOCII,
NATURÆ CURIOSORUM CENSURÆ SUBJICIT
NICOLAUS E. DAHLBERG,
OSTROGOTHUS.
IN AUDIT. [illegible] JULII, AN. 1755.
H. A. M. S.
HOLMIÆ, E TYPOGRAPHIA REGIA.

9. Dahlberg, N.E. 1755. Metamorphoses plantarum. Holmiae. Typographia Regia. [3] + 6-26.

Dissertation defended on June 3, 1755[434], and reprinted in the fourth volume of *Amoenitates Academicae* (Dahlberg, 1759). In the introduction, Dahlberg (1755) recounts the changes observed in various animal species, and it is interesting to transcribe what he says of amphibians (pp. 6-7):

••••

In not a few amphibians, we see the same transformation[435]. The frog, which deposits its eggs in water, initially appears almost like a fish, as a *Cottus*[436], with a large head, a compressed tail, and no legs; eventually, it acquires legs, resembling a lizard; then it loses its tail, leaves the water, and finally attains its proper form.

Rana piscis, from Suriname[437], about which Vincent, Merian, and Seba[438], and others have tried to show transforms from a frog into a fish, in the opposite way that is often seen in frogs; but the metamorphosis of this animal is not so, but the same, which at first with its compressed tail was very similar to a Lizard, finally after it becomes adult and reaches its proper size, it changes into a frog with four split toes on its hands, five palmate toes on its feet, and thighs obliquely striated on the back. Thus, also many animals are compared.

[432]Footnote in the original. (d) Amphibia Gyllenborgiana ibidem p. 107

[433]Footnote in the original. (e) Surinamensia Grilliana ibidem p. 483

[434]Data completed by hand on the cover of the dissertation, overwriting "Julii". Date explicited as June 3, 1755 in Dahlberg (1759).

[435]In the preceding paragraph, the metamorphosis in insects was discussed.

[436]*Cottus* Linnæus, 1758, genus of Perciformes that includes species known as sculpins.

[437]*Rana piscis, Surinamensis incola* inthe original.

[438]Refers to Levinus Vincent (1726b), Maria Sibylla Merian (1719) and Albertus Seba (1734).

I. N. S. S. T.
CALENDARIUM
FLORÆ,
QUOD,
VENIA EXPERIENTISSIMÆ FACULT. MED.
IN ALMO UPSALIENSI LYCÆO,
PRÆSIDE,
VIRO NOBILISSIMO ET EXPERIENTISSIMO
DN. DOCT. CAROLO
LINNÆO,
EQUIT. DE STELLA POLARI, ARCHIATR.
REG. MED. ET BOTAN. PROFESS. ACAD. UPS. HOLMIENS. PETROPOL. BEROL. IMPERIAL. LONDIN.
MONSPEL. TOLOS. FLORENT. SOCIO,
PUBLICÆ VENTILATIONI SUBJICIT
STIPENDIARIUS REGIUS,
ALEXANDER. MAL. BERGER,
VERMELANDUS.
IN AUDIT. CAROL. MAJ. D. XXXI. MARTII,
ANNI MDCCLVI.
H. A. M. S.
UPSALIÆ, Excud. L. M. HÖJER, Reg. Acad. Typogr.

10. Berger, A. M. 1756. Calendarium florae. Upsaliae. L.M. Höjer, Reg. Acad. Typogr. [3] + 1-5 + [19].

Dissertation defended on March 31, 1756, and reprinted in the fourth volume of *Amoenitates Academicae* (Berger, 1759). Explores seasonal changes through the flowering times of plant species, suggesting these natural patterns could better guide agricultural activities than traditional methods.

••••

s/n.

II. THAWING[439].

From the first melting of snow to the breaking up of river ice.

IV: (...)

6. Hills are exposed with half-melted snow.

Snakes emerge.

10. Inundation of snow water.

Rivers thaw and ice floats away.

Esox Lucius spawns.

———

Footnote: 10 *Esox Lucius* stops spawning when *Rana temporaria* begins.

s/n

III. GERMINATION[440].

From the arrival of the white wagtail to the arrival of the swallow.

From the first flower to the budding of the first tree.

through the entire blooming of Leucojum vernum.

IV. 12 (...)

Rana temporaria croaks.

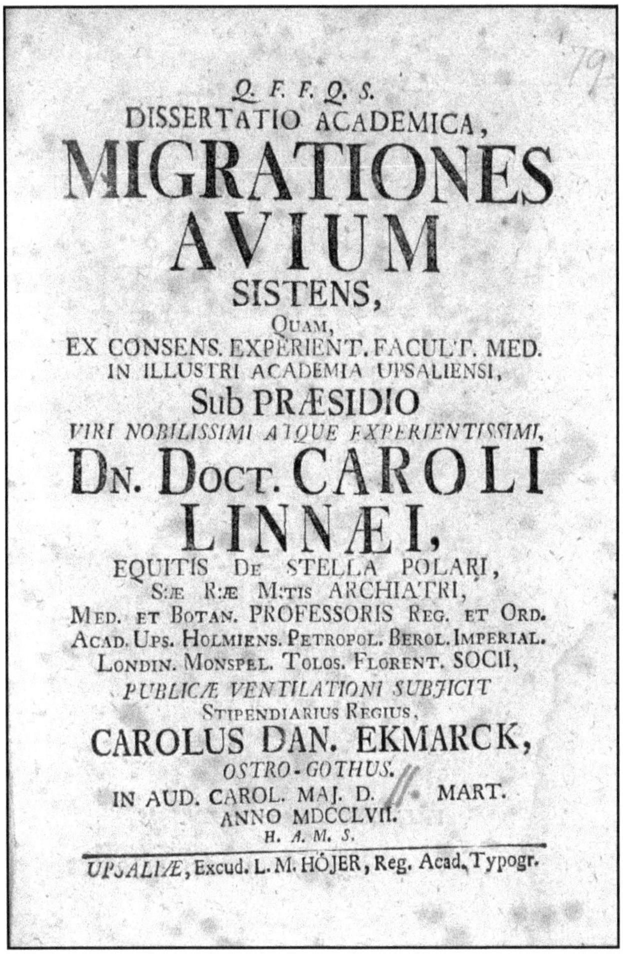

Q. F. F. Q. S.
DISSERTATIO ACADEMICA,
MIGRATIONES
AVIUM
SISTENS,
QUAM,
EX CONSENS. EXPERIENT. FACULT. MED.
IN ILLUSTRI ACADEMIA UPSALIENSI,
Sub PRÆSIDIO
VIRI NOBILISSIMI ATQUE EXPERIENTISSIMI,
DN. DOCT. CAROLI
LINNÆI,
EQUITIS DE STELLA POLARI,
S:Æ R:Æ M:TIS ARCHIATRI,
MED. ET BOTAN. PROFESSORIS REG. ET ORD.
ACAD. UPS. HOLMIENS. PETROPOL. BEROL. IMPERIAL.
LONDIN. MONSPEL. TOLOS. FLORENT. SOCII,
PUBLICÆ VENTILATIONI SUBJICIT
STIPENDIARIUS REGIUS,
CAROLUS DAN. EKMARCK,
OSTRO-GOTHUS.
IN AUD. CAROL. MAJ. D. MART.
ANNO MDCCLVII.
H. A. M. S.
UPSALIÆ, Excud. L. M. HÖJER, Reg. Acad. Typogr.

11. Ekmarck. C.D. 1757. Migrationes avium. Upsaliae. L.M. Höjer, Reg. Acad. Typogr. [4] + 6-38.

Dissertation defended on March 2, 1757, and reprinted in the fourth volume of Amoenitates Academicae (Ekmarck, 1759). Ekmarck provides an extensive examination of bird migration, focusing on the reasons behind their movements and their destinations.

••••

[p. 12]

[439]REGELATIONIS. KJÄLDRAGET in the original.

[440]GERMINATIONIS. SANJNGSTIDEN in the original.

We see, therefore, that one of the primary causes of migration is related to matters of survival, namely, the scarcity of food in certain places that birds leave and the abundance in other places that they seek.

They also consider this in choosing their summer locations. Storks, which mainly feed on snakes, hardly ever visit us, as we are freed from their multitude. Instead, they typically summer in Scania, where *Coluber Natrix*, which is quite rare in the northern provinces of Sweden, is very common. (...)

p. 26-27

Ciconia alba Fn. 136, emerging beyond Scania in the spring, scarcely proceeds further, so that its sighting in the upper provinces during the summer is considered a rare omen of some future event. They depart around the end of August and return at the beginning of April. Whether, as Cl. Klein conjectures, they submerge themselves in the autumn is not at all known; but according to Bellon, it seems more consistent with the analogy that they migrate to southern regions (n). (...)

———

footnote:

(n) *Ciconia* come to Egypt in such great numbers that the fields and meadows turn white with them. Nevertheless, the Egyptians love them, as frogs breed there in such abundance that if the storks did not devour them, nothing would be more common. Additionally, the storks capture and devour snakes.[441] (...)

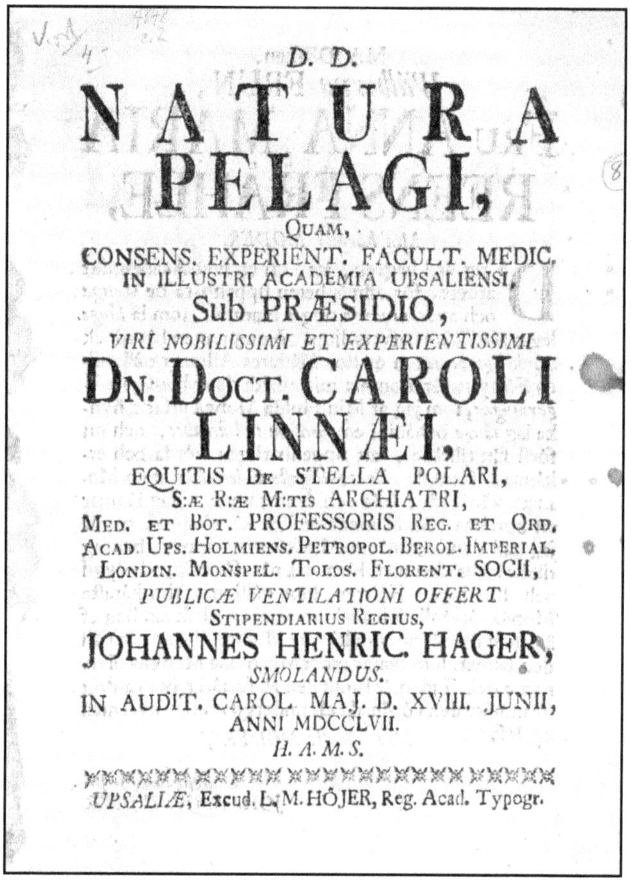

D. D.
NATURA PELAGI,
QUAM,
CONSENS. EXPERIENT. FACULT. MEDIC.
IN ILLUSTRI ACADEMIA UPSALIENSI,
Sub PRÆSIDIO,
VIRI NOBILISSIMI ET EXPERIENTISSIMI
DN: DOCT. CAROLI LINNÆI,
EQUITIS DE STELLA POLARI,
S:Æ R:Æ M:TIS ARCHIATRI,
MED. ET BOT. PROFESSORIS REG. ET ORD.
ACAD. UPS. HOLMIENS. PETROPOL. BEROL. IMPERIAL.
LONDIN. MONSPEL. TOLOS. FLORENT. SOCII,
PUBLICÆ VENTILATIONI OFFERT
STIPENDIARIUS REGIUS,
JOHANNES HENRIC. HAGER,
SMOLANDUS.
IN AUDIT. CAROL. MAJ. D. XVIII. JUNII,
ANNI MDCCLVII.
H. A. M. S.
UPSALIÆ, Excud. L. M. HÖJER, Reg. Acad. Typogr.

12. Hager, J. H. 1757. Natura pelagi. Upsliae. L.M. Höjer. [4] + 6-15 + [1].

Dissertation defended on June 18, 1757, and reprinted in the fifth volume of Amoenitates Academicae (Hager, 1760). Showcases the ocean's diversity, from *Sargassum* algae covering vast sea areas to coral reefs forming islands.

••••

AMPHIBIA.

Amphibians that dwell in the deep sea are very different in appearance from terrestrial ones.

Testudo Midas (sic) seeks the surface of the sea to sleep and floats on its back (Osb. 66). At night, it ascends to the shores, where, among uncultivated rocks, it lays its eggs in the sandy, shell-laden soil (Osb. 293). Voracious sharks[442] (Osb. it. 293, 78, 70, Kalm. it. 2. p. 147)[443], resembling tigers, emulate the swiftest course of ships, devouring anything they encounter.

Lophius histrionius (Osb. it. 305), living among floating seaweed, hunts and consumes Zoopterygia.

[441]See Belon, 1605: 102), and notes 176 and 424.

[442]The inclusion of sharks here foreshadows the incorporation of NANTES among the Amphibia, a fact formalized by Linnæus (1758), in the tenth edition of the Systema Naturæ. This is repeated in the theses of Wännman (1759) and Wilcke (1760).

[443]Both comments are based on Osbeck (1757) and will be later reproduced in SN 10.

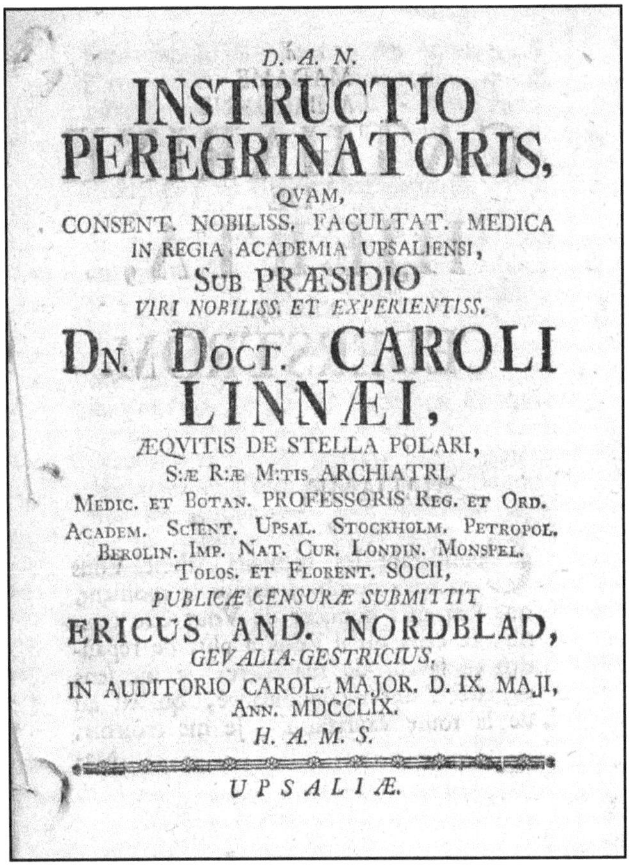

D. A. N.
INSTRUCTIO
PEREGRINATORIS,
QVAM,
CONSENT. NOBILISS. FACULTAT. MEDICA
IN REGIA ACADEMIA UPSALIENSI,
SUB PRÆSIDIO
VIRI NOBILISS. ET EXPERIENTISS.
DN. DOCT. CAROLI
LINNÆI,
ÆQVITIS DE STELLA POLARI,
S:Æ R:Æ M:TIS ARCHIATRI,
MEDIC. ET BOTAN. PROFESSORIS REG. ET ORD.
ACADEM. SCIENT. UPSAL. STOCKHOLM. PETROPOL.
BEROLIN. IMP. NAT. CUR. LONDIN. MONSPEL.
TOLOS. ET FLORENT. SOCII,
PUBLICÆ CENSURÆ SUBMITTIT
ERICUS AND. NORDBLAD,
GEVALIA-GESTRICIUS,
IN AUDITORIO CAROL. MAJOR. D. IX. MAJI,
ANN. MDCCLIX.
H. A. M. S.
UPSALIÆ.

13. Nordblad, E.A. 1759. Instructio Peregrinatoris. Upsaliae. S. n. [3] + 1-15.

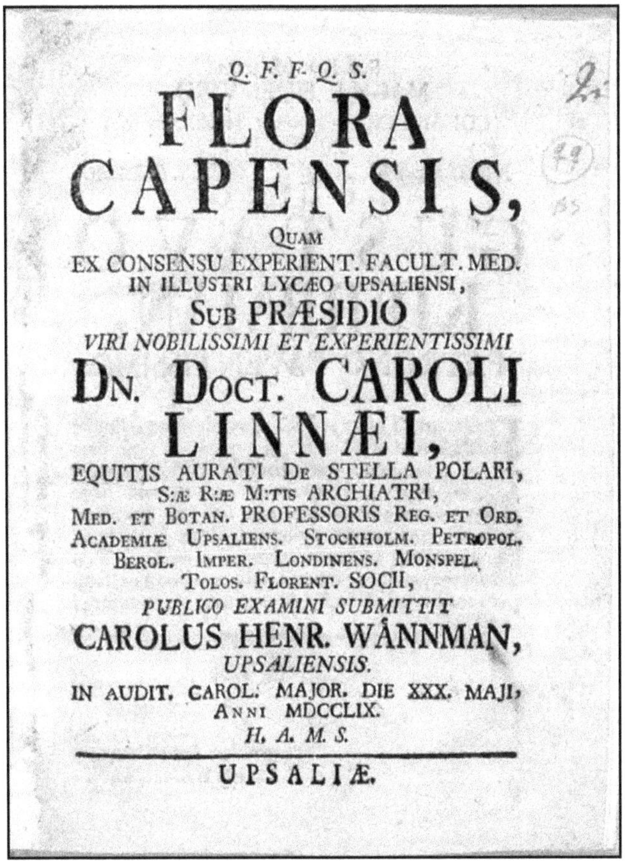

Q. F. F. Q. S.
FLORA
CAPENSIS,
QUAM
EX CONSENSU EXPERIENT. FACULT. MED.
IN ILLUSTRI LYCÆO UPSALIENSI,
SUB PRÆSIDIO
VIRI NOBILISSIMI ET EXPERIENTISSIMI
DN. DOCT. CAROLI
LINNÆI,
EQUITIS AURATI DE STELLA POLARI,
S:Æ R:Æ M:TIS ARCHIATRI,
MED. ET BOTAN. PROFESSORIS REG. ET ORD.
ACADEMIÆ UPSALIENS. STOCKHOLM. PETROPOL.
BEROL. IMPER. LONDINENS. MONSPEL.
TOLOS. FLORENT. SOCII,
PUBLICO EXAMINI SUBMITTIT
CAROLUS HENR. WÅNNMAN,
UPSALIENSIS.
IN AUDIT. CAROL. MAJOR. DIE XXX. MAJI,
ANNI MDCCLIX.
H. A. M. S.
UPSALIÆ.

14. Wännman, C.H. 1759. Flora capensis. Upsaliae. S. n. [3] + 1-19 + [1].

Dissertation defended on May 9, 1759, and reprinted in the fifth volume of *Amoenitates Academicae* (Nordblad, 1760). This is a guide on how to prepare a journey, to behave abroad, keep a journal and collect voucher specimens, which also contains the intellectual requirements that the traveler must possess. On Amphibia (XI. Zoologica, page 8) says:

••••

XI.

Zoology.

Exotic amphibians, less known, should be preserved in alcohol; for snakes, the structure of their fangs and the number of scales should be recorded.

Dissertation defended on May 30, 1759, and reprinted in the fifth volume of *Amoenitates Academicae* (Wännman, 1760). In the introduction Wännman provides a list of animals that inhabit the jungles of Africa, mentioning *Najae* and *Crocodili.*

••••

§. I.

It was a very common proverb among the ancient Romans that something new always comes out of Africa, and to this day it is well proven. For no land is as abundant in so many unusual and wondrous natural phenomena as this one. Here in the forests are found roaring lions, incredibly fast tigers (sic), the largest elephants and rhinoceroses, and hippopotamuses, countless golden jackals, as well as ichneumons, civets, crested porcupines, Cape hares, gazelles, zebras, cobras, electric rays, ostriches, peacock herons, and hummingbirds; along the shores, crocodiles; in the sea, sharks and chimeras. Even humans here have an entirely different external appearance. Thus, the black Moors display huge eyes, thicker lips, a flattened nose, pendulous breasts, and a loincloth for modesty, not to mention the *Troglodytas, Satyros, Sylvanos, Sphinges, Dianas, Hamadryades, Cynomolgos*[444], and the rest.

[444]On these improbable creatures, see the dissertation of Hoppius (1760).

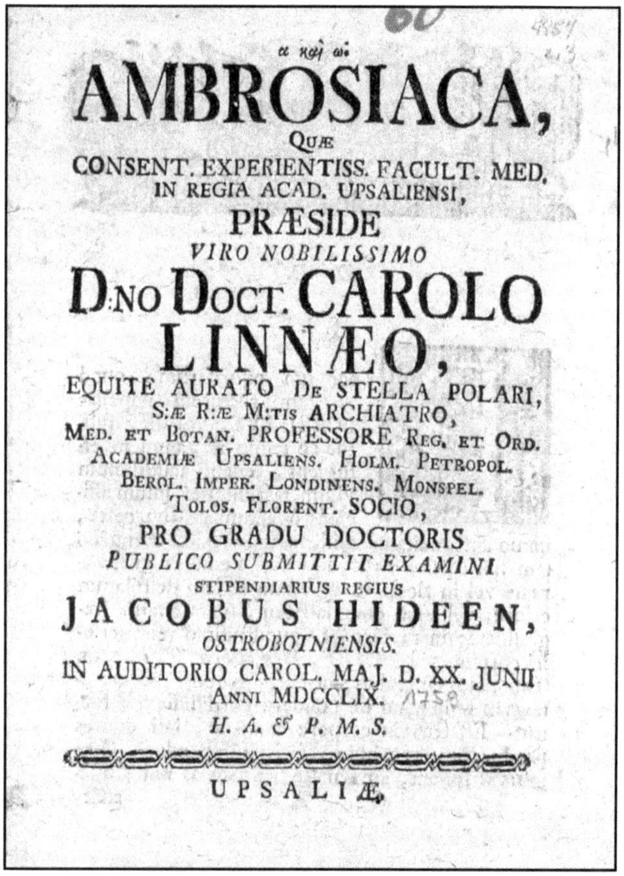

AMBROSIACA,
QUÆ
CONSENT. EXPERIENTISS. FACULT. MED.
IN REGIA ACAD. UPSALIENSI,
PRÆSIDE
VIRO NOBILISSIMO
D:NO DOCT. CAROLO
LINNÆO,
EQUITE AURATO DE STELLA POLARI,
S:Æ R:Æ M:TIS ARCHIATRO,
MED. ET BOTAN. PROFESSORE REG. ET ORD.
ACADEMIÆ UPSALIENS. HOLM. PETROPOL.
BEROL. IMPER. LONDINENS. MONSPEL.
TOLOS. FLORENT. SOCIO,
PRO GRADU DOCTORIS
PUBLICO SUBMITTIT EXAMINI
STIPENDIARIUS REGIUS
JACOBUS HIDEEN,
OSTROBOTNIENSIS.
IN AUDITORIO CAROL. MAJ. D. XX. JUNII
ANNI MDCCLIX.
H. A. & P. M. S.
UPSALIÆ.

15. Hideen, J. 1759. Ambrosiaca. Upsaliae. S. n. [1] + 3-14.

Dissertation defended on June 20, 1759, and reprinted in the ninth volume of *Amoenitates Academicae* (Hideen, 1785). In section V of this thesis on aromatic plants (*Ambrosiaca*), Hideen highlights the various attributes that the animals have to protect themselves against possible assaults and injuries.

••••

[p. 8]

§. V.

Animals are endowed with various faculties and are differently armed so that they can better sustain themselves or defend against the injuries of aggressors. *Crotalus horridus* and *Coluber Naja* possess venom, nothing in nature being more potent, and *Raja Torpedo* has a tremulous motion, which is extremely bothersome even to those who touch it lightly. The capuchin monkey emits a cruel scream, as if it is breaking bones, whenever someone harasses it. *Mustela Ermina*, when terrified, collapses in convulsions, thus becoming easy prey for its enemies; but at the same time, it emits a foul odor that drives them away. *Viverra Memphitis* and *Putorius* are equipped with a peculiar defense mechanism: when someone pursues or attacks these animals, they expel a stench through their spray that almost suffocates other animals. See the works of Noble D: ni Praes. S. Nat. editi. X[445]. and Cel. Prof. Kalm It. Am. T. II. p. 378.[446]

[445]Linnæus, 1758: 44: "...Irritata fundit cum ventris crepitu halitum fœtidissimum, ut nihil cum eo comparari queat..." [Irritated, it pours out the foulest breath with a rumbling of its belly, so that nothing can be compared to it].

[446]"...*Polcat* tallades af Angelsmännerna hårstås des et fyrfotadt djur, hwilket finnes nog allmånt ej allenast i Pennsylvanien, utan ock längre bås de til Norr och Söder i Norra America. Uti New Yorks Province fallas det gemenligen *Skunk*. De Swenske hår i landet kallade det *Fiskarta*, af den oliderliga stank, den ibland gifwer från sig, hwarom strar nedanföre. Fransoserne i Canada *kalla* det för samma orsak skuldels *Béte puant*, dels *Enfant du diable*. En del Fransoser kalla det ock *Pekan*. Det är beskrifs wit och med lefwande fårgor afritadt af Herr Catesby uti deß Natural History of Carolina Vol. II. p. 62. f. 62. under namn af *Putorius Americanus Striatus*..." [The polecat is described by the English as a four-legged animal which is found quite commonly not only in Pennsylvania but also further both to the North and South in North America. In New York's Province, it is commonly called a skunk. The Swedish inhabitants of the country call it fiskarta due to the unbearable stench it sometimes emits. For the same reason, the French in Canada also call it béte puant and enfant du diable. Some French people also call it pékan. It is described as white and with living characteristics, illustrated by Mr. Catesby in his Natural History of Carolina, Volume 11, p. 62, under the name Putorius Americanus Striatus].

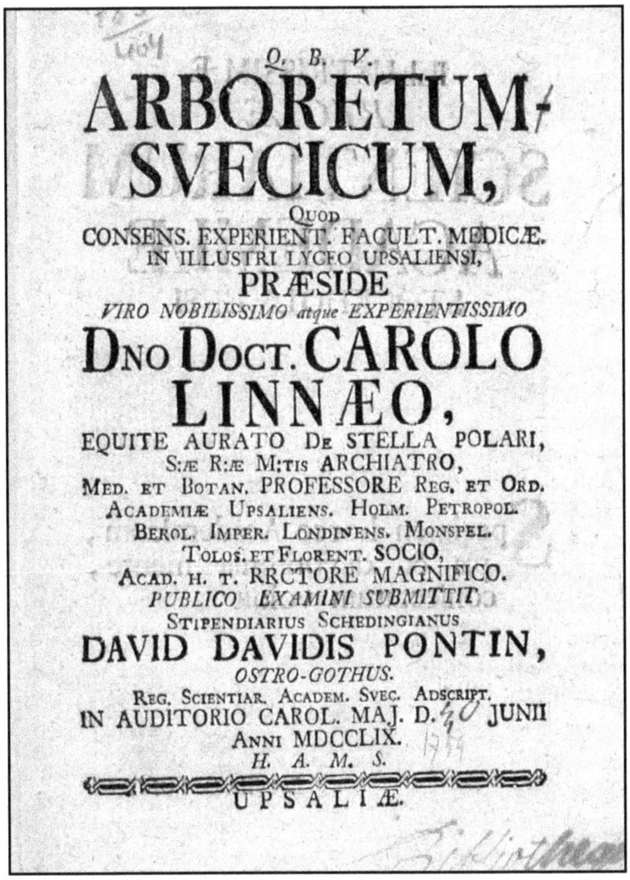

Q. B. V.
ARBORETUM SVECICUM,
QUOD
CONSENS. EXPERIENT. FACULT. MEDICÆ.
IN ILLUSTRI LYCEO UPSALIENSI,
PRÆSIDE
VIRO NOBILISSIMO atque EXPERIENTISSIMO
DNO DOCT. CAROLO
LINNÆO,
EQUITE AURATO DE STELLA POLARI,
S:Æ R:Æ M:TIS ARCHIATRO,
MED. ET BOTAN. PROFESSORE REG. ET ORD.
ACADEMIÆ UPSALIENS. HOLM. PETROPOL.
BEROL. IMPER. LONDINENS. MONSPEL.
TOLOS. ET FLORENT. SOCIO,
ACAD. H. T. RRCTORE MAGNIFICO.
PUBLICO EXAMINI SUBMITTIT
STIPENDIARIUS SCHEDINGIANUS
DAVID DAVIDIS PONTIN,
OSTRO-GOTHUS.
REG. SCIENTIAR. ACADEM. SVEC. ADSCRIPT.
IN AUDITORIO CAROL. MAJ. D. 30 JUNII
ANNI MDCCLIX.
H. A. M. S.
UPSALIÆ.

16. Pontin, D.D. 1759. Arboretum svecicum. Upsaliae. S. n. [2] + 4-30.

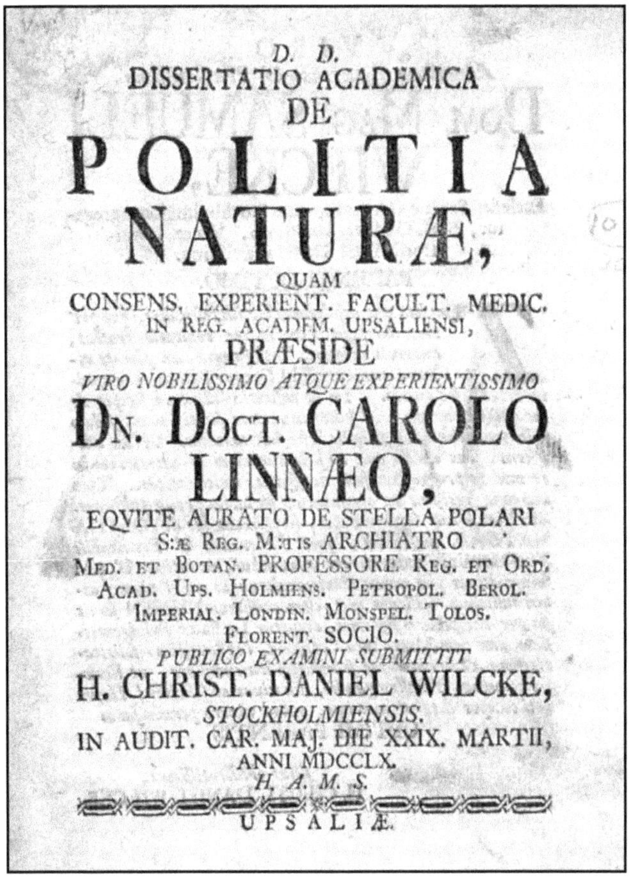

D. D.
DISSERTATIO ACADEMICA
DE
POLITIA
NATURÆ,
QUAM
CONSENS. EXPERIENT. FACULT. MEDIC.
IN REG. ACADEM. UPSALIENSI,
PRÆSIDE
VIRO NOBILISSIMO ATQUE EXPERIENTISSIMO
DN. DOCT. CAROLO
LINNÆO,
EQVITE AURATO DE STELLA POLARI
S:Æ REG. M:TIS ARCHIATRO
MED. ET BOTAN. PROFESSORE REG. ET ORD.
ACAD. UPS. HOLMIENS. PETROPOL. BEROL.
IMPERIAL. LONDIN. MONSPEL. TOLOS.
FLORENT. SOCIO.
PUBLICO EXAMINI SUBMITTIT
H. CHRIST. DANIEL WILCKE,
STOCKHOLMIENSIS.
IN AUDIT. CAR. MAJ. DIE XXIX. MARTII,
ANNI MDCCLX.
H. A. M. S.
UPSALIÆ.

17. Wilcke, H.C.D. 1760. De Politia naturae. Upsaliae. S. n. [1] + 1-22.

Dissertation defended on June 30, 1759, and reprinted in the fifth volume of *Amoenitates Academicae* (Pontin, 1760). Referring to the common aspen, comments that "*Ranas bufones* & *Colubros*" hide beneath its deciduous leaves.

••••

[p. 25]

XX. POPULUS tremula Sl. Sv. 909. [Asp.] does not avoid hills, valleys, plains, or forests within the borders of our country, but tends to grow in any soil or place, from the highest Lapland mountains to the farthest field of Scania, although it thrives less in clayey soil. It impoverishes the land, suffocates the grass with its fallen leaves, and covers the surface of the ground with countless spreading shoots. Although clearings among the rustics often consist of these trembling poplars, they are not very suitable for this purpose; for not only do the leaves constantly tremble, producing an unpleasant rustling sound, but the fallen leaves also conceal frogs, toads, and snakes beneath them. (...)

Dissertation defended on March 29, 1760, and reprinted in the sixth volume of *Amoenitates Academicae* (Wilcke, 1763); there is an abridged English translation in Brand (1781). When discussing the balance that exists in nature, Wilcke highlights the benefits that amphibians and reptiles bring in their role as predators and prey.

••••

[p. 12]

19. The number of amphibians is indeed smaller, but each has its own duties. Frogs, which swim in ponds, marshes, and cisterns, consume any excess such as duckweed[447] and especially the numerous larvae of mosquitoes. Snakes and lizards devour the insects they encounter everywhere. Sharks and rays scavenge carcasses on the sea floor.

[p. 13]

24. Wading birds (*Grallae*) primarily carry out their duties and tasks in marshes and swamps, where numerous threadworms, earthworms, and insects work to remove impurities; however, with the help of these birds, balance is maintained. When, in warmer lands, the marshes dry up during the summer, they migrate to us at that time; but because the winter ice in our lands denies them sustenance,

[447] *Lemna* in the original.

they are forced to return to more southern lands. When the Nile, which had flooded Egypt, recedes, many amphibians are left on solid ground; at that time, storks (*Ciconia*) are present to devour the surplus[448].

[p.18]

VIII. From what is known, we gather how essential each element of Nature's design is, to the extent that if even a single earthworm were missing, the soil would be corrupted by stagnant water, and mold would infect everything. If even one superior minister among the animals were lacking, a great disaster would be feared for the world. The blackbird[449], which is destined to maintain the balance of the pea beetle[450] population, being destroyed by weapons, has caused a ruin of peas in northern America. If all the sparrows in our lands were to perish, our plantations would fall prey to crickets and other insects. If America were devoid of pigs, it would be overrun by snakes to the point of destruction; likewise, in our homeland, homes and possessions would suffer great damage from mice if the family of cats were to completely die out. The same must be said of other servants of nature, down to the smallest insect, since the wisest Author of things has never acted without cause or reason.

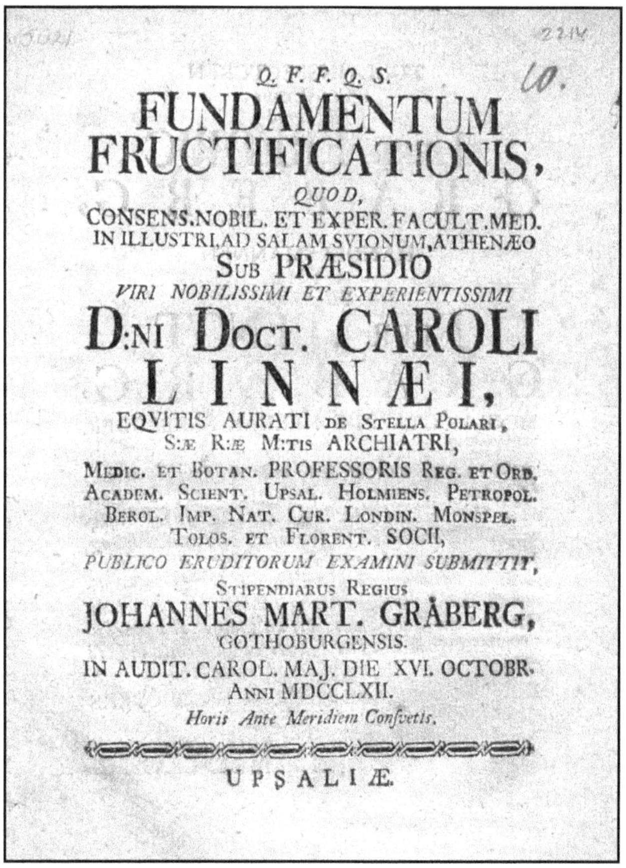

Q. F. F. Q. S.
FUNDAMENTUM FRUCTIFICATIONIS,
QUOD,
CONSENS. NOBIL. ET EXPER. FACULT. MED.
IN ILLUSTRI AD SALAM SVIONUM, ATHENÆO
SUB PRÆSIDIO
VIRI NOBILISSIMI ET EXPERIENTISSIMI
D:NI DOCT. CAROLI
LINNÆI,
EQVITIS AURATI DE STELLA POLARI,
S:Æ R:Æ M:TIS ARCHIATRI,
MEDIC. ET BOTAN. PROFESSORIS REG. ET ORD.
ACADEM. SCIENT. UPSAL. HOLMIENS. PETROPOL.
BEROL. IMP. NAT. CUR. LONDIN. MONSPEL.
TOLOS. ET FLORENT. SOCII,
PUBLICO ERUDITORUM EXAMINI SUBMITTIT,
STIPENDIARUS REGIUS
JOHANNES MART. GRÅBERG,
GOTHOBURGENSIS.
IN AUDIT. CAROL. MAJ. DIE XVI. OCTOBR.
ANNI MDCCLXII.
Horis Ante Meridiem Consvetis.
UPSALIÆ.

18. Gråberg, J.M. 1762. Fundamentum fructificationis. Upsaliae. S. n. [3] + 1-24.

Dissertation defended on October 16, 1762, and reprinted in the sixth volume of *Amoenitates Academicae* (Gråberg, 1763). The fifth section of Gråberg's dissertation is a synthesis of Swammerdam (1693: 56-57), and Gråberg compares there the transformations observed in insects and in *Rana temporaria* with those of plants.

••••

[p. 8]

§. V.

(...) *Rana temporaria* is first a larva, exactly resembling a fish; afterwards, it is equipped with two and soon four legs and becomes a pupa. Finally, it loses its tail, and only then is it considered complete and declared. Therefore, to understand the frog, one must recognize it in its perfect state, after metamorphosis: the same judgment applies to plants.

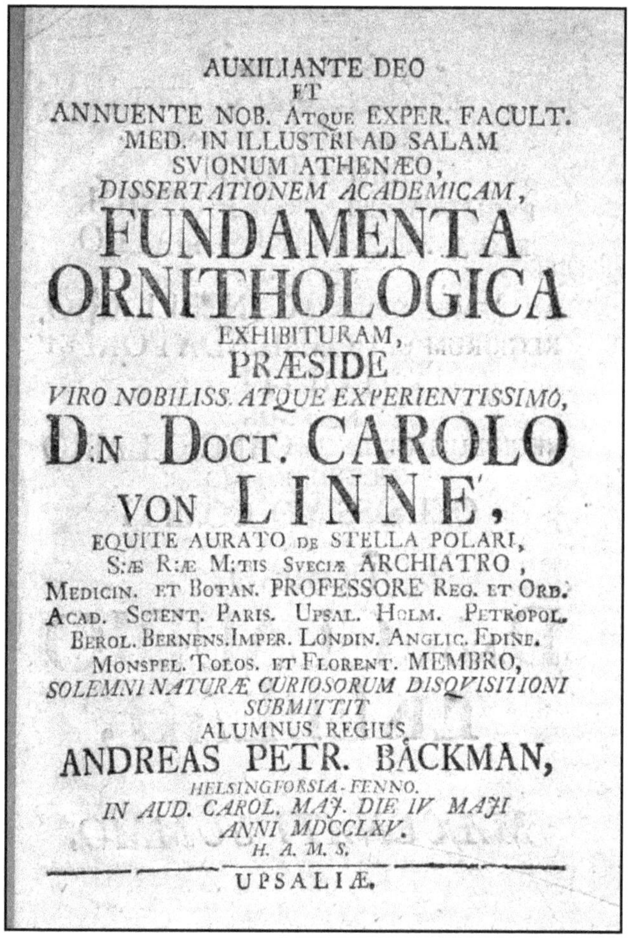

AUXILIANTE DEO
ET
ANNUENTE NOB. ATQUE EXPER. FACULT.
MED. IN ILLUSTRI AD SALAM
SVIONUM ATHENÆO,
DISSERTATIONEM ACADEMICAM,
FUNDAMENTA ORNITHOLOGICA
EXHIBITURAM,
PRÆSIDE
VIRO NOBILISS. ATQUE EXPERIENTISSIMO,
D:N DOCT. CAROLO
VON LINNÉ,
EQUITE AURATO DE STELLA POLARI,
S:Æ R:Æ M:TIS SVECIÆ ARCHIATRO,
MEDICIN. ET BOTAN. PROFESSORE REG. ET ORD.
ACAD. SCIENT. PARIS. UPSAL. HOLM. PETROPOL.
BEROL. BERNENS. IMPER. LONDIN. ANGLIC. EDINB.
MONSPEL. TOLOS. ET FLORENT. MEMBRO,
SOLEMNI NATURÆ CURIOSORUM DISQVISITIONI
SUBMITTIT
ALUMNUS REGIUS
ANDREAS PETR. BÄCKMAN,
HELSINGFORSIA-FENNO.
IN AUD. CAROL. MAJ. DIE IV MAJI
ANNI MDCCLXV.
H. A. M. S.
UPSALIÆ.

[448]A similar idea is cited in Ekmarck (1757).

[449]*Quiscula* in the original.

[450]*Dermestis pisorum* in the original.

19. Bäckman, A.P. 1765. Fundamenta ornithologica. Upsaliae. S. n. [5] + 6-28 + 1 plate.

Dissertation defended on May 4, 1765, and reprinted in the seventh volume of *Amoenitates Academicae* (Bäckman, 1769). In Chapter 2, when dealing with the diagnostic characters of birds, Bäckman (1765: 10) questions the flight, because this ability was recorded in other classes, like *Vespertiliones* and *Noctilionem* among mammals, *Draconis* and *Pegassus* among amphibians and *Mugilis, Triglae* and *Gasterosteis* among fish.

••••

CAP. II.

On the Diagnostic Characteristics of Birds

§ 4.

All animals, from the largest to the smallest, are conveniently classified by nature into six classes: Mammalia, Aves, Amphibia, Pisces, Insecta & Vermes. Among these, Mammals and Birds are most closely related in affinity and are very similar in internal characteristics: both have a heart with two ventricles, warm blood, and lungs that breathe reciprocally. However, Mammals and Birds are distinguished by their covering: Mammals have simple hair, while Birds are adorned with branched or feathered plumes. To say nothing of the beak, which in Birds is extended into a bare horny structure without teeth, nor that they stand only on two feet, nor that their arms are equipped with fingers modified into wings that allow them to climb the air, nor that they all lay eggs covered with a calcareous shell and do not carry their young in the womb. The venerable antiquity has designated birds with the name *Volatiles* to express their characteristic with the name itself: I confess it is a beautiful characteristic denomination, provided it were based on such certain foundations as the ancients believed. However, the inadequacy of this characteristic, given the flight of bats among Mammals, *Draco* and *Pegasus*[451] among Amphibians, *Mugil, Trigla*, and *Gasterosteus* among Fish, and many Insects (except the wingless ones), is clearer than midday light. On the other hand, not all birds can soar high: for example, the flightless ostriches, the *Dodo*, the submerged *Diomedea*, the flightless *Alca*, and the submerged *Phaeton*.

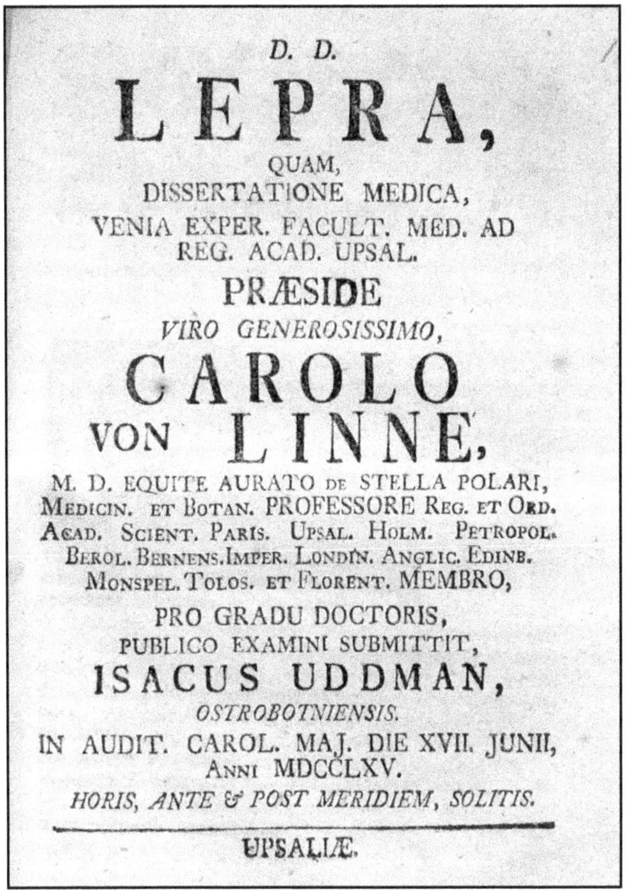
D. D.
LEPRA,
QUAM,
DISSERTATIONE MEDICA,
VENIA EXPER. FACULT. MED. AD
REG. ACAD. UPSAL.
PRÆSIDE
VIRO GENEROSISSIMO,
CAROLO
VON LINNE,
M. D. EQUITE AURATO DE STELLA POLARI,
MEDICIN. ET BOTAN. PROFESSORE REG. ET ORD.
ACAD. SCIENT. PARIS. UPSAL. HOLM. PETROPOL.
BEROL. BERNENS. IMPER. LONDIN. ANGLIC. EDINB.
MONSPEL. TOLOS. ET FLORENT. MEMBRO,
PRO GRADU DOCTORIS,
PUBLICO EXAMINI SUBMITTIT,
ISACUS UDDMAN,
OSTROBOTNIENSIS.
IN AUDIT. CAROL. MAJ. DIE XVII. JUNII,
ANNI MDCCLXV.
HORIS, ANTE & POST MERIDIEM, SOLITIS.
UPSALIÆ.

20. Uddman, I. 1765. Lepra. Upsaliae. S. n. 1-14.

Dissertation defended on June 17, 1765, and reprinted in the seventh volume of *Amoenitates Academicae* (Uddman, 1769).

••••

§. IX.

Worms and insects were comprehended under one natural class by the ancients, yet they differ entirely, as much as birds and mammals. Consequently, they concluded that what harms insects would also harm worms, which is now clearly unfounded. Worms of various kinds inhabit human bodies, three of which are commonly found in the intestinal tract: *Taenia*, *Lumbricus*, and *Ascaris*. Besides these, other animals occasionally descend into the stomach. For instance, snakes, frogs, and lizards, for which one can refer to the observations of Barthol.[452] and others.

[451]Bäckman (1765) pre-dates Linnæus's (1766) formal inclusion of *Pegasus* among the Amphibia: Nantes in the twelfth edition of the Systema Naturæ. In the tenth (Linnæus, 1758), this genus was classified among the Pisces Branchiostegi.

[452]Probably refers to Bartholini (1622), who in page 51, and referring to the orifices of the stomach, wrote about pylorus: "*...fibras transversas habet, circulumque crassiorem circumjectum (alij vocant tubercula glandulosa) instar musculi orbicularis vel Sphyncteris. Dicitur piloros & janitor, quias chylum emittit. Potest valdè dilatari, quéadmodum & sinistrum. Hinc varia exempla testantur, sæpè deglutita fuisse per vomitum & secessum rejecta, quæ oppido magna fuere; ut annuli aurei, nuclei, cultelli, lapilli, ferramenta, ranæ, lacertæ, serpentes, anguilla integra, &c...*". ["...It has transverse fibers and a thicker circular band around it (some call it glandular tubercles), resembling an orbicular muscle or sphincter. It is called the pylorus and gatekeeper because it emits chyme. It can be greatly dilated, just like the left one. Hence, various examples testify that often things swallowed have been expelled through vomiting and excretion, which were exceedingly large; such as gold rings, nuts, knives, small stones, tools, frogs, lizards, snakes, whole eels, etc..."].

§. XI.

TREATMENT. To ward off leprosy, the ancients used a broth made from cooked vipers, which, however, is less effective in our time. But it should be noted that the viper of the ancients was the *Coluber* of S.N. 140[453], whereas our contemporaries use the *Coluber Bero*. We know that true vipers were once dried and shipped in full loads from Egypt to Venice; now, however, pharmacists substitute our common *Bero*, considering it sufficient to administer any snake, as long as it is a serpent. However, the fact that the result is not the same is sufficiently demonstrated by the different smell of the various serpents, as well as the equally different effects of their bites.

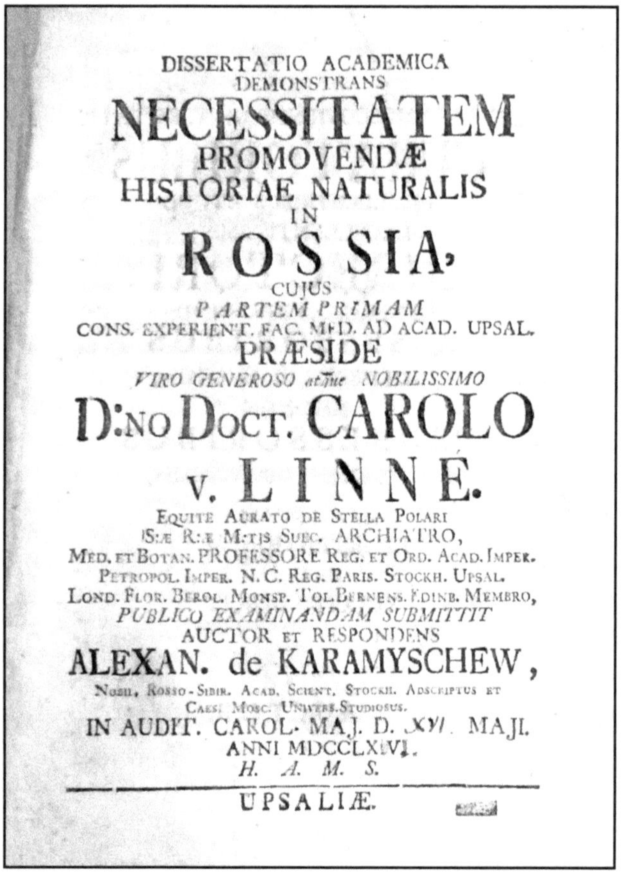

DISSERTATIO ACADEMICA
DEMONSTRANS
NECESSITATEM
PROMOVENDÆ
HISTORIAE NATURALIS
IN
ROSSIA,
CUJUS
PARTEM PRIMAM
CONS. EXPERIENT. FAC. MED. AD ACAD. UPSAL.
PRÆSIDE
VIRO GENEROSO atque NOBILISSIMO
D:NO DOCT. CAROLO
v. LINNÉ.
EQUITE AURATO DE STELLA POLARI
S:Æ R:Æ M:TIS SUEC. ARCHIATRO,
MED. ET BOTAN. PROFESSORE REG. ET ORD. ACAD. IMPER.
PETROPOL. IMPER. N. C. REG. PARIS. STOCKH. UPSAL.
LOND. FLOR. BEROL. MONSP. TOL. BERNENS. EDINB. MEMBRO,
PUBLICO EXAMINANDAM SUBMITTIT
AUCTOR ET RESPONDENS
ALEXAN. de KARAMYSCHEW,
NOBIL. ROSSO-SIBIR. ACAD. SCIENT. STOCKH. ADSCRIPTUS ET
CAES. MOSC. UNIVERS. STUDIOSUS.
IN AUDIT. CAROL. MAJ. D. XVI. MAJI.
ANNI MDCCLXVI.
H. A. M. S.
UPSALIÆ.

21. Karamyschew, A. de. 1766. Necessitas promovendae historiae naturalis in Rossia. Upsaliae. S. n. [4] + 1-34 + [10] + 1 plate.

Dissertation defended on May 16, 1766, and reprinted in the seventh volume of *Amoenitates Academicae* (Karamyschew, 1769). The basic hypothesis of this dissertation is that the development of natural science studies in Russia will allow for the development and progress of the country (more precisely, Siberia), as happened in other European nations[454].

••••

§. VIII.

Numerous new and still undescribed ANIMALS are believed to be found in our homeland for many reasons. While perusing the esteemed Museum of the Imperial Academy of Sciences in St. Petersburg in §:pho 3:tia[455], I observed with my own eyes among the numerous Indian objects of the animal's kingdom indigenous to Siberia, such as Mammals, Birds, Amphibians, Fish, and Insects, which have never been seen or described by any mortal and still lie shrouded in darkness. If I am fortunate enough to return to my homeland, nothing will be more in my wishes than to have the opportunity to organize and present these to the learned world. Who, pray tell, can understand and recognize from an unknown name alone the imperfectly determined animals? Tell me, and you will be my great Apollo, what kind of mammals and birds are described in our museum under names such as: *Furunculus myodes* p. 343. *Mustela Daurica cauda carens*, *Cuniculus Dauricus caudatus*, pag. 344. *Mus cauda carens* pag. 347. *Glis montanus* pag. 347. *Talpa aurea* pag. 349. *Tinunculus fuliginosus* pag. 357. *Noctua Daurica* pag. 363. *Pica Daurica* pag. 363. *Columba rupicola* pag. 373. *Alauda Tungusica* pag. 376. *Hirundo Daurica* pag. 377. *Motacilla capite ventre flavo* pag. 339. *Unaria longicauda* pag. 396. *Tschiptirra Tungusica Messerschmidii* pag. 397. *Cirlus Tungusicus* pag. 397. *Grus Daurica* pag. 400. *Tringa Samoedica* pag. 402. *Ergulus Lutorlslka* pag. 405. *Anser Leucorinchos* pag. 406., and others? These names are confusing, presented without descriptions, without specific notes, without illustrations. Therefore, to promote the study of natural history in Russia, it is necessary for our own people to recognize their own wealth.

§. IX.

There is no doubt that Amphibians exist in our land, although in my homeland—speaking of Siberia—there are very few Snakes, and indeed, I suspect, due to the climate, they are similar to those in Sweden. Whether there are toads escapes me, as they are particularly recognized in their homeland of Ukraine, where they are plentiful, much like *Anthemis cotula*, which abounds there[456]. I am convinced that we have various species of FROGS, which, however, will remain forever untouched in natural science. Gmelin, the best investigator of Siberian plants, could

[453] *Coluber vipera.*

[454] This is an idea taken from Linnæus (1741; 1743b; 1751) and his extensive network of travelers and collectors; see, among others, Frangsmyr (1994) y Kœrner (1999).

[455] See Anonymous (1742).

[456] The reference on the interactions with *Cotula*, was previously noted in Linnæus (1746a) *Fauna svecica* and in the tenth and twelfth editions of the Systema Naturæ (Linnæus, 1758; 1766).

hardly extend himself to describe animals, although he collected numerous specimens and introduced them to the St. Petersburg Museum, many of which perished in an unfortunate fire; thus, they remain unknown to this day in the icy region. Among the Swimming animals, the *Acipenser Huso* (our Beluga) claims its place as the largest amphibian in the world, considered a holy fish from the beginning of the world until this century, and has been transferred to a new class by the power of true science; the same must be said of the *Acipenser Rutheno* (Sterlet), which primarily recognizes Siberia as its homeland. I have no doubt that a curious investigator of natural phenomena will discover many new objects of this class if they visit our shores.

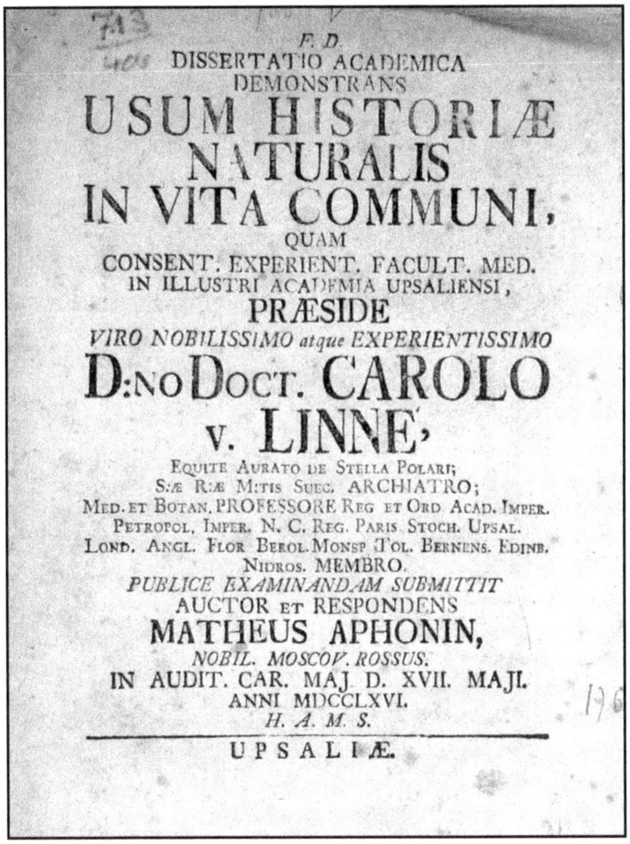
F. D.
DISSERTATIO ACADEMICA
DEMONSTRANS
USUM HISTORIÆ
NATURALIS
IN VITA COMMUNI,
QUAM
CONSENT. EXPERIENT. FACULT. MED.
IN ILLUSTRI ACADEMIA UPSALIENSI,
PRÆSIDE
VIRO NOBILISSIMO atque EXPERIENTISSIMO
D:NO DOCT. CAROLO
v. LINNÉ,
EQUITE AURATO DE STELLA POLARI;
S:Æ R:Æ M:TIS SUEC. ARCHIATRO;
MED. ET BOTAN. PROFESSORE REG ET ORD ACAD. IMPER.
PETROPOL. IMPER. N. C. REG. PARIS. STOCH. UPSAL.
LOND. ANGL. FLOR BEROL. MONSP. TOL. BERNENS. EDINB.
NIDROS. MEMBRO.
PUBLICE EXAMINANDAM SUBMITTIT
AUCTOR ET RESPONDENS
MATHEUS APHONIN,
NOBIL. MOSCOV. ROSSUS.
IN AUDIT. CAR. MAJ. D. XVII. MAJI.
ANNI MDCCLXVI.
H. A. M. S.
UPSALIÆ.

22. Aphonin, M.I. 1766. Usum historiae naturalis in vita communi.
Upsaliae. S. n. [6] + 3-30 + [2] + 1 plate.

Dissertation defended on May 17, 1766, and reprinted in the seventh volume of *Amoenitates Academicae* (Aphonin, 1769). Mentions to Amphibians are related to a series of tips to combat some species considered harmful.

••••

[p. 25-26]

CHAPTER THREE.

§. I.

Lacerta aquatica[457], a terrible inhabitant of fishponds and the fiercest enemy of freshwater creatures, is best expelled with muriatic salt[458]. Snakes are devoured by pigs and storks; they are mesmerized by Aristolochia, as Jaquin noted[459]. In my homeland[460], my fellow countrymen rub their hands with growing *Ligusticum* or *Levisticum*[461], which we call "Sorja" in our native tongue, and it is so common in Russia that there is hardly a garden where it does not grow. They handle the worst snake without harm, often placing it on their hand, rendering it inert and seemingly mesmerized. I have found the same to be true in Siberia, as confirmed by my dearest comrade, the most noble Lord Karamyschew, whose companionship I have enjoyed among foreigners for six years, and who today will excellently discuss the advancement of natural history in our beloved homeland[462].

[457]*Lissotriton vulgaris* (Linnæus, 1758).

[458]Sodium or magnesium chloride.

[459]Se comments on Jacquin in Acrel (1762) and SN 12 (Linnæus, 1766).

[460]The mention of "*mea patria*" (my country, i.e., Russia) is a clue that points to Aphonin, not Linnæus, as the author of the dissertation.

[461]Ligustico or levistico, common names of *Levisticum officinale*.

[462]Refers to Karamyschew (1766) dissertation (see above).

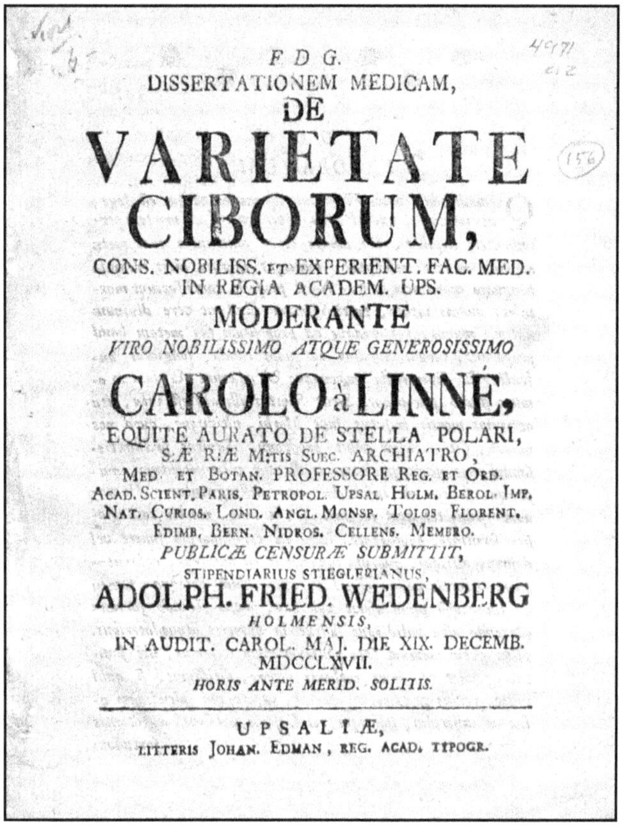

F D G.
DISSERTATIONEM MEDICAM,
DE
VARIETATE
CIBORUM,
CONS. NOBILISS. ET EXPERIENT. FAC. MED.
IN REGIA ACADEM. UPS.
MODERANTE
VIRO NOBILISSIMO ATQUE GENEROSISSIMO
CAROLO à LINNÉ,
EQUITE AURATO DE STELLA POLARI,
S:Æ R:Æ M:TIS SUEC. ARCHIATRO,
MED. ET BOTAN. PROFESSORE REG. ET ORD.
ACAD. SCIENT. PARIS. PETROPOL. UPSAL. HOLM. BEROL. IMP.
NAT. CURIOS. LOND. ANGL. MONSP. TOLOS. FLORENT.
EDIMB. BERN. NIDROS. CELLENS. MEMBRO.
PUBLICÆ CENSURÆ SUBMITTIT,
STIPENDIARIUS STIEGLERIANUS,
ADOLPH. FRIED. WEDENBERG
HOLMENSIS,
IN AUDIT. CAROL. MAJ. DIE XIX. DECEMB.
MDCCLXVII.
HORIS ANTE MERID. SOLITIS.

UPSALIÆ,
LITTERIS JOHAN. EDMAN, REG. ACAD. TYPOGR.

23. Wedenberg, A.F. 1767. De Varietate ciborum. Upsaliae. Johan. Edman, Reg. Acad. Tipogr. [2] + 4-20.

Dissertation defended on December 19, 1767, and reprinted in the seventh volume of *Amoenitates Academicae* (Wedenberg, 1769).

••••

[p. 4-5]

Life is in continuous motion; nerves, fibers, and humors of the human body are perpetually agitated. Since all motion wears down and deteriorates the body, it would easily be destroyed unless it acquired new particles daily to replace those that were worn away and constantly restored its own strength. Hence the necessity of food and drink; hence the use of exercise, which applies nourishment to the body, and of sleep, which assimilates it. Food is varied in every way, and very different nutrients are provided. Therefore, the selection of food recommends itself by necessity. It makes a great difference whether we repair a crumbling wall with lime and clay or with any other material. The wisest Creator has provided each animal with its own suitable food according to its nature. Thus, He has given distinct insects a unique food, for example, the mulberry leaves for the silkworm, nettle leaves for the urticating butterfly, and cabbage leaves for the cabbage butterfly, and so on for others[463]. For the larger and more complex animals, the same providential power has granted several types of food. To cattle, He has given vegetation, and to wild beasts, flesh. To man, above all others, He has wonderfully provided and generously bestowed a rich and varied diet. Man takes nourishment from fleshy pericarps and seeds, from roots, stalks, leaves, and flowers, from animals, birds, and finally fish. Indeed, the inhabitants of America even eat boa snakes, just as some southerners eat frogs. We eat insects and crabs, while the arabs locusts; indeed, even worms, razor clams, oysters, mussels, and snails, as is well known.

Not air, not sea, not even all the earth's bounty,
can barely satisfy our insatiable appetite.

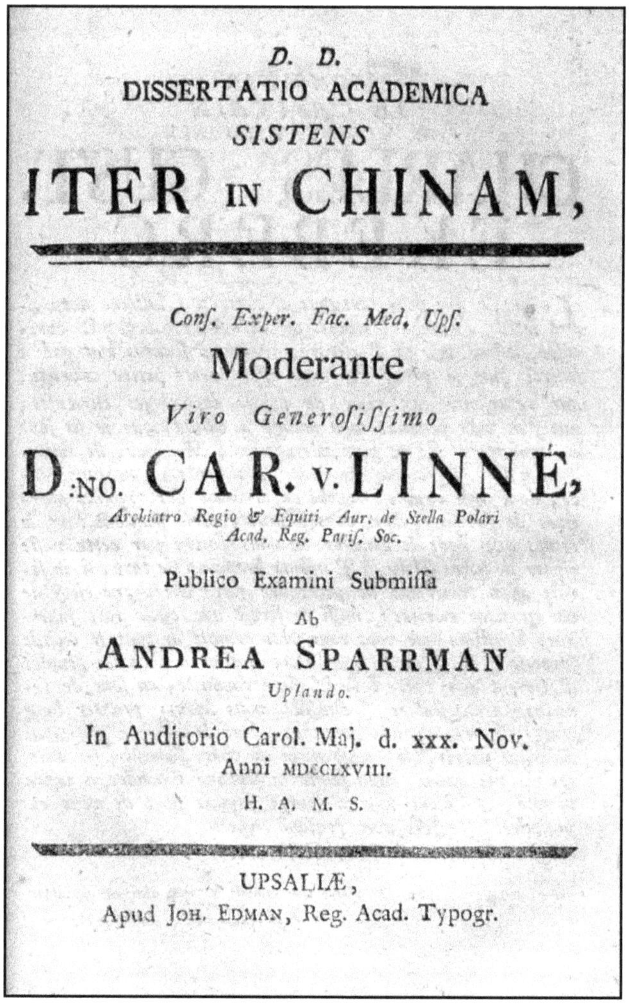

D. D.
DISSERTATIO ACADEMICA
SISTENS
ITER IN CHINAM,
Conf. Exper. Fac. Med. Upf.
Moderante
Viro Generofiffimo
D:NO. CAR. v. LINNÉ,
Archiatro Regio & Equiti Aur. de Stella Polari
Acad. Reg. Parif. Soc.
Publico Examini Submiffa
Ab
ANDREA SPARRMAN
Uplando.
In Auditorio Carol. Maj. d. xxx. Nov.
Anni MDCCLXVIII.
H. A. M. S.

UPSALIÆ,
Apud JOH. EDMAN, Reg. Acad. Typogr.

[463]"...Phalene Bombyci folia Mori, Papilioni urticario folia Urtica dioicæ, Papilioni Brassicae folia Brassicæ, & sic in ceteris..." in the original.

[464]Testudinis midas in the original.

[465]Portulacam olera in the original.

[466]Balistes Vetula in the original.

24. Sparrman, A. 1768. Iter in Chinam. Upsaliae. Johan. Edman, Reg. Acad. Tipogr. [3] + 5-16.

Dissertation defended on November 30, 1768, and reprinted in the seventh volume of *Amoenitates Academicae* (Sparrman, 1769).

••••

[p. 15]

On May 25th we arrived at Ascension Island where we captured 79 turtles[464] on the shore and gathered purslane[465]. Here, we observed the western pelican and caught a triggerfish[466] with beautiful colors but with a dusky hue.

On July 19th, I saw a turtle named Midas sleeping in the water.

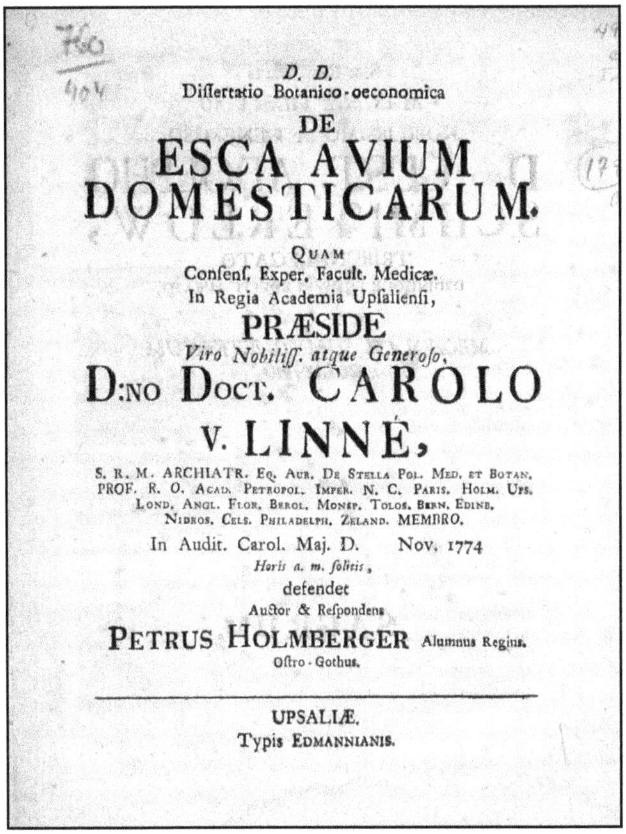

D. D.
Dissertatio Botanico-oeconomica
DE
ESCA AVIUM DOMESTICARUM.
QUAM
Consens. Exper. Facult. Medicæ.
In Regia Academia Upsaliensi,
PRÆSIDE
Viro Nobiliss. atque Generoso,
D:NO DOCT. CAROLO V. LINNÉ,
S. R. M. ARCHIATR. EQ. AUR. DE STELLA POL. MED. ET BOTAN. PROF. R. O. ACAD. PETROPOL. IMPER. N. C. PARIS. HOLM. UPS. LOND. ANGL. FLOR. BEROL. MONSP. TOLOS. BERN. EDINB. NIDROS. CELS. PHILADELPH. ZELAND. MEMBRO.
In Audit. Carol. Maj. D. Nov. 1774
Horis a. m. solitis,
defendet
Auctor & Respondens
PETRUS HOLMBERGER Alumnus Regius.
Ostro-Gothus.
UPSALIÆ.
Typis EDMANNIANIS.

25. Holmberger, P. 1774. Esca avium domesticarum. Upsaliae. Typis Edmannianis. [1] + 1-15.

Dissertation defended on November 26, 1774, and reprinted in the eighth volume of *Amoenitates Academicae* (Holmberger, 1785).

••••

§. II.

Experience, founded on true principles, provides the safest guidance in Natural Science, offering the most solid arguments that both Insects and the fruits of plants are numerous for these reasons: primarily for propagation, and secondly to serve as food for many quadrupeds, birds, amphibians, and countless insects. For if these animals were deprived of this food, the entire cohort would perish from hunger, which contradicts the purpose of creation. (...)

I observed that *Lacerta agili*, dissected with a scalpel, had consumed *Cicindelam campestrem, Gryllum biguttatum* & *Phalangium opilionem*.

In the open stomach of *Rana Bufonis*, I saw a *Dermestem pedicularium, Chrysomelam politam, Carabum latum, Carabum vulgarem, Elaterem sangvineum, Elaterem murinum, Vespam vulgarem, Araneam labyrinthicam, Scolopendram Laguram* and *Julum terrestrem*.

The female *Rana temporaria* swallows aquatic plants. *Asilus crabroniformis* devours *Gryllus buguttulum*, just as *Asilus gibbosus* kills *Vespa vulgaris* with its strong jaws, to mention just a few examples. Ultimately, all animals feed on a specific diet suited to their nature, which they are not allowed to exceed at will, under penalty of disease or death. For if each were permitted to eat anything, the rest would perish from hunger, and the entire order of things would be disrupted. Behold divine providence!

Literature Cited

Entries in ***bold italics*** correspond to citations from the original dissertations, followed by their respective references in regular typeface. Occasionally these abbreviated texts do not follow strict alphabetical order; this liberty has been taken to facilitate their location with the corresponding citation.

Acosta, C. 1578. *Tractado de las drogas, y medicinas de las Indias Orientales, con sus Plantas debuxadas al biuo por Christoual Acosta medico y cirujano que las vio ocularmente. En el cual se verifica mucho de lo que escriuio el Doctor Garcia de Orta. Dirigido a la muy noble y muy mas leal ciudad de Burgos cabeça de Castilla y camara de su Magestad.* Burgos. Martin de Victoria. [23] + 1-448 + 1-38 + [1].

Acrel, J. G. 1796. *Tal, om läkare-vetenskapens grundläggning och tilväxt vid rikets älsta lärosäte i Upsala, hållet vid praesidii nedläggande uti Kongl. Vetenskaps Academien den 10 augusti 1796.* Stockholm. Joh. P. Lindh. [i] + 1-176.

Acrell, J. G. 1762. *Disputatio medica, de morsura serpentum.* Upsaliae. S. n. [4] + 2-19 + 1 fig. + [1].

Acrell, J. G. 1763. Morsura serpentum. *Amoenitates academicae,* 6: 197-216. (Holmiae edition).

Act. Londin. Tyson, 1683.

Act. paris. 1737. Geoffroy & Hunauld. 1740.

Act. stockholmens. 1740. Linnaeus, 1740c.

Act. Ups. 1736. Linnaeus, 1742.

Acta Anglicana. Mortimer, 1741.

Aelian. 1958. *On the characteristics of animals. With an English translation by A. F. Scholfield. In three volumes. I.* Books I-V. London. William Heinemann. Cambridge, Mass. Harvard University Press. i-xix + 1-359.

Aëtius. Aëtius. 1534.

Aëtius. 1534. *Aetii Amideni medici clarissimi Libri sexdecim nunc primum Latinitate donati, in quibus cuncta quae ad artem curandi pertinent sunt congesta: ex omnibus qui usque ad eius tempora scripserant diligentissime excerpta. Amideum autem qua ratione appellaverimus, cum nonnulli Antiochenum nominaverint, facile quilibet intelligere poterit, qui ea legerit quae in praefatione sunt dicta. Totum vero opus summa diligentia excusum, cum INDICE omnium quae adnotatione digna visa sunt copiossisimo, in tres divisum est tomos. Quorum primus, septem libros continet a Ioanne Baptista Montano medico veronensi eleganter quidem Latinitate donatos. Secundus autem, libros sex Iano Cornario Zuiccaviensi interprete. Ac tertius, libros tres habet ab eodem Ioanne Baptista Montano felicissime latinos factos* (...). Venetiis. Luceantonii Iunte. [x] + 1-303 + [viii] + 1-304 + [1] + 1-156 + [xlvi].

Agardh, C. A. 1817-1826. *Aphorismi botanici ... praeside Carolo Ad. Agardh, . pro gradu philosophico p. p..., smolandus in Lyceo Carolino ... I: J. Forsander; pt. II: P. Ahlquist; pt. III: C. E. Hallström; pt. IV: J. Bruzelius; pt. V: N. Kihlgren; pt. VI: L. G. Starck; pt. VII: A. Rahnberg; pt. VIII: J. Dahl; pt. IX: S. Lyckberg; pt. XIII: G. C. Ljungstedt; pt. XIV: L. A. Wadstein; pt. XV: G. J. Tönningh; pt. XVI: B. E. Hildebran, respondents).* Lundae. Literis Berlingianis. 1-246.

Aldrov. ins. Aldrovandi, 1623.

Aldrovandi, U. 1623. *Historiam Natvralem in patria olim profitentis. De Animalibvs Insectis Libri Septem. In Qvibvs Omnia Illa Animalia accuratißime describuntur, eorum Icones ad viuum ob oculos ponuntur, tandemque etiam natura, mores ac proprietates ita declarantur, vt quicquid de iis dici queat, facile inde innotescat. His Adivnctvs Est Index Geminvs Capitvm scil. & verborum copiosissimus.* Francofvrti. Typis Ioannis Hoferi, Impensis Ioannis Treudelii. [10] + 1-299 + [22] + 15 pl.

Aldrov. quadr. Aldrovandi, 1637.

Aldrovandi, U. 1637. *De quadrupedib'. digitatis viviparis Libri Tres, et de quadripedib'. digitatis oviparis Libri Dvo.* Bonon. M. Antoni Berniae. Nicolau Tebaldinum. [5] + 718 + [16] pp.

Aldrovand. serp. Aldrovandi, 1640.

Aldrovandi, U. 1640. *Serpentum, et draconũ historiae libri duo. Bartholomaevs Ambrosinvs In Patrio Bonon Gymnasio Simplicium med. Professor ordinarius. Horti publici nec non Musęi Ill. Senatus Bonon. Prefectus. Summo labore opus concinnauit ad Illvstrissimvm Reverendissimvm et Excellentissimv Virvm D. Franciscv Perettvm Abbatem Venafri Principem Nomenti Marchione et Celani Comitem Meritissimvm. Cum Indice memorabilium nec non variarũ linguarũ locupletissimo.* Bononiae. M. Antonii Bernia. [6] + 1-427 + [30].

Alpin.; Alpin. Aegypt. Alpin, 1735

Alpin, P. 1735. *Historiae Aegypti Naturalis pars prima. Qua continentur Rerum Aegyptiarum libri quatuor. Opus postumum nunc primum ex auctoris autographo, diligentissime recognito, editum; atque ex eodm tabellis aeneis XXV, illustratum & uberrimo indice auctum.* Lugduni Batavorum. Gerardum Potvliet. [18] + 1-248 + XXV plates + [12].

Amoen. Acad. 112. Hast, 1749.

Amoen. acad. 2. Darelius, 1751.

Amoenitat. Kaempferus. Kaempfer, 1712.

Amph. [ib]. Gyllenb. Hast, 1745.

An Italian. Redi, 1664; 1670.

Anonymous. 1700. *Musaei Indici Index Exhibens Varia exotica animalia vegetabilia Nativâ facie liquori Basamico innatantia. Conspicienda in Horti Academici Ambulacro.* Lugduni Batavorum. s.n. 11 unnumbered pages.

Anonymous. 1742. *Musei Imperialis Petropolitani. Volumen I. Pars Prima qua continentur res naturales ex regno animali.* Petropolitanae. Typis Academiae Scientiarum: i-viii + 1-755.

Anonymous. 1896. [Linnaean Society of London]. *Catalog of the library of the Linnean Society of London. New edition.* London. Taylor and Francis. 1-727 pp.

Aphonin, M.I. 1766. *Usum historiae naturalis in vita communi.* Upsaliae. S. n. [6] + 3-30 + [2] + 1 plate.

Aphonin, M. I. 1769. Usus historiae naturalis in vita communi. *Amoenitates Academicae*, 7: 409-437 + 1 plate. (Holmiae edition).

Aristoteles. 1533. *De historia animalium libri IX. De partibus animalium & earum causis libri IIII. De generatione libri V. Theodoro Gaza Interprete. De communi animalium gressu liber I. De communi animalium motu liber I. Petro Alcyono Interprete.* Parisiis. Simonis Colinaei. [70] + 232 folios numbered in recto.

Aristotle. 1910. *Historia Animalium (History of Animals).* In J. A. Smith & W. D. Ross (eds.) *The works of Aristotle translated into English.* vol. 4. London. Oxford University Press. i-xv + paragraphs 486a - 633b + [30].

Arted. Syn. Artedi, 1738b.

Artedi, P. 1738a. *Genera Piscium. In quibus systema totum Ichtyologiae proponitur cum classibus, ordinibus, generum characteribus, specierum differentiis, observationibus plurimis. Redactis speciebus 242 ad genera 52. Ichthyologiae pars III.* Lugduni Batavorum. Conradum Wishoff. [6] + 1-84 + [4].

Artedi, P. 1738b. *Synonymia Nominum Piscium fere omnium; in qua recensio fit nominum piscium, omnium facile authorum, qui umquam de piscibus scripsere: uti graecorum, romanorum, barbarorum, nec non omnium insequentium ichthyologorum, una cum nominibus inquilinis variarum nationum. Opus sine pari. Ichthyologiae Pars IV.* Lugduni Batavorum. Conradum Wishoff. [2] + 1-66 + [2].

Artedio.; Arted. Gen. Artedi, 1738a.

Avellan, N. 1775. *Dissertatio Physiologica De Perspiratione Insensibili.* Upsaliae. Joh. Edman, Regiae Acad. Typogr. 12 pp.

Bäckman, A. P. 1765. *Fundamenta ornithologica.* Upsaliae. S. n. [5] + 6-28 + 1 plate.

Bäckman, A. P. 1769. Fundamenta ornithologica. *Amoenitates Academicae*, 7: 109-128 + 1 plate. (Holmiae edition).

Bails, B. 1790. *Arismetica para negociantes.* Madrid. Viuda de Ibarra. i-xiv + 1-407.

Baldaeus, P. 1672. *Wahrhaftige Ausführliche Beschreibung der Berühmten Ost-Indischen Kusten Malabar und Coromandel, als auch der Insel Zeylon samt dero angräntzenden und untergehörigen Reichen /Fürstentühmen / Ländern / Städten / vornehmsten Hafen / Bedäuden / Pagoden / Gewäschen / Thieren; der Eimvohner Gestalt / Sitten / Kleidertacht / haushaltung / Ceremonien: So wol auch der merkwürdigsten Kriegshändet / Belägerungen / Feld- und Seeschlachten / sonderlich zwischen den Portugesen und Holländern / Handel- und Kaufmannschaften: Durchgehends verzieret mit neuen Landkahrten und Abbildungen der vornehmsten Städte / Festungen / Trachten / Thiere / Früchte / &c. so in India selbst nach dem Leben gezeichnet / uns folgends mit Fleiß zu Kupfer gebracht: Benebst einer Umständlichen und Gründlichen Entdeckung der Abgötterey der Ost-Indischen Heyden, Malabaren, Benjanen, Gentiven, Bramines &c. So wol aus ihrem eigenen mit anhero gebrachten Vedam oder Gesetzbuch / und urkundlichen Handgeschriften / als Gespräch und Beywohnung ihrer vornehmsten Priester und Schriftgelehrten / nachgespüret / erforschet / und widerleget: zusamt den Abbildungen ihrer Götzen nach dero eigenen Bildern gezeichnet und fürgestellet. (...) Anitzo aber aus dem Niederländischen ins Hochteutsche mit Fleiß übergesetzt und mit einem vollständigen Register versehen.* Amsterdam. Johannes Janssonius von Waesberge und Johannes Someren. [3] + 1 plate + 1 map + 1-610 + [26] + mapas and plates.

Balk, L. 1746. *Museum Adolpho-Fridericianum.* Holmiae. Laurentii Salvii. [8] + 2-48 + 2 pl. + [2].

Balk, L. 1749. Museum Adolpho-Fridericianum. *Amoenitates Academicae*, 1: 277-326. (Holmiae & Lipsiae edition).

Barchewitz, E. C. 1730. *Allerneueste und wahrhaffte Ost-Indianische Reise-Beschreibung: Darinnen I. Seine durch Teutsch- und Holland nach Jndien gethane Reise; II. Sein eilff-jähriger Aufenthalt auf Java, Banda und den Südwester-Jnsuln, Glücks- und Unglücks-Fälle, seltsame Begebenheiten, auch remarquirte rare Gewächse, Bäume, Früchte, Thiere, Fische, Insecten, Berge, Vestungen, Nationen, Gewohnheiten, Aberglauben der Wilden, und viele andere Denckwürdigkeiten mehr; III. Seine Rück-Reise, der dabey erlittene grausame Sturm, und endlich glücklich erfolgte Ankunft in sein Vaterland, umständliche erzählet wird; Benebst einer ausführlichen Land-Charte der Sudwester- und Bandanesischen Insulen, welche in anderen Land-Charten nicht gefunden, noch den Geographien beschreiben werden; und einem vollständigen Register.* Chemnitz. Johann Christoph und Johann David Stößeln. i-xix + 1-657 + i-xv.

Bardon, H. 1956. *La littérature Latine inconnue. Tome II. L'Époque Impériale.* Paris. C. Klincksieck. [11] + 12-338.

Barthol. hist. anat. Bartholini, 1654.

Bartholin. acta haffniens. v. 5. Obs. 97. Jacobaeus [Jacobai], 1680.

Bartholin, G. 1622. *Anatomicae Institutiones Corporis Humani, Utriusq' sexus Historiam & Declarationem exhibentes, Cum plurimis novis Observationibus & Opinionibus, Necnon Illustriorum, quae in anthropologia occurrunt, controversiarum decisionibus, cum indice Capitum & Rerum locupletismo.* Rostochi. Ferberi Impens. & Litt. [23] + 1-442 + [51].

Bartholini, T. 1654. *Historiarum anatomicarum rariorum Centuria I et II.* Hafniae. Typis Academicis Martzani. Sumptibus Petri Hauboldt. [14] + 1-360 + [8].

Bauh. pin. Bauhin, 1671.

Bauhin, C. 1671. *ΠΙΝΑΞ Theatri Botanici sive index in Theophrasti Dioscoridis Plinii et botanicorum qui a seculo scripserunt opera plantarum circiter sex millium ab ipsis exhibitarum nomina cum earundem Synonymiis & differentiis methodice secundum genera & species proponens. Opus XL. annorum.* Basileae. Impensis Joannis Regis. [22] + 1-518 + [21] + [6] + 1-160 + [12].

Behren diaet. 225. Behrens, 1710.

Behrens, C.B. 1710. *Selecta diaetetica, seu, De recta ac conveniente ad sanitatem vivendi ratione tractatus: quo non solùm quicquid circa äerem, alimenta, motiones, somnum, aliaque ad sanitatem conservandam, cuicunq. aetati, sexui, vitae generi observatu necessarium, speciatim evolvitur; sed & origines ac occasiones morborum, praecavendi rationes, & remedia domestica explicantur, varisque observationibus illustrantur; ac simul de methodico acidularum, thermarumque usu agitur: cum indice copioso.* Francofurti & Lipsiae. Ludolphum Schröder. [13] + 2-548 + [29].

Bellon. itin. Bell. itiner. Belon, 1605.

Belon, P. 1553. *De aquatilibus, libri duo cum eiconibus ad viuam ipsorum effigiem, quoad eius fieri potuit, expressis. Ad amplissimum Cardinalem Castillionaeum.* Parisiis. Carolum Stephanus. [30] + 1-448.

Belon, P. 1554. *Les observations de plusieurs singularitez & choses memorables, trouvées en Grece, Asie, Iudée, Egypte, Arabie, & autres pays estranges, redigées en trois livres.* Paris. Gilles Corrozet. [24] + 1-211 folios numbered in recto + [1].

Belon, P. 1605. *Plurimarum singularium & memorabilium rerum in Graecia, Asia, Aegypto, Iudaea, Arabia, alisque exteris Provinciis ab ipso conspectarum observationes tribus libris expressae. Carolus Clusius Atrebas è Gallicis Latinas faciebat, & denuò recensebat. Altera editio, longè castigatior, & quibusdam Scholiis illustrata.* Pp. [10] + 1-208 in C. Clusius, *Exoticorvm Libri Decem: quibus Animalium, Plantarum, Aromatum, aliorumque Peregrinorum Fructuum Historiae Describuntur: Item Petri Bellonii Observationes, eodem Carolo Clusio Interprete. Series totius Operis post Praefationem Indicabitur.* s. l. [Leiden?] Officinâ Plantinianâ Raphelengii. [12] + 3-208.

Belonius. Bell. aquat. Bellon. aquat. Belon, 1553.

Berger, A. M. 1756. *Calendarium florae.* Upsaliae. L. M. Höjer, Reg. Acad. Typogr. [3] + 1-5 + [19].

Berger, A. M. 1759. Calendarium florae. *Amoenitates Academicae,* 4: 387-414. (Holmiae edition).

Beverly, R. 1705. *The History and Present State of Virginia, In Four Parts. I. The History of the First Settlement of Virginia, and the Government Thereof, to the Present Time. II. The Natural Productions and Conveniencies of the Country, Suited to Trade and Improvement. III. The Native Indians, Their Religion, Laws, and Customs, in War and Peace. IV. The Present State of the Country, as to the Polity of the Government, and the Improvements of the Land. By a Native and Inhabitant of the Place.* London. R. Parker. [1-16] + 1-104 + 1-40 +1-64 + 1-83 + 14 pl. + 1 fold. tab.

Biberg, I. J. 1749. *Oeconomia naturae.* Upsaliae. S. n. [7] + 1-48.

Biberg, I. J. 1751. Oeconomia naturae. *Amoenitates Academicae,* 2: 1-58. (Holmiae edition).

Biberg, S. 1759. *The Oeconomy of Nature.* Pp. 31-108 *in* B. Stillingfleet, *Miscellaneous Tracts Relating to Natural History, Husbandry and Physick,* translated from the Latin with Notes by (...). London. R. and J. Dodsley, S. Baker, and M. Cooper.

Biberg, S. 1789. Économie de la nature. Pp. 217-287 *in* R. Pulteney, *Revue Générale des Ecrits de Linné; Ouvrage dans lequel on trouve les Anecdotes les plus intéressantes de sa Vie privée, un Abrégé de ses Systêmes et de ses Ouvrages, un Extrait de ses Aménités Académiques, &c. &c. &c. Traduit de l'Anglois Par L. A. Millin de Grandmaison; Avec des Notes et des Additions du Traducteur. Tome Second.* A Londres & se trouve Paris. Chez Buisson.

Bjerkén, P. af. 1757. *Morbi Expeditionis Classicae.* Upsaliae. L. M. Hojer. [3] + 4-22.

Blunt. W. 2001. *The compleat naturalist. A life of Linnaeus.* London. Frances Lincoln Ltd. 1-264.

Bochart. hieroz. Bochart. [o]. S. 1663.

Bochart. [o]. S. 1663. *Hierozoicon, sive bipertitum opus de animalibus Sacrae Scripturae. Pars prior agit libris quatuor De animalibus in genere et de quadrupedibus viviparis et oviparis Pars posterior agit libris sex De avibus, serpentibus, insectis, aquaticis, et fabulosis animalibus. Cum indice septuplici I. Locorum Scripturae. II. Authorum qui citantur. III. Vocum Hebraici, Chaldaici, Syriaci, Samaritanici, & Aethiopici sermonis. IV. Vocum Arabicarum. V. Graecarum. VI. Romanarum, quae passim explicantur. VII. Rerum.* Londini. Ex. Tho. Roycroft. Impensis Jo. Martyn & Jac. Allerstry. [95] + 1-1094 columns + [125].

Boecler, C.E. 1772. *D. A. G. De Suturis Vulnerum in Genere.* Upsaliae. Typis Edmannianis. 22 pp.

Boerhaav. Boerhaave, 1728.

Boerhaave, H. 1728. *Commentationes novae de Lue Venerea, et huius curatione omnimoda, praefatae operi magno, Aphrodisiaco.* Clarissimo Hermanno Boerhaave. Ubi posuit huius Mali sedem in membrana adiposa, ejusque propaginibus: Curamque Gonorrhoeae virulentae cujusque speciei tradidit. Londini. Abraham Vandenhoeck. 1-68.

Boerhaave, H. 1738. *Praxis medica, sive commentarium in aphorismos Hermanni Boerhaave de cognoscendis et curandis morbis. Pars secunda. Editio tertia, aucta & accuratissima.* Londini. Sumptibus Societatis. 1-400 + [8].

Boerhaave, H. & A. Haller. 1743. *Hermani Boerhaave (...) Praelectiones Academicae. In proprias institutiones rei medicae edidit, et notas addidit Albertus Haller. Vol. IV. Sensus externi, interni, somnus.* Gottingae. Abram Vandenhoeck. 1-616.

Bontius.; Bont. Jav. Bontius, 1658.

Bontius, I. 1658. *Historiae naturalis & medicae Indiae Orientalis Libri Sex. I. De conservanda valetudine. II. Methodus medendi. III. Observationes e cadaveribus. IV. Notae in Garciam ab Orta. V. Historia Animalium. VI. Historia Plantarum. In quorum libro penultimo, Naturae Animalium, Avium, & Piscium: In ultimo autem, Arborum & Plantarum species mirae, Europaeis incognita, ac ad vivum delineatae explicantur. Commentari, quos Auctor, morte in Indiis praeventus, indigestos reliquit, a Gulielmo Pisone, in ordinem redacti & illustrati, atque annotationibus & additionibus rerum & iconum necessariis adaucti.* Independent pagination (1-160) in Piso, G. 1658. *De Indiae Utriusque re naturali et medica libri quatuordecim. Quorum contenta pagina sequens exhibet.* Amstelaedami. Ludovicum et Daniele Elzevierum.

Borelli de mot. Anim. 2. propr. 219. Borelli, 1681.

Borelli, J.A. 1681. *De Motu Animalium Pars Altera. Svb Iisdem Favstissimis Reginae Christinae Avspiciis.* Roma. Angeli Bernabò. [3} + 1-520.

Brand, F. J. 1781. *Select dissertations from the Amoenitates Academicae, a supplement to Mr. Stillingfleet's Tracts relating to Natural History. In two volumes.* Volume I. London. G. Robinson / J. Roeson. [i-xv] + 1-483.

Breyn. prod 2. Breyn, 1739.

Breyn, J. 1739. *Prodromi fasciculi rariorum plantarum primus et secundus, quondam separatim, nunc nova hac editione multum desiderata coniunctim editi, notulisque illustrati. Accedunt icones rariorum et exoticarum plantarum aeri incisae, fasciculo olim promisso destinatae: adiectis nominibus et succinctis descriptionibus. Quibus praemittuntur vita et effigies auctoris. Cura et studio Johannis Philippi Breynii, Jac. fil. [...] Huius ad calcem annectitur Dissertatio botanico-medica, de radice Gim-sem., seu Nisi et herba Acmella, cum additamentis.* Gedani. Sumptibus Editoris. Typis Thom. Joh. Schreberi. [7] + 2-108 + [3] + 4-54 + 34 pl.

Burm. afr.; Burm. afric. Burmann, 1738.

Burmann, J. 1737. *Catalogi plantarum Africanarum duo, quorum prior complectitur plantas ab Hermanno observatas, posterior vero illas continet, quas Oldenlandus & Hartogius indagarunt.* Amstelaedami. Janssonio-Waesbergios [2] + 1-33 + [1].

Burmann, J. 1738. *Rariorum Africanarum Plantarum, Ad vivum delineatarum, Iconibus ac descriptionibus illustratarum. Decas Prima (-decima).* Amstelaedami. Henricum Boussiere. i-viii + 1-268 + [2] + 50 pl.

Burman, J. 1739. *Rariorum Africanarum plantarum: ad vivum delineatarum, iconibus ac descriptionibus illustratarum decas octava.* Amstelaedami. Henricum Boussiere. [2] + 195-228 + [10]. [Tab. LXXI-LXXX].

Camel, J.J. 1704. *Herbarium Aliarumque Stirpium in Insulâ Luzone Philippinarum.* Pp. 1-135 in J. Ray, *Historiae plantarum tomus tertius: Qui est supplementum duorum praecedentium: Species omnes vel omissas, vel post voluminia illa evulgata editas, praeter innumeras fere novas & indictas ab amicis communicatas complectens: cum synonymis necessariis, et usibus in cibo, medicina & mechanicis (...). Accesit Historia stirpium Ins. Luzonis & reliquarum Philippinarum a R. P. Geo. Jos. Camello (...). Item D. Jos. Pitton Tournefort (...).* Londini. Sam. Smith & Benj. Waldorf.

Carniolica. Scopoli, 1763.

Catesby carol. 2. Catesby, 1743.

Catesby, M. 1743. *The natural history of Carolina, Florida and the Bahama islands: containing the figures of birds, beasts, fishes, serpents, insects, and plants: particularly, the forest-trees, shrubs, and other plants, not hitherto described, or very incorrectly figured by authors. Together with their descriptions in English and French. To which, are added observations on the air, soil, and waters: with remarks upon agriculture, grain, pulse, roots, &c. To the whole, is prefixed a new and correct map of the countries treated of. Vol. 2. -Histoire naturelle de la Caroline, la Floride, & les Isles Bahama: contenant les desseins des oiseaux, animaux, poissons, serpents, insectes, & plantes. Et en particulier, des arbres des forets, arbrisseaux, & autres plantes, qui n'ont point été decrits, jusques à present par les auteurs, ou peu exactement dessinés. Avec leur descriptions en françois et en anglois. A quoi on a adjouté, des observations sur l'air, le sol, & les eaux, avec des remarques sur l'agriculture, les grains, les legumes, les racines, &c. Le tout est precedé d'une carte nouvelle & exacte des païs dont ils s'agit. Tome II.* London. Sold by Innys, Manby and the Author. [12] + 1-100 + 100 pl.

Charas. Charas, 1670.

Charas, M. 1670. *Nouvelles experiences sur La Vipere, ou l'on verra une description exacte de toutes ses parties, la source de son venin, ses divers effets et les remedes exquis que les artistes peuvent tirer de la Vipere, tant pour la guerison de ses morsures, que pour celle de plusieurs autres maladies.* Paris. Auteur / Olivier de Varennes. [10] + 1-245 + 3 plates + [6].

Charas, M. 1684. *Operum tomus tertius, continens Historiam Naturalem animalium, plantarum et mineralium, Theriacae Andromachi compositionem ingredientium; cum experimentis circa viperam, addita serie novorum experimentorum.* Genevae. Joannis Ludovici Du-Four. [2] + 1-275.

Charlet. excercit. Charleton, 1677.

Charleton, G. 1677. *Exercitationes de Differentiis & Nominibus Animalium quibus accedunt Mantissa Anatomica, et quaedam de variis fossilium generibus, deque differentiis*

& nominibus colorum. Editio secunda, duplo fere auctior priori, novisque iconibus ornata. Oxoniae, E. Theatro Sheldoniano. [12] + 1-119 + 1-106 + 1 plate + 1-78 + [19].

Clas.* (sic) *posth. Clus. post. Clusius, 1611.

Cleijero. Cleyerus in ephemeridibus germanicis. Cleyer. [Cleijer], 1684.

Cleyer. [Cleijer], A. 1684. Observatio 7, De Serpente magno Indiae Orientalis, Urobubalum deglutiente. *Miscellanea curiosa medico-physica Academiae naturae curiosorum sive Ephemerides Germanicae, Decuria II, Annus II*: 18-20 + 1 pl.

Clus. exot. Clusius, 1605.

Clusius, C. 1605. *Exoticorvm libri decem: quibus animalium, plantarum, aromatum, aliorumque peregrinorum fructuum historiae describuntur: Item Petri Bellonii observationes, eodem Carolo Clusio interprete. Series totius operis post praefationem indicabitur.* s.l. [Leiden?]. Officinâ Plantinianâ Raphelengii. [15] + 1-378 + [9].

Clusius, C. 1611. *Caroli Clvsii Atrebatis Cvrae Posteriores, sev Plurimarum non antè cognitarum, aut descriptarum stirpium, peregrinorumque aliquot animalium. Novae descriptiones: Quibus & omnia ipsius Opera, aliaque ab eo versa augentur, aut illustrantur. Accessit seorsim Everardi Vorstii, Medicinae professoris Clarissimi, de eiusdem Caroli Clvsi Vita & Obitu Oratio, aliorumque de eisdem Epicedia.* s.l. [Leiden?] Officinâ Plantinianâ Raphelengii. [4] + 1-71.

Colbe cap. bon. spei. 3. p. 127. Kolbe, 1741.

***Comm. Hort.* I.** Commelin, 1697.

Commelin, L. 1697. *Horti Medici Amstelodamensis Rariorum Tam Orientalis, quàm Occidentalis Indiae, aliarumque Peregrinarum Plantarum Magno studio ac labore, sumptibus Civitatis Amstelodamensis, longâ annorum Serie collectarum, Descriptio et Icones ad vivum aeri incisae. Auctore Joanne Commelino Urbis Amstelodamensis (dum viveret) Senatore. Opus Posthumum. Latinitate donatum, Notisque & Observationibus Illustratum, à Frederico Ruyschio M.D. Botanices Professore, &c. & Francisco Kiggelario.* Amstelodami. P. & J. Blaeu nec non Abrahamum à Someren. [20] + 1-220 + 112 pl.

Comp. Med. Chym. Grim, 1684.

Cuba, J. de. [Johann Wonnecke von Kaub]. 1485. *Gart der Gesundheit.* Straßburg. Johann Grüninger. 224 unnumbered leaves.

Cuba, J. de. 1491. *Ortus sanitatis.* Moguntiae. Jakob Meydenbach. 454 unnumbered folios.

Curiosi isti observatores Angli. Jones, 1700.

Dahlberg, N. E. 1755. *I. N. D. Dissertatio botanica Metamorphoses plantarum sistens.* Holmiae. Typographia Regia. [3] + 6-26.

Dahlberg, N. E. 1759. Metamorphosis plantarum. *Amoenitates Academicae*, 4: 368-386. (Holmiae edition).

Dal. Pharm. Dale, 1739.

Dale, S. 1739. *Pharmacologia, seu manuductio ad materiam medicam; In qua Medicamenta Officinalia Simplicia, hoc est Mineralia, Vegetabilia, Animalia eorumque partes in Medicina: Officinis usitata, in Methodum naturalem digesta succincte & accurate describuntur. Cum Notis Generum Characteristicis Specierum Synonymis, Differentiis & Viribus. Opus Medicis, Philosophis, Pharmacopoeis, Chirurgis, &c. utilissimum. Ad calcem adjiciuntur indices necessarii. Quarta Editio, Indice Gallico, Germanico, Belgico aucta.* Lugduni Batavorum. Joh. Arnold. Langerak. [3] + ii-vii + [6] + 2-459 + [51].

Dalech. hist. Dalechamps, 1587.

Dalechamps, J. 1587. *Historia Generalis Plantarvm, In libros XVIII per certas classes artificiose digesta, Hec, plusquam mille imaginibus plantarum locupletior superioribus, omnes propemodum que ab antiquis scriptoribus, Graecis, Latinis, Arabibus, nominantur: necnon eas quae in Orientis atque Occidentis partibus, ante seculum nostrum incognitis, reperta fuerunt, tibi exhibet. Habeş etiam earundem plantarum peculiaria diuersis nationibus nomina: habes amplas descriptiones, e quibus singularum genus, formam, vbi crescant & quo tempore vigeant, natiuum temperamentum, vires denique in medicina proprias cognosces. Adiecti sunt Indices, non solum Graeci & Latini, sed aliarum quoque linguarum, locupletissimi.* Lugduni. Gvlielmvm Rovillivm. [10] + 1-1095.

Darellis Dissertatio. Darelius, 1751.

Darelius, J.A. 1749. *Lignum Colubrinum leviter delineatum.* Upsaliae. S. n. [2] + 3-22.

Darelius, J. A. 1751. Lignum colubrinum. *Amoenitates Academicae*, 2: 100-125. (Holmiae edition).

De Quadrupedibus. Ray, 1693.

Desallier d'Argenville, M. 1780. *La Conchyliologie, ou histoire naturelle des coquilles de mer, d'eau douce, terrestres et fossiles ; Avec un Traité de la Zoomorphose, ou représentation des Animaux qui les habitent: Ouvrage dans lequel on trouve une nouvelle Méthode de les diviser. Ouvrage considérablement augmenté de Planches en taille-douce, qui représentent les figures de plus de deux mille Testacées, dessinées d'après nature, accompagnées de Descriptions étendues & de Remarques sur chaque Famille, avec plusieurs Tables systématiques & des Nomenclatures étrangeres. Troisieme édition. Tome Premier.* [*second*]. Paris. Guillaume de Bure. i-iv + [4] + i-lx + 1-878 + pls.

Diss. de China. Lönqvist, 1744.

Dissertation of April 9, 1743. Linnaeus, 1743a.

Dubois, G. 1748. Taenia. Upsaliae. S. n. [3] + 1-30 + 1 plate.

Dubois, G. 1751. Taenia. *Amoenitates Academicae*, 2: 59-99, plate I. (Holmiae edition).

Dufay, M. 1741. Abstracts of Two Letters from M. Dufay, F. R. S. &c. to Dr. Mortimer, Secr. R. S. concerning the Efficacy of Oil of Olives in Curing the Bite of Vipers. *Philosophical Transactions*, 40: 444-445.

Duméril, A. M. C. & G. Bibron. 1837. *Erpétologie Générale ou Histoire Naturelle Complète des Reptiles. Tome second. Contenant l'histoire de toutes les espèces de l'ordre des Tortues ou Cheloniens, et les généralités de celui des Lézards ou Sauriens.* Paris. Roret. i-ii + 1-680.

Ekmarck. C. D. 1757. *Q. F. F. Q. S.Dissertatio academica,*

Migrationes avium. Upsaliae. L. M. Höjer, Reg. Acad. Typogr. [4] + 6-38.

Ekmarck, C. D. 1759. Migrationes avium. *Amoenitates Academicae*, 4: 565-600. (Holmiae edition).

Elff, E.E. 1749. *D. D. Dissertatio Medica De Haemorrhagiis Uteri Sub Statu Graviditatis*. Upsaliae. s.n. 24 pp.

Ellis, J. 1766. XXII. An account of an Amphibious Bipes. *Philosophical Transactions of the Royal Society*, 56: 189–192.

F. Zeyl. Fl. Zeyl. Linnaeus, 1748b.

Faun. Svec. 1281. Fn. Sv. Linnaeus, 1746.

Faun. Svecica 331. t. 2. Linnaeus, 1761.

Fernández de Oviedo y Valdéz, G. 1535. *Primera parte de la historia natural y general de las indias yslas y tierra firme del mar oceano: escripta por el capitan gonçalo hernandez de Ouiedo y valdes: alcayde de la fortaleza de la ciudad de sancto Domingo de la ysla Española/y cronista dela sacra cesarea y catholicas majestades del emperador don carlos quinto de tal nõbre: rey de españa: y de la serenissima y muy poderosa reyna doña Juana su madre nuestros señores. Por cuyo mandado el auctor escriuio las cosas maravillosas que ay en diuersas yslas y partes destas indias y imperio de la corona real de castilla: segũ lo vido y supo en veyte y dos años y mas que ha que biue y reside en aquellas partes. La qual historia comiença en el primero descubrimiento destas indias: y se contiene en veynte libros este primero volumen*. Sevilla. Juam Cronberger. [6] + CCXCII folios numbered in recto.

Fl. Svec. Linnaeus, 1745b.

Fragoso, J. 1572. *Discursos de las Cosas Aromaticas, arboles y frutales, y de otras muchas medicinas simples que se traen de la India Oriental, y siruen al uso de medicina*. Madrid. Francisca Sanchez. [17] + 211 folios numberd in recto + [34].

Fragosus. Fragoso, J. 1601.

Fragoso, J. 1601. *Aromatum, Fructuum et Simplicium Aliquot Medicamentorum ex India Utraque, et Orientali et Occidentali, In Europam delatorum, quorum iam est usus plurimus, Historia breuis, utilis, et iucunda. Conscripta Primum Hispanice a Ioanne Fragoso Philippi II, Hispan. Regis Medico et Chirurgo. Nunc Latine Edita Opera ac Studio Israelis Spachio Med. D. & Pro.f Argentinensis cum Notis marginalibus atque indice. Plura indicat ad lectorem Praefatio*. Argentinae. Iodocus Martinus. [14] + 1-115 fols, numbered in recto + [3].

Frangsmyr, T. (ed.). 1994. *Linnaeus, the man and his work. Revised edition.* Canton, MA. Science History Publications. i-xiv + 1-206.

Frisch, J. L. 1730. *Beschreibung von allerley Insecten in Teutschland, nebst nützlichen Anmerckungen und nöthigen Abbildungen von diesem kriechenden und fliegenden inländischen Gewürme, zur Bestätigung und Fortsetzung der gründlichen Entdeckung, so einige von der Natur dieser Creaturen heraus gegeben, und zur Ergäntzung und Verbesserung der andern. Neunter Theil. Samt einer Vorrede, worinnen ein Auszug aus des Herrn Francisci Redi Buch von Erzeugung der Insecten enthalten ist. Wie auch einer Beschreibung der Strich-Heuschrecken, welche in diesem Jahr grossen Schaden gethan. Im übrigen wird theils zu folge der Schwammerdamischen Ordnung, wie im achten Theil geschehen ist, theils auch ausser derselben fortgefahren*. Berlin. Christof Gottlieb Nicolai. [9], 1-37, [1], Pl. I-XXII.

Frisch. insect. Frisch, 1730.

Garzia. Orta [Horto], 1563.

Garzias; Garz. Arom. Orta [Horto], 1593.

Gaztambide-Géigel, A. 2004. The Invention of the Caribbean in the 20th Century (The Definitions of the Caribbean as a Historical and Methodological Problem). *Social and Economic Studies*, 53 (3): 127-157.

Gen. Plant. Linnaeus, 1747b.

Geoff. Mat. 2. Geoffroy, 1741.

Geoffroy, E.F. 1741. *Tractatus de Materia Medica sive de Medicamentorum Simplicium Historiâ, virtute, delectu & usu. Tomus Secundus. De Vegetabilibus Exoticis*. Parisiis. Joannis Desaint & Caroli Saillant. 1-794 + [6].

Geoffroy, C.-J. & F.J. Hunauld. 1740. Mémoire dans lequel on examine si l'huile d'olive est un spécifique contre la morsure des viperes. *Histoire de l'Académie royale des sciences. Année M.DCCXXXVII. Avec les mémoires de mathématique et de physique*: 183-205.

Gesner. quadr. Gesner, 1586.

Gesner, C. 1586. *Historiae animalium liber II. qui est de Quadrupedibus Oviparis. Nunc denuo recognitus ac pluribus in locis ab ipso authore ante obitum emendatus & actus, atq; aliquot novis iconibus & descriptionibus locupletatus, ac denique brevibus in margine annotationibus illustratus. Adiecti sunt indices alphabetici undecim super nominibus oviparorum quadrupedum in totidem linguis diversis: & ante illos enumeratio eo ordine quo in hoc volumine continentur.* Francofurdi, Typographica Ioannis Wecheli, Imprensis Roberti Cambieri. [4] + 1-125.

Gesner. pisc. Gesner, 1604.

Gesner, C. 1604. *Historiae animalium liber IV. Qui est de piscium & aquatilium animantium natura. Cum iconibus singulorum ad viuum expressis fere omnibus DCCVI. Editio secunda; nouis iconibus, necnon observationibus non paucis auctior, atque etiam multis in locis emendatior. Continentur in hoc volumine, Gvlielmi Rondeletii (...) & Petri Bellonii (...) de aquatilium singulis scripta. Paralipomena quadam ad finem adiecta sunt.* Francofurti. Andreae Cambieri. [38] + 1-1052 + 1-38.

Gesner. serpent. Gesner, 1621.

Gesner, C. 1621. *Historiae animalium liber V. Qui est de Serpentium natura. Ex variis schedis et collectaneis eiusdem compositus per Jacobum Carronum Francofurtensem. Adiecta est ad calcem, Scorpionis insecti historia a D. Casparo VVolphio Tigurino Medico, ex eiusdem Paralipomenis conscripta. Accesservnt indices nominvm serpentivm secundum diuersas linguas: & ante illos enumeratio eo ordine quo in hoc volumine continentur.* Francofurti. Henrici Laurenti. [14] + 1-170.

Godofr. Mat. Med. vol. 2. Geoffroy, 1741.

Gråberg, J. M. 1762. *Fundamentum fructificationis*. Upsaliae. S. n. [3] + 1-24.

Gråberg, J. M. 1763. Fundamentum fructificationis. *Amoenitates Academicae*, 6: 279-304. (Holmiae edition).

Gråberg, J.M. 1767. *D. D. Dissertatio medica inauguralis De Haemoptysi*. Upsaliae. Joh. Edman. 14 pp.

Grew. mus. Grew, 1681.

Grew, N. 1681. *Musaeum Regalis Societatis. Or a catalog and description of the natural and artificial rarities belonging to the Royal Society, and preserved at Gresham Colledge. Whereunto is subjoyned the comparative anatomy of stomachs and guts*. London. W. Rawlins. [10] + 1-386 + 1-43 + 31 plates.

Grim. Labor. Zeyl. Grim, 1677.

Grim, H.N. 1677. *Laboratorium Chymicum, Gehouden op het voortreffelycke Eylands Ceylon, Soo in 't Animalische, Vegetabilische, als Mineralische Ryck. Wordende de Liefhebberen niet alleen de Praeparatien der selve trouwelyck op-geteyckent, maer oock desselver gebruyck, en hoedanigh sy moeten geadhibeert worden. Dienende tot een bewys, hoe dat men de swaere Eysschen te vooren gedaen, grootelyoks verminderen, en van betere bedienen kan*. Batavia. Abraham van den Eede. Boeck drucker der E. Compagnie. [14] + 1-108 + [7].

Grim, H.N. 1684. *Compendium Medico Chymicum, Seu Accurata medendi Methodus, quae excellentissimis Medicamentis, tam Europae quàm Indiae Orientali proficuis, repleta, rariores Observationes, & curiosam optimorum Medicamentorum, in libelli hujus formulis contentorum, praeparationem exhibet, Auctior & Emendatior*. Augustae Vindelicorum. Impensis Theophili Göbelii. Typis Johan Jacob Schönigii. [12] + 2-475 + [11].

Gron. Virg.; Gronov. Flor. Virgin. Gronovius, 1743.

Gronovius, J. F. 1743. *Flora Virginica exhibens plantas quas V. C. Johannes Clayton in Virginia observavit atque collegit. Easdem methodo sexuali disposuit, ad genera propria retulit, nominibus specificis insignivit, & minus cognitas descripsit. Pars Prima*. Lugduni Batavorum. Cornelium Haak. [4] + 3-128 + [8] + 129-206 + [4].

Gronovius, L. T. 1754. *Museum Ichthyologicum, sistens piscium indigenorum & quorundam exoticorum, qui in Museo Laurentii Theodori Gronovii, J. U. D. adservantur, descriptiones ordine systematico, accedunt nonnullorum exiticorum piscium icones aeri incisae*. Lugduni Batavorum. Theodorum Haak. [8] + 1-70.

Gronovius, L. T. 1756. *Musei Ichthyologici tomus secundus sistens piscium indigenorum & nonnullorum exoticorum, quorum maxima pars in Museo Laurentii Theodori Gronovii, J. U. D. adservantur, nec non quorumdam in aliis museis observatorum descriptiones. Accedunt nonnullorum exoticorum piscium icones aeri incisae, et amphibiorum animalium historia zoologica*. Lugduni Batavorum. Theodorum Haak. [6] + 1-88 + 7 plates.

Hager, J. H. 1757. *Natura pelagi*. Upsaliae. L. M. Höjer, Reg. Acad. Typogr. [4] + 6-15 + [1].

Hager, J. H. 1760. Natura pelagi. *Amoenitates Academicae*, 5: 68-77. (Holmiae edition).

Hall. Helv. Haller, 1742.

Haller, A. von. 1742. *Enumeratio Methodica Stirpium Helvetiae Indigenarum. Qua Omnium Brevis Descriptio Et Synonymia Compendium Virium Medicarum Dubiarum Declaratio Novarum Et Rariorum Vberior Historia et Icones Continentur. Tomus I*. Gottingae. Abrami Vandenhoek. [3] + 2-794 + 24 pl.

Haller in Boerhav. vol. 4. P. 13. Boerhaave & Haller, 1743.

Hasaei, T. 1731. *Dissertationum et observationum philologicarum sylloge*. Bremae. Hermanni Jaegeri. [12] + 1-650 + [19].

Hasaeus' dissert. Philolog. Hasaei, 1731.

Hasselquist, F. 1747. *Vires Plantarum*. Upsaliae. S. n. [2] + 4-37 + [3].

Hasselquist, F. 1749. Vires plantarum. *Amoenitates Academicae*, 1: 418-453. (Holmiae & Lipsiae edition).

Hasselq. itin. Hasselquist, 1757.

Hasselquist, F. 1757. *Iter Palaestinum eller resa til heliga landet, förrättad ifrån år 1749 til 1752, med beskrifningar, rön, anmärkningar, öfver de märkvärdigaste naturalier, på hennes Kongliga Maj:ts befallning. Utgifven af Carl Linnaeus*. Stockholm. Lars Salvii. [14] + 1-619 + [1].

Hast, B. R. 1745. *Amphibia Gyllenborgiana*. Upsaliae. (s. n.). [6] + 1-34 + [2].

Hast, B. R. 1749. Amphibia Gyllenborgiana. *Amoenitates Academicae* 1: 107-140. (Holmiae & Lipsiae edition).

Hedin, S.A. 1775. *D. D. Canones Medici*. Upsaliae. Joh. Edman, Regiae Acad. Typogr. 14 pp.

Heide. Heide, 1684.

Heide, A. 1684. *Anatome Mytuli, Belgicè Mossel, Structuram elegantem ejusque motum mirandum exponens. Subjecta est Centuria Observationum Medicarum*. Amstelodami. Janssonio-Waesbergios. [14] + 1-199 + [1] + 16 pl.

Heide, A. 1686. *Experimenta Circa Sanguinis Missionem, Fibras Motrices, Urticam Marinam &c. Accedunt Ejusdem Auctoris Observationes Medicae nec non Anatome Mytuli, Editio Nova*. Amstelodami. Janssonio Waesbergios. [6] + 1-205 + 1-48 + 21 pl.

Heidenstam, E.J.M., 1772. *D. F. Dissertatio inauguralis Medica, De Haemorrhagiis ex Plethora*. Upsaliae. Typis Edmannianis. 36 pp.

Hellenius, C. N. 1776. D. D. *Hypericum*. Upsaliae. Typis Edmannianis. [5] + 3-14 + 1 pl.

Heller, J. L. 1983. Notes on the titulature of Linnaean dissertations. *Taxon*, 32 (2): 218-251.

Herm. Leyd. Hermann 1687.

Hermann P. 1687. *Horti academici Lugduno-Batavi catalogus exhibens plantarum omnium nomina, quibus ab anno MDCLXXXI ad annum MDCLXXXVI hortus fuit instructus ut & plurimarum in eodem cultarum & à nemine hucusque editarum descriptiones & icones*. Leiden, C. Boutesteyn. [17] + 1-699 + [1].

Herm. prodr. Hermann, 1689.

Hermann, P. 1689. Pauli Hermani Paradisi Batavi Prodromus sive Plantarum Exoticarum in Batavorum Hortis Observatarum. Index. pp. 303-386 + [4]. *In S. Warton, Schola botanica sive Catalogus plantarum, quas ab aliquot annis in horto regio Parisiensi studiosis indigitavit vir clarissimus Joseph Pitton Tournefort, d.m. ut et Pauli Hermanni p.p. Paradisi Batavi prodromus, in quo plantae rariores omnes, in Batavorum hortis hactenus cultae, & plurimam partem à nemine antea descriptae recensentur. Edente in lucem S.* [*imon*]. *W.* [*arton*]. *A.* [*nglo*]. Amstelaedami. Henricum Wetstenium.

Herm. Zeyl. Hermann, 1717.

Hermann, P. 1717. *Musaeum Zeylanicum, sive Catalogus Plantarum, In Zeylana Nascentium, observatarum & descriptarum a Viro Celeberrimo Paulo Hermanno, In Academia Lugduno-Batava quondam Medicinae & Botanices Professore.* Lugduni Batavorum. Isaacum Severinum. [2] + 1-71.

Hermann, P. 1726. *Cynosura Materiae Medicae, ante sedecim annos in lucem emissa, brevibusque annotatis exornata à Dn. Joh. Sigismundo Henningero, Med. Doct. & Prof. Publ. Ord. in Universitate Argentoratensi, nunc diffusius explanata et compositorum medicamentorum recensione aucta, Curante Johanne Boeclero Philos. & Medic. Doct. Botan. atque Chym. apud Argentinenses Prof. Publ. Ord. Capit. Thom. Canon.* Argentorati. Johannis Beckil. Pars prima: [15] + 2-728 + Pars secunda: 1-153 + [7].

Hermannus in Mat. Med. Hermann, 1726.

Hern. mex.; Hernand. mex. Hernandez, 1651.

Hernandez, F. 1651. Historiae animalium et mineralium Novae Hispaniae. Liber unicus in sex tractatus divisus. pp. 1-90, in N. A. Recchi, *Rerum medicarvm Novae Hispaniae thesaurus seu plantarum animalium mineralium Mexicanorum historia ex Francisci Hernandez novi orbis medici primarii relationibus in ipsa Mexicana urbe conscriptis a Nardo Antonio Reccho (...). Nunc primũ in naturaliũ rer. studiosor. gratiã lucubrationibus Lynceorũ publici iuris facta, quibus jam excussis accessere demum alia quor. omnium synopsis sequenti pagina ponitur opus duobus voluminibus diuisum.* Romae. V. Mascardi. [26] + 1-950 + [2] + 1-90 + [14].

Hideen, J. 1759. *Ambrosiaca.* Upsaliae. S. n. [1] + 3-14.

Hideen, J. 1785. Ambrosiaca. *Amoenitates Academicae*, 9: 106-117. (Earlangen edition).

Hiortzberg, L. 1754. *D. D. Dissertatio Chemico Medica De Methodo Investigandi Vires Medicamentorum Chemica.* Upsaliae. L.M. Höjer, Reg. Acad. Typ. 16 pp.

Historical-Political Geography London 1729. Paschoud, 1729.

Hodaporico India Orientalis 1730 8vo. Barchewitz, 1730.

Holmberger, P. 1774. *Esca avium domesticarum.* Upsaliae. Typis Edmannianis. [1] + 1-15.

Holmberger, P. 1785. Esca avium domesticarum. *Amoenitates Academicae*, 8: 205-220. Erlangae edition).

Hoppius, C. E. 1760. *Anthropomorpha.* Upsaliae. S. n. [4] + 1-16 + 1 plate.

Hort. Cliff. Linnaeus, 1737a.

Hortus sanitatis. Cuba, 1491.

Hultman, D. 1753. *Instructio musei rerum naturalium.* Upsaliae. Laur. Magnus Höjer. [3] + 1-19 + [1].

Hultman, D. 1756. Instructio musei rerum naturalium. *Amoenitates Academicae*, 3: 446-464. (Holmiae edition).

It. Gothl. Linnaeus, 1745a.

It. Oeland. Linnaeus, 1745a.

Itinere helvetico. Scheuchzer, 1723.

Jackson, B. D. 1913. Authorship in the "*Amoenitates Academicae.*" *Journal of Botany*, 51: 101–103.

Jacob. mus. Jacobaeus, 1696.

Jacobaeus, O. 1696. *Museum Regium seu Catalogus Rerum tam naturalium, quàm artificialium, quae in Basilica Bibliothecae Augustissimi Daniae Norvegiaeq; Monarchae Christiani Quinti Hafniae asservantur, Descriptus ab Oligero Jacobaeo Med. & Phil. Prof. Regio cum Auctoris cum Privilegio.* Hafniae. Joachim Schmetgen. [17] + 1-201 + 38 plates + [4].

Jacobaeus [Jacobai], O. 1680. XCVII. Anatome Piscis Torpedinis motusq' tremuli examen. In *Thomae Bartholini Acta Medica et Philosophica Hafniensia*, 5: 253-259 + 7 figs.

Jacquin. Refers to a letter from Jacquin to Linnaeus, probably the one dated in Wien on August 28, 1759 (identified as L2562 in linnaeus.c18.net).

Jacquin, N. J. 1763. *Selectarum stirpium americanum historia, in qua ad Linnaeanum Systema determinatae descriptaeque sistuntur plantae illae, quas in insulis Martinica, Jamaica, Domingo, aliisque, et in vicinae continentis parte, observavit rariores; adjectis iconibus in solo natali delineatis.* Vindobonae. Officina Krausiana. [5] + I-VII + [5] + 1-284 + [14] + 183 pl.

Jobi. Petri, 1541.

Jones, H. 1700. Part of a letter from the Reverend Mr. Hugh Jones to the Reverend Dr. Benjamin Woodrose, F.R.S. concerning several observables in Maryland. *Philosophical Transactions*, XXI: 436-442.

Jonst. serp.; Jonst. serpent. Jonston, 1653.

Jonston, J. 1653. *Historiae naturalis de serpentibus, libri duo.* Francofurti ad Moenum. Haeredum Merianeorum. 1-40 + 12 plates.

Jonst. pisc. Jonston, 1657.

Jonston, J. 1657. *Historiae naturalis de piscibus et cetis libri V. Cum aeneis figuris.* Amstelodami. Ioannem Iacobi Fil. Schipper. [6] + 1-160 + 47 pl.

Jonst. quadr. Jonst. quadrup. Jonston, 1657.

Jonston, J. 1657. *Historiae naturalis de quadrupetibus libri cum aeneis figuris.* Amstelodami. Ioannem Iacobi Fil. Schipper. [3] + 4-163 + [1] + 80 plates.

Kaempf. amoen.; Kaempferus; Kaempf. Kaempfer, 1712.

Kaempfer, E. 1712. *Amoenitatum exoticarum politico-physico-medicarum fasciculi V, quibus continentur variae relationes, observationes & descriptiones rerum Persicarum & ulterioris Asiae, multâ attentione in peregrinationibus per universum orientem, collectae.* Lemgoviae. Henrici Wilhelmi Meyeri. [18] + 1-912 + [32].

Karamyschew, A. de. 1766. *Necessitas promovendae historiae naturalis in Rossia.* Upsaliae. S. n. [4] + 1-34 + [10] + 1 plate.

Karamyschew, A. de. 1769. Necessitas promovendae historia naturalis in Rossia. *Amoenitates Academicae*, 7: 438-465 + 1 plate. (Holmiae edition).

Kiernander, J. 1749. *Radix Senega*. Holmiae. Laurentii Salvii. [3] + 4-32 + 1 pl.

Kiernander, J. 1751. Radix Senega. *Amoenitates Academicae*, 2: 126-153. (Holmiae edition).

Kircher, A. 1664. *Mundus subterraneus in XII libros digestus; quo Divinum Subterrestris Mundi Opificium, mira ergasteriorum Naturae in eo distributio, verbo παντάμορφον Protei Regnum, Universe denique Naturae Majestas & divitiae summa rerum varietate exponuntur. Abditorum effectum causae acri indagine inquisitae demonstrantur; cognitae per Artis & Naturae conjugium ad humanae vitae necessarium usum vario experimentorum apparatu, necnon novo modo, & ratione applicantur. Tomus I*. Amstelodami, apud Joannem Janssonium à Waesberge, & Viduam Elizaei Weyerstraet. [29] + 1-346 + [6].

Kircher, A. 1665. [1664]. *Mundi Subterranei Tomus II. us In v. libros digestus quibus Mundi Subterranei fructus exponuntur, et quidquid tandem rarum, insolitum, et portentosum in foecundo Naturae utero continetur, ante oculos ponitur curiosi Lectoris*. Amstelodami. Joannem Janssonium & Elizeum Weyerstraten. [12] + 1-487 + [9].

Kircher, A. 1667a. *China monumentis qua sacris qua profanis, nec non varius naturae & artis spectaculis, aliarumque rerum memorabilium argumentis illustrata (...)*. Amstelodami. Joannem Janssonium a Waesberge & Elizeum Weyerstraet.

Kircher, A: 1667b. *Magneticum naturae regnum sive disceptatio physiologica | De triplici in natura rerum Magnete, iuxta triplicem eiusdem Naturae gradum digesto | Inanimato | Animato | Sensitivo | Qua occultae prodigiosarum quarundam motionum vires & proprietates, quae in triplici Naturae Oeconomia nonnullis in corporibus noviter detectis observantur, in apertam lucem eruuntur, & luculentis argumentis, experientia duce, demostrantur. Ad inclytum, & eximium virum Alexandrum Fabianum Novi orbis Indigenam*. Amstelodami. Joannem Janssonium a Waesberge & Viduae Elizeum Weyerstraet. 1-136.

Koerner, L. 1999. *Linnaeus: Nature and Nation*. Cambridge, Massachusetts & London, England. 1-298.

Kolbe, P. 1741. *Description du Cap de Bonne Esperance; Où l on trouve tout ce qui concerne l'Histoire-naturelle du pays; la religion, les moeurs & les usages des Hottentots; et l' établissement des Hollandois. Tirée des mémoires de Mr. Pierre Kolbe, Maitre ès Arts, dressés pendant un séjour de dix années dans cette Colonie, où il avoit été envoyé pour faire des Observations Astronomiques & Physiques. Tome Troisieme*. Amsterdam. Jean Catuffe. [16] + 1-280.

Krecsák, L. & A.M. Bauer. 2025. Carl Linnaeus the herpetologist: a review of his complete herpetological activity and legacy. *Zootaxa*, 5594 (2): 201-257.

Krecsák, L. & R. Wahlgren. 2008. A survey of the Linnean type material of *Coluber berus*, *Coluber chersea* and *Coluber prestes* (Serpentes, Viperidae). *Journal of Natural History*, 42 (35–36): 2343–2377.

Krok, T. O. B. N. 1925. [Entries for Linnaeus] In: Bibliotheca botanica suecana ab antiquissimis temporibus ad finem anni MCMXVIII. *Svensk botanisk litteratur från äldsta tider t. o. m. 1918*: 422–465. Uppsala. Almqvist & Wiksell. xvi + 799 pp.

Laet, J. de. 1633. *Novus Orbis seu descriptionis Indiae Occidentalis libri XVIII. Novis tabulis geographicis et variis animantium, plantarum, fructuumque iconibus illustrati*. Lugd. Batav. Elzevirios. [32] + 1-690 + [10] + 14 maps.

Laet. Amer. Marcgravius, 1648.

Laet. occid. Laet, J. de. 1633.

Lavilla, E.O. 2012. Skepticism and gullibility in Linnaeus' Herpetological Contributions. *Herpetologia Brasileira*, 1 (2): 65-81.

Leeuwenhoek, A. A. 1719. *Epistolae ad Societatem Regiam Anglicam, et Alios Illustres Viros Seu Continuatio mirandorum Arcanorum Naturae detectorum, Quadraginta Epistolis contentorum, Quae ex Belgica in Latinam Linguam translatae sunt, Cum figuris aeneis & Indicibus locupletissimis*. Lugduni Batavorum. Joh. Arnold Langerak. [15] + 1 pl. + 1-426 + [10] + 23 pl.

Linder, J. 1705. *Dissertatio medica De foeda lue dicta venerea, quam consensu ampliss. fac. med. in reg. Ups. acad. praeside Dn. Laurentio Roberg ... censurae publicae ventilandam subjicit regius stipendiarius, Johannes Linder Carlstadio-Wermlandus, ad diem 2. Decembr. anno 1705. horis & loco solitis*. Upsaliae. Joh. Henr. Werner. [4] + 2-47 + 1.

Linnaeus, C. 1735. *Systema Naturae, sive Regna Tria Naturae Systematice Proposita per Classes, Ordines, Genera, & Species*. Lugduni Batavorum. Theodorum Haak. Ex Typographia Joannis Wilhelmi de Groot. 14 unnumbered pages.

Linnaeus, C. 1737a. *Hortus cliffortianus Plantas exhibens quas In Hortis tam Vivis quam Siccis Hartecampi in Hollandia, Coluit Vir Nobilissimus & Generosissimus Georgius Clifford Juris Utriusque Doctor, Reductis Varietatibus ad Species, Speciebus ad Genera, Generibus ad Classes, Adjectis Locis Plantarum natalibus Differentiisque Specierum. Cum Tabulis Aeneis*. Amsteledami. 1 pl. + [28] + 1-501 + [17] + XXXII pl.

Linnaeus, C. 1737b. *Viridarium Cliffortianum, in quo exhibentur Plantae omnes, quas Vivas aluit Hortus Hartecampensis Annis 1735, 1736, 1737. Indicatae nominibus Ex Horto Cliffortiano Depromtis*. Amsteledami. S.L. [2] + 1-104 + [8].

Linnaeus, C. 1740a. *Caroli Linnaei Naturae Curiosorum Dioscorides Secundi. Systema Naturae in quo Regna Tria, secundum Classes, Ordines, Genera, Species, Systematice proponuntur. Editio Secunda, Auctior*. Stockholmiae. Gottfr. Kiesewwetter. [3] + 2-80.

Linnaeus, C. 1740b. *Systema Naturae, sive regna tria naturae systematice proposita per Classes, Ordines, Genera et Species. Natur Systema, oder die in ordentlichem Zusammenhange vorgetragene Drey Reiche der Natur, nach ihren Classen, Ordnungen, Geschlechtern und Arten, in die deutsche Sprache übersetzet und mit einer Vorrede herausgegeben von Johann Joachim Langen, &c*. Halle. Gebauerischen Schriften. [8] + 2-70 + [1].

Linnaeus, C. 1740c. Beskrifning om guld-fisken och silfwerfisken, *Cyprinus pinna ani duplici, caudae trifurcae*. *Kungl. Svenska vetenskapsakademiens handlingar*, 1 (1): 403-410 + Pl. I, figs. 3-8.

Linnaeus, C. 1741. *Oratio, qua peregrinationum intra patriam asseritur necessitas, habita Upsaliae, in auditorio carolino majori MDCCXLI octobr. XVII. quum medicinae professionem regiam & ordinariam susciperet*. Upsaliae. Typis Academ. Reg. [2] + 18 unnumbered pages.

Linnaeus, C. 1742. Animalia per Sveciam observata. *Acta Literaria et Scientiarum Svecicae, Volumen Quartum. Continens annos 1735. 1736. 1737. 1738 & 1739*. Anni MDCCXXXVI: 97-138.

Linnaeus, C. 1743a. *Patronis, hospitibus, patribus civibusque academicis et urbicis, S. P. D. Carolus Linnaeus, Medicinae & Botanices Professor Reg. & Ord. Nec non in solenni hac inauguratione medica constitutus promotor*. Upsaliae. Typis Viduae R. Höjeri. Frontis + 3 unnumbered pages.

Linnaeus, C. 1743b. *Oratio de necessitate peregrinationum intra patriam, ejusque Elenchus animalium per Sueciam observatorum: Accedunt Johannis Browallii Examen epicriseos Siegesbeckianae in systema plantarum sexuale, et Johannis Gesneri... Dissertationes de partium vegetationis et fructificationis structura, differentia et usu, In quibus Elementa Botanica dilucide explicantur*. Lugduni Batavorum. Cornelium Haak. [2] + 3-28.

Linnaeus, C. 1744. *Systema Naturae. In quo proponuntur Naturae regna tria secundum Classes, Ordines, Genera, & Species. Editio quarta ab Auctore emendata & aucta. Accesserunt nomina Gallica*. Parisiis. Michaelis-Antonii David. 1-108.

Linnaeus, C. 1745a. *Öländska och Gothländska resa på riksens högloflige ständers befallning förrättad åhr 1741. Med anmärkningar uti oeconomien, natural-historien, antiquiteter &c. med åtskillige figurer*. Stockholm, Upsala. G. Kiesewetter. [13] + 1-344 + [27] + [2].

Linnaeus, C. 1745b. *Flora Svecica exhibens plantas per Regnum Sveciae crescentes, systematice cum differentis specierum / synonymis auctorum / nominibus incolarum / solo locorum / usu pharmacopaeorum*. Stockholmiae. Laurentii Salvii. [16] + 1-418 + [1] + 1 pl.

Linnaeus, C. 1746. *Fauna Suecica. Sistens animalia Sueciae Regni: Quadrupedia, Aves, Amphibia, Pisces, Insecta, Vermes. Distributa per Classes & Ordines, Genera & Species. Cum differentiis specierum, synonymis auctorum, nominibus incolarum, locis habitationum, descriptionibus insectorum*. Stockholmiae. Laurentii Salvii. [27] + 1-411.

Linnaeus, C. 1747a. *Caroli Linnaei Naturae Curiosorum Dioscorides Secundi. Systema Naturae in quo Naturae Regna Tria, secundum Classes, Ordines, Genera, Species, Systematice proponuntur. Recvsvm et Societatis, qvae impensas contvlit, vsvi accommodatvm. Cvrante Mich. Gottl. Agnethlero, Saxone Transilvano. Editio Altera Avctior et Emendator*. Hale Magdebvrgicae. S. n. [5] + 6-87.

Linnaeus, C. 1747b. *Genera plantarum, Eorumque characteres naturales secundum numerum, figuram, situm, & proportionem. Omnium fructificationis partium. Editio secunda, nominibus Plantarum Gallicis locupletata*. Parisiis. Michaelis Antonii David. i-xxxii + [3] + 1-413 + [59] + 2 pl.

Linnaeus, C. 1748a. *Systema naturae sistens regna tria naturae, in classes et ordines genera et species redacta tabulisque aeneis illustrata. Editio sexta, emendata et aucta*. Stockholmiae. Godofr. Kiesewetteri. [3] + 2-224 + [28] + 8 plates.

Linnaeus, C. 1748b. *Flora Zeylanica sistens Plantas Indicas Zeylone Insulae; Quae Olim 1670-1677 lectae fuere a Paulo Hermanno, Prof. Bot. Leydensi; Demum, Post 70 annos Ab Augusto Gunthero, Pharmacop. Hafniensi. Orbi redditae; Hoc Vero Opere Revisae, Examinatae, Determinatae, & Illustratae Generibus Certis, Differentiis Specificis, Synonymis Propriis, Descriptionibus, Compendiosis Iconibus Paucis*. Amstelodami. J. Wetstenium. [5] + 6-240 + [24] + 1-14 + [1] + 4 pl.

Linn. Mat. Med. Linnaeus, 1749.

Linnaeus, C. 1749. *Materia medica. Liber I. De Plantis. Secundum genera, differentias, synonyma, loca, durationes, culturas, nomina, simplicia, praeparata, qualitates, modos, potentias, vires, usus, composita, digestus*. Holmiae. Laurentii Salvii. [33] + 1-252.

Linnaeus, C. (ed.). 1749b. Hypothesis Nova de Febrium Intermittentium Causa. *Amoenitates Academicae* Volumen Primum. [3] + 1-610 + [9] + 14 pl. Lugduni Batavorum. Cornelium Haak.

Linnaeus, C. 1751. *Skånska resa, på höga öfwerhetens befallning förrättad år 1749. Med rön och anmärkningar uti oeconomien, naturalier, antiquiteter, seder, lefnads-sätt. Med tilhörige figurer*. Stockholm. Laurentii Salvii. [8] + i-xiv + 1 mapa + 1-434 + [34].

Linnaeus, C. 1752. Anmärkning Om Ormarnas Skilje-märken. *Kongliga Svenska Vetenskaps Academiens handlingar*, 3: 206-207.

Linnaeus, C. 1753a. *Species plantarum, exhibentes plantae rite cognitas, ad genera relatas, cum differentiis specificis, nominibus trivialibus, synonymis selectis, locis natalibus. Secundum systema sexuale digestas. Tomus I*. Holmiae. Laurentii Salvii. [12] + 1-560.

Linnaeus, C. 1753b. *Species plantarum, exhibentes plantae rite cognitas, ad genera relatas, cum differentiis specificis, nominibus trivialibus, synonymis selectis, locis natalibus. Secundum systema sexuale digestas. Tomus II*. Holmiae. Laurentii Salvii. 561-1200 + [31].

Linnaeus, C. 1756. *Systema Naturae. Sistens Regna Tria Naturae in Classes et Ordines, Genera et Species redacta. Tabulisque aeneis illustrata. Accedunt vocabula Gallica. Editio multo auctior & emendatior*. Lugdunum Batavorum. Theodorum Haak. [6] + 1-227 + [27] + 8 plates.

Linnaeus, C. 1758. *Systema naturae per regna tria naturae, secundum classes, ordines, genera, species, cum characteribus, differentiis, synonymis, locis*. Tomus I. Editio decima, reformata. Holmiae. Laurentii Salvii. [8] + 6-823 + [1].

Linnaeus, C. 1761. *Fauna Suecica. Sistens animalia Sueciae Regni: Mammalia, Aves, Amphibia, Pisces, Insecta, Vermes. Distributa per Classis & Ordines, Genera & Species, cum differentis specierum, synonymis auctorum, nominibus incolarum, locis natalium, descriptionibus insectorum. Editio Altera, Auctior*. Stockholmiae, Laurentii Salvii. [47] + 1-578 + 2 plates.

Linnaeus, C. 1766. *Systema naturae per regna tria naturae, secundum classes, ordines, genera, species, cum characteribus, differentiis, synonymis, locis. Tomus I. Editio duodecima, reformata*. Holmiae. Laurentii Salvii. 1-532.

Linnaeus, C. 1767. *Systema naturae per regna tria naturae, secundum classes, ordines, genera, species, cum characteribus, differentiis, synonymis, locis. Tomus I. Pars II. Editio duodecima reformata*. Holmiae. Laur. Salvii. 533-1327 + [37].

Lochnerus* - *Mungos Animalculum & Radix. Lochner, 1715.

Lochner, M.F. 1715. Mungos Animalculum et Radix Descripta. *Academiae Caesareo Leopoldinae-Carolinae Naturae Curiosorum Ephemerides. Centuria III & IV Cum Appendice*, [2] + 59-88.

Loevvenhaek. epist. Leeuwenhoek, 1719.

Löfling, P. 1749. *Gemmae Arborum*. Upsaliae. SN. [4] + 2-32.

Lönnberg, E. 1896. Linnean type-specimens of birds, reptiles, batrachians and fishes in the Zoological Museum of the R. University in Upsala. *Bihang till Kungliga Svenska Vetenskapsakademiens Handlingar*, 22. [4 (1)]: 1-45.

Lönqvist, N.J. 1744. *D.D. Dissertatio Medico Practica De Cortice Peruviano Vulgo China dicto, Ejusque Amplissimo in Medicina usu; Quam, Ex consensu Ampliss. Facultatis Medicae in Reg. Acad. Gothor. Carol. Ad publicam disquisitionem modeste deferunt, Auctor Eberhard Rosen Med. Doctor, Regiique Collegii Med. Assessor; et Respondens Nicolaus Johannes Lönqvist Scanus. Ad diem XXVIII Aprilis Anni MDCCXLIV. Loco, horisque solitis*. Typis Ludovici DECREAUX Directoris Officinae Typographica Londini Gothorum. [4] + 2-60 + 1 pl. + 1 tab.

Lucan, M. 1962. *The civil war* (Pharsalia); *with an English translation by J. D. Duff. 6th edition*. Cambridge, Mass. - London. Harvard University Press. I-XV + 1-638.

Lucano, M.A. 1984. *Farsalia. Introducción, traducción y notas de Antonio Holgado Redondo. Biblioteca Clásica Gredos, 71*. Madrid. Gredos. 1-344.

Lucanus. Lucanus, 1469.

Lucanus, M.A. 1469. *M. Annei Lucani Pharsaliae. Pomponii infortunati M. Annaeus Lucanus*. Romae. Konrad Sweynheym et Arnold Pannartz. 217 unnumbered pages.

Lucretius Lib. 9. Lucano, 1469.

Mag. monsp. Magnol, 1686.

Magnol, P. 1686. *Botanicum Monspeliense, sive Plantarum circa Monspelium nascentium index. In quo plantarum nomina meliora seliguntur, loca in quibus plantae sponte adolescunt, tum a prioribus botanicis, tum ab authore observata indicantur, & praecipuae facultates traduntur. Adduntur variarum plantarum descriptiones et icones. Cum appendice quae plantas de novo repertas continet & errata emendat*. Monspelii. Ex Officinâ Danielis Pauli. Impensis Marret Bibliopolae Monspeliensis. [12] + 1-309 + [2].

Marcgr. bras.; Marcgrav. Marcgravius, G. 1648.

Marcgravius, G. 1648. Historiae rerum naturalium Brasiliae, libri octo: quorum tres priores agunt de plantis. Quartus de piscibus. Quintus de avibus. Sextus de quadrupedibus & serpentibus. Septimus de insectis. Octavus de ipsa regione, & illivs incolis. Cum appendice de Tapuyis, et Chilensibus. Pp. [3] + 1-293 + [7] in. [*De Laet, I.*], *Historia naturalis Brasiliae, auspicio et beneficio illustriss. J. Mauritii Com. Nassav illius provinciae et maris summi praefecti adornata: in qua non tantum plantae et animalia, sed et indigenarum morbi, ingenia et mores describuntur et iconibus supra quingentas illustrantur*. Lugdun. Batavorum: Apud Franciscum Hackium; et Amstelodami: Lud. Elzevirium.

Marian. in folio. Merian, 1726.

Mathiol. dioscorid. Mathiolus. Matthioli, 1570.

Matthioli, P. A. 1570. *Petri Andreae Mattioli senensis medici, Commentarii in sex libros Pedacij Dioscoridis Anazarbei de Medica materia, iam denvo ab ipso avtore recogniti, et locis plvs mille avcti. Adiectis plantarum, & animalium Iconibus, supra priores editiones longè pluribus, ad uiuum delineatis. Accesserunt quoq; ad margines Graeci contextus quàm plurimi, ex antiquissimis codicibus desumpti, qui Dioscoridis ipsius deprauatam lectionem restituunt. Cvm locupletissimis indicibvs, tvm ad rem Herbariam, tvm Medicamentariam pertinentibus. Cum privilegiis amplissimis, vt videre est statim post Praefationem ad Lectores*. Venetiis. Officina Valgrisiana. [159] + 1-956.

Merian, M. S. 1719. *Over de voortteeling en wonderbaerlyke veranderingen der Surinaemsche insecten, waer in de Surinaemsche rupsen en wormen, met alle derzelfer veranderingen, naer het leeven afgebeeldt, en beschreeven worden; zijnde elk geplaest op dezelfde gewassen, bloemen en vruchten, daer ze op gevonden zijn; beneffens de beschrijving dier gewassen. Waer in ook de wonderbare padden, hagedissen, slangen, spinnen, en andere zeldzame gediertens worden vertoond, en beschreven. Alles in Amerika door den zelve M. S. Meriaen naer het leeven, en leevensgrootte geschilderd, en nu in't koper overgebracht. Benevens een aanhangsel van de veranderingen van visschen in kikvorschen, en van kikvorschen in visschen*. Amsterdam. Joannes Oosterwyk. [6] + 1-72 + 72 plates.

Merian, M. S. 1726. *Dissertatio de generatione et metamorphosibus insectorum Surinamensium: In quá, praeter vermes & erucas surinamenses, earumque admirandarum metamorphosin, plantae, flores & fructus, quibus vescuntur, & in quibus fuerunt inventae, exhibentur. His adjunguntur bufones, lacerti, serpentes, araneae, aliaque admiranda istius regionis animalcula; omnia manu ejus-

dem Matronae in America ad vivum accurate depicta, & nunc aeri incisa. Accedit Appendix Transformationum Piscium in Ranas, & Ranarum in Pisces. Dissertation sur la Generation et les Transformations des Insectes de Surinam: Dans laquelle on traite des vers et des chenilles de Surinam, des Plantes, fleurs, & fruits dont ils vivent & dans lesquels on les a trouvez. On en parle aussi des crapaux, lézards, serpens, aragnées, & autres petits animaux du même païs, peints sur les lieux d'après nature, & depuis graves avec beaucoup de soin. On a joint un Traité des changemens des Poissons en Grenouilles & des Grenouilles en poissons. Hagae Comitum / La Haye. Pierre Gosse. 1-72 + 72 plates.

Merret. pin. Merrett, 1667.

Merrett, C. 1667. *Pinax rerum naturalium Britannicarum, continens vegetabilia, animalia, et fossilia, in hac insula reperta inchoatus.* Londini. T. Roycroft. [32] + 1-223 + [1].

***Mill. lexic.* (a).** Miller, 1740.

Miller, P. 1740. *The Second Volume Of The Gardeners Dictionary: Which Completes the Work. The Second Edition.* London. Printed for the author and sold by Charles Rivington. Pages and plates unnumbered.

***Mill. lexic.* (b).** Miller, van Royen & van Eems. 1745.

Miller, P., A. van Royen & J. van Eems. 1745. *Groot en algemeen kruidkundig, hoveniers, en bloemisten woordenboek, behelzende de Manier om Moes-, Bloem-, Vrugt-, Kruid-Tuinen, Wildbossen, Wijngaarden, Oranje-Huizen, Stook-Kassen, enz. Aanteleggen, enz. en allerlei Gewassen te kweeken. Alles volgens de Ondervinding van de Ervarenste Hoveniers der tegenwoordige Eeuw; Doormengd met de Historien der Planten, de Kentekens van elk Geslagt en de Naamen van alle de bijzondere Soorten in het Latijn en Duits; en eene Verklaaring van alle de Kunst-woorden in de Kruidkunde en het Tuinieren gebruikelijk, waarby ook komt Een Berigt van de natuur en het gebruik der Barometers, Thermometers, en Hygrometers, welke den Hoveniers te pas komen, als meede van den Oorsprong, Oorzaaken, en Natuur der Verhevelingen, en den bijzonderen invloed van Aarde, Lugt, Vuur en Water op de Planten, volgens de beste Natuurkundigen. Met koperen platen versierd.* Leyden. Pieter van der Eyk & Jakob van der Kluis. 595-1238 + several plates with mixed numeration.

Mortimer, C. 1741. An abstract by C. Mortimer, M.D. Secr. R.S. of an Inaugural Dissertation published at Wittenberg, 1736. by Dr. Abraham Vater, F.R.S. concerning the Cure of the Bite of a Viper, cured by Sallad-oil. *Philosophical transactions*, vol. XL. for the years 1737, 1738. With a supplement, being the Croonean Lectures on muscular motion, for the year 1738: 441-444.

Mundo subterraneo. Kircher, 1664.

Mus. Petr., Mus. Petrop., Mus. Petropol., Musaeo Petrop., Anonymous. 1742.

Mus. Princip. Balk, 1746.

Nauclér, S. 1745. *Hortus Upsaliensis.* Upsaliae. S. n. [3] + 1 pl. + 1 map + [2] + 3-45 + [3].

Nauclér, S. 1749. Hortus Upsaliensis. *Amoenitates Academicae*, 1: 172-210, plate N°6. (Holmiae & Lipsiae edition).

Nieremb. hist.; Nieremb. nat.; Niremb. natur. Nieremberg, 1635.

Nieremberg, J.E. 1635. *Historiae Naturae, maxima peregrinae, Libris XVI distincta. In quibus rarissima Naturae arcana, etiam astronomica, & ignota Indiarum animalia, quadrupedes, aves, pisces, reptilia, insecta, zoophyta, plantae, metalla, lapides, & alia mineralia, fluviorumque & elementorum conditiones, etiam cum proprietatibus medicinalibus, describuntur; novae & curiosissimae quaestiones disputantur, ac plura sacrae Scripturae loca erudite enodantur. Accedunt de iris & miraculosis Naturis in Europâ Libri duo: item de iisdem in Terrâ Hebraeis promissa Liber unus.* Antverpiae. Balthasaris Moreti. [6] + 1-502 + [102].

Nordblad, E. A. 1759. *Instructio Peregrinatoris.* Upsaliae. S. n. [3] + 1-15.

Nordblad, E. A. 1760. Instructio Peregrinatoris. *Amoenitates Academicae*, 5: 298-313. (Holmiae edition).

Old. afr. Burman, 1737.

Olear. mus. Olearius, 1666.

Olearius, A. 1666. *Gottorffische Kunst-Cammer/ Worinnen Allerhand ungemeine Sachen/ So theils die Natur/ theils künstliche Hände hervor gebracht und bereitet; Vor diesem Aus allen vier Theilen der Welt zusammen getragen.* Schleßwig in der Fürstl. Johan Holwein. [12] + 1-88 + 36 plates.

Orta [Horto], G. ab. 1563. *Coloquios dos simples, e drogas he cousas mediçinais da India, e assi dalgu. [m]as frutas achadas nella onde se tratam algu. [m]as cousas tocantes amediçina, pratica e outras cousas boas, pera saber / cõpostos pello Doutor garçia dorta: fisico del Rey nosso senhor, vistos pello muyto Reuerendo senhor, ho liçençiado Alexos diaz : falcam desenbargador da casa da supricaçã inquisidor nestas partes.* Goa. Ioannes de Endem. [7] + 1-217. [i.e. 257].

Orta [Horto], G. ab. 1593. *Aromatum, et Simplicium Aliquot Medicamentorum Apud Indos Nascentium Historia: Primùm Quidem Lusitanica Lingua διαλογικῶς Conscripta, à D. Garcia Ab Horto, Proregis India Medico: Deinde Latino Sermone in Epitomen Contracta, & Iconibus Ad Vivum Expreßis, Locupletioribusq' Annotatiunculis Illustrata À Carolo Clusió Atrebate. Quarta Editio, Castigatior, & Aliquot Locis Auctior.* Antverpiae. Officina Plantiniana, Apud Viduam, & Ioannem Moretum. 3-217 + [6]. Ill.

Osbeck, P. 1757. *Dagbok öfwer en Ostindisk resa åren 1750. 1751. 1752. Med anmärkningar uti naturkunnigheten, främmande folkslags språk, seder, hushållning, m. m. (...) Afkene Skepps-Predikanten Toréns Bref.* Stockholm. Lor. Ludv. Grefing. [6] + 1-276 + [16] + 12 plates.

Österdam, A. 1766. *Siren lacertina.* Upsaliae. S. n. [3] + 1-15 + [1] + 1 plate.

Ovid. 1939. *Ovid, with an English translation. Tristia - Ex Ponto. By Arthur Leslie Wheeler.* Cambridge, Massachussetts. Harvard University Press + London. William

Heinemann Ltd. i-xliv + 1-511.

Ovied. amer. Fernandez de Oviedo y Valdéz, G. 1535.

Paschoud, Rev. Mr. 1729. *Historico-political geography: or, a particular description of the several countries in the world. The second edition, with additions.* London. William France. [2] + i-xiv + 1-395 + [5].

Pennant, T. 1769. *British Zoology. Class III. Reptiles. IV. Fish. Vol. III.* Chester. Eliz. Adams for Benjamin White. xii + 1-358.

Peregrinantium per Indiam. Baldaeus, 1672.

Peters, J. L. 2009. The history of central venous access. Pp. 1-12 *en H. Hamilton & A. R. Bodenham (eds.), Central Venous Catheters.* Wiley-Blackwell, Oxford, UK.

Petersen, J.C.P. 1758. *D. F. Dissertatio De Cortice Peruviano.* Upsaliae. L. M. Hojer, Reg. Acad. Typogr. 38 pp.

Pet. mus. Petiver, 1695.

Pet. sicc. Petiver, 1693.

Peti. Gazoph. Petiver, 1702.

Petiv. mus. Petiver, 1698.

Petiver, J. 1693. *Hortus siccus pharmaceuticus sive, collectio curiosissima plantarum officinalium; tàm exoticarum quàm indigenarum, alphabetico ordine diligenter digesta. Cum synonymis necessariis, locis natalibus & tempore vigendi. Accesserunt etìam illarum partes ac preparationes quae maximè sunt in usu.* No data. 5 sheets.

Petiver, J. 1695. *Musei Petiveriani Centuria Quarta & Quinta, Rariora Naturae continens: viz. Animalia, Vegetabilia, &c. ex variis Mundi Plagis advecta, Ordine digesta, & Nominibus propriis signata.* Londini. S. Smith & B. Walford. 33-42.

Petiver, J. 1698. *Musei Petiveriani centuria secunda & tertia, rariora naturae continens: viz. animalia, fossilia, plantas, ex variis mundi plagis advecta, ordine digesta, & nominibus propriis signata.* Londini. Smith & Walford. [2] + 1 pl. + 3-93 + [3].

Petiver, J. 1702. *Gazophylacii naturae & artis decas prima-quinta. In quâ animalia, quadrupeda, aves, pisces, reptilia, insecta, vegetabilia; item fossilia, corpora marina & stirpes minerales è terra eruta, lapides figurâ insignes &c. Descriptionibus brevibus & iconibus illustrantur. Hisce annexa erit supellex antiquaria, numismata, gemmae excisae, & sculpturae, opera figulina, lucernae, urnae, instrumenta varia, inscriptiones, busta, reliquaque ad rem priscam spectantia: item machinae, effigies clarorum virorum, omniaque arte producta.* Londini. Christ. Bateman. 2 plates + 1-96 + plates I-L.

Petri, L. 1541 (translator). *Biblia / Thet är / All then Helgha Scrifft / på Swensko.* Uppsala: Georg Richolff.

Piso, W. [Pisonis, G.]. 1648. De medicina Brasiliensi libri quatuor: I. De äere, aquis, & locis. II. De morbis endemiis. III. De venenatis & antidotis. IV. De facultatibus simplicium. Pp. [8] + 1-122 + [2] in [De Laet, I.], *Historia naturalis Brasiliae, auspicio et beneficio illustriss. J. Mauritii Com. Nassav illius provinciae et maris summi praefecti adornata: in qua non tantum plantae et animalia, sed et indigenarum morbi, ingenia et mores describuntur et iconibus supra quingentas illustrantur.* Lugdun. Batavorum: Apud Franciscum Hackium; et Amstelodami: Lud. Elzevirium.

Pis. bras.; Pison. bras.; Pis. natural.; Pison. americ. Piso, [Pisonis]. 1658.

Piso, W. [Pisonis, G.]. 1658. *De Indiae utriusque re naturali et medica libri quatuordecim: quorum contenta pagina sequens exhibet.* Amstelaedami. Ludovicum et Danielem Elzevirios. [21] + 1-327 + [5].

Plin. Plinii, 1582.

Plinii, C. 1582. *C. Plinii Secundi Historiae Mundi Libri XXXVII, A Sigismundo Galenio, diligenter castigati, maioréque studio & fide, quàm hactenus unquam a vitiis quibus multiplici olim impressione contaminati fuerant, vindicati. Accesserunt ad marginem variae lectionis ac Notae ex Fer. Pintiani, Adr. Tvrnebi, Ios. Scaligeri, Ivsti Lipsi, & aliorum doctißimorum virorum scriptis diligenter excerptae, quorum nomina sequenti pagina indicantur. Vnà cum duplici Indice totius operis copiosissimo, non poenitenda rursus accessione locupletato, locisque propemodum innumeris, quae cum autores sensu non satis congruebant, quàm accuratissimè restituto.* Lugdunum. Petrum Santandreanum. [36] + 1-679 + [230].

Pliny. 1855. *The Natural History of Pliny. Translated, with copious notes and illustrations by the late John Bostock, M.D., F.R.S., and H. T. Riley, Esqu. B.A., (...).* Vol. II. London. Henry G. Bohn. i-xii + 1-555.

Plukenet, L. 1691. *Phytographia. Seu Stirpium Εἰκόνες fidelitèr & exquisitè ad vivum delineatae. Aeneis 93 tabulis illustrata in quibus novae plures & exoticae plantae a nemine hactenus animadversae nunc primum proponuntur. Necnon aliae quaedam apud Neotericos dubiae atq' obscurae (mutatis aliquando nominibus prout res ipsa postulat) perspicua brevitate dilucidantur. Pars Altera et sic Meminisse Juvabit.* Londini. Sumptibus Autoris. [6] + Plates LX-CXX.

Pluck. Alm.; Pluk. Almag. Plukenet, 1696.

Plukenet, L. 1696. *Almagestum Botanicum: Sive Phytographiae Pluc'netianae Onomasticon Methodo Syntheticâ digestum Exhibens Stirpium exoticarum, rariorum, novarumque Nomina, quae Descriptionis locum supplere possunt. Cui (ad ampliandum regnum Vegetabilium) accessere Plantae circitèr Quingentae suis Nominibus similiter insignitae; Quae nullibi nisi in hoc opere (sex ferè Plantarum Chiliadas complectente) memorantur. Adjiciuntur & aliquot novarum Plantarum ICONES In gratiam Phylophytosophorum in lucem nùnc editae.* Londini. Sumptibus Autoris. [3] + 1-402 + [2].

Pluk. mant. Plukenet, 1700.

Plukenet, L. 1700. *Almagesti Botanici Mantissa Plantarum novissimè detectarum ultrà Millenarium numerum Complectens. Cui, tanquam pedi jam stantis Columnae, Plus Ultra Inscribere fas est. Cum Indice Totius Operis ad calcem adjecto.* Londini. Sumptibus Autoris. [5] + 1-191 + [1] + CCCXXIX - CCCL pl.

Plum. Spec. Plumier, 1703.

Plumier, C. 1703. *Nova Plantarum Americanarum Genera.* Parisiis. Joannem Boudot. [5] + 2-52 + [4] + 1-27 + [1]

+ 40 pl.

Pontin, D. D. 1759. *Arboretum svecicum*. Upsaliae. S. n. [2] + 4-30.

Pontin, D. D. 1760. Arboretum svecicum. *Amoenitates Academicae*, 5: 170-203. (Holmiae edition).

Posthius, J. 1563. *Iohan. Posthii Germershemii Tetrasticha in Ovidii Metam. lib. XV. qvibus accesserunt Vergilij Solis figurae elegantiss. & iam primùm in lucem editae = Schoene Figuren auss dem fuertrefflichen Poeten Ouidio, allen Malern, Goldtschmiden, und Bildthauwern, zü nutz unnd gütem mit fleiss gerissen durch Vergilium Solis, und mit teutschen Reimen kürtzlich erkläret, dergleichen vormals im Truck nie ; aussgangen durch Johan. Posthium von Germerssheim.* Francofurti. Georgium Coruinum, Sigismundum Feyrabent, & haeredes VVigandi Galli. [16] + 1-178 + [14].

Pulteney, R. 1781. *A general view of the writings of Linnaeus.* London. T. Payne & B. White. i-iv + 1-425 + [1].

Pulteney, R. 1805. *A general view of the writings of Linnaeus. The second edition; with corrections, considerable additions, and memoirs of the author, by William George Maton, M.D. F.R.S. F.S.A. Fellow of The Royal College of Physicians and a Vice President of the Linnean Society of London. To which is annexed the Diary of Linnaeus written by himself and now translated into English, from the Swedish manuscript in the possession of the Editor.* London. Printed for J. Mawman. R. Taylor and Co. i-xv + 1-595 + 3 pl.

Radloff, F. W. 1787. *D. D. Museum naturalium Academiae Upsaliensis cujus partem primam, Consensu Exp. Fac. Med. Upsal. praeside Carol. Pet. Thunberg (...) publico examini proponit, Fridericus Wilhelm. Radloff, Stockholmiensis. In auditorio Gust. Maj. D. XIV. April. MDCCLXXXVII. [1787]. H. A. M. S. -pp. 1-16.* Upsaliae. Edmannianis. Raj. [pis]

Rai quadr.; Raj. Anim.; Raj. Quadr.; Raj. quadrup.; Raj. syn. Ray, J. 1693.

Raj. app.; Rajus hist. Ray, J. 1704.

Raj. Hist. 2. Ray, J. 1688.

Ramsbottom, J. 1959. Caroli Linnaei Pan Suecicus. *Botanical Society of Edinburgh Transactions* 38: 151–167.

Ray, J. 1688. *Historiae Plantarum Tomus Secundus cum Duplici Indice; Generali Altero Nominum & Synonymorum praecipuorum Altero Affectuum & Remediorum: Accessit Nomenclator Botanicus Anglo-Latinus.* Londini. Typis Marie Clark: Prostant apud Henricum Faithorne Regia Societatis Typographum, [6] + 985-1944 (missnumbered as 1940) + [35].

Ray, J. 1693. *Synopsis methodica animalium quadrupedum et serpentini generis. Vulgarium notas characteristicas, rariorum descriptiones integras exhibens: cum historiis & observationibus anatomicis perquam curiosis. Praemittuntur nonnulla de animalium in genere, sensu, generatione, divisione, &c.* Londini. S. Smith & B. Walford. [14] + 1-336 + [8].

Ray, J. 1704. *Historiae Plantarum Tomus Tertius: Qui est Supplementum Duorum praecedentium: Species omnes vel omissas, vel post Volumina illa evulgata editas, praeter innumeras fere novas & indictas ab Amicis communicatas complectens: Cum Synonymis necessariis, et Usibus in Cibo, Medicina, & Mechanicis: Addito ad Opus consummandum Generum Indice copioso. Accessit Historia Stirpium Ins. Luzonis & reliquarum Philippinarum. A R.P. Geo. Jos. Camello, Moravo-Brunensi, S. J. conscripta. Item D. Jos. Pitton Tournefort, M. D. Parisiensis & in Horto Reg. Botanices Professoris, Corollarium Institutionum Rei Herbaria.* Londini. Sam. Smith & Benj. Walford. i-ix 1-666 + 1-135 + 1-255 + [7].

Reamur acta. gallic. Ao 1714. Réaumur, 1714.

Réaumur, R.-A.F. de. 1714. Des effets que produit le poisson appellé en François Torpille, ou Tremble, sur ceux que le touchent; et de la cause dont ils dépendent. *Histoire de l'Académie royale des sciences avec les mémoires de mathématique et de physique pour la même année tirez des registres de cette Académie*: 344-362.

Reaumur. Réaumur, 1939.

Réaumur, R.A.F. de. 1939. [1736]. XVI - Mémoire sur les grenouilles. L'accouplement des grenouilles. Pp. 247-253 in Réaumur. *Morceaux Choisis. Présentés et annotés par Jean Torlais.* Paris. Gallimard.

Redi, F. 1664. *Osservazioni intorno alle vipere fatte da Francesco Redi Gentiluomo Aretino, Accademico della Crusca. E da lui scritte in una lettera all' illustrissimo signor Lorenzo Magalotti gentilhuomo della camera del Ser. mo G. Duca di Tosc. na.* Firenze. All'Insegna della Stella. 1-91 + [4].

Redi, F. 1666. Some observations of vipers. *Philosophical Transactions*, 1665-66 (1): 160-162.

Redi. Redi, 1670a.

Redi, F. 1670a. *Observationes de Viperis Francisci Redi Nobilis Aretini Academici Furfurariorum sive della Crusca scriptae in litteris ad generosissimum dominum Laurentium Magalotti magni ducis Hetruriae camerarium ex italica in latinam translatae. 1-40. Appendix with own numbering after page 344 of Miscellanea curiosa medico-physica academiae naturae curiosorum sive ephemeridum medico-physicarum germanicarum curiosarum. Annus primus. Anni scilicet M. DC. LXXmi.* Lipsiae. Viti Jacobi Trescheri / Johannis Baueri.

Redi, F. 1670b. *Lettera di Francesco Redi (...) sopra alcune opposizioni fatte alle sve Osservazioni in torno alle vipere. Scritta alli signori abate Bourdelot (...), e Alessandro Moro. Firenze. Stamperia della Stella. 1-47 + [1]*

Redi, F. 1671. *Esperienze intorno a diverse cose natvrali e particolarmente a qvelle, che ci son portate dall'Indie fatte da Francesco Redi e scritte in una lettera al Reverendissimo Padre Atanasio Chircher della Compagnia di Giesu.* Firenze. All'Insegna della Nave. 1-152 + 6 plates.

Redi experiment. nat. Redi, 1675.

Redi, F. 1675. *Experimenta circa res diversas naturales, speciatim illas, quae ex Indiis adferentur. Ex Italico Latinate donata.* Amstelodami. Andreae Frisii. [1] + 1-193 + [15].

Redo. Rhedi. Redi, 1675.

Reed Malab. Rhede, 1686.

Rhede, H. van. 1686. *Horti Malabarici Pars Septima, De Varii Generis Fruticibus Scandentibus; Latinis, Malabaricis, Arabicis, Belgicis, Bramannum characteribus nominibusque expressis, Adjecta Florum, Fructuum, Seminumque nativae magnitudinis vera delineatione, colorum viriumque accurata descriptione, Adornata per Nobilissimum ac Generosissimum D. D. Henricum van Rhede tot Draekestein, Toparcham in Mydrecht, quondam Malabarici Regni Gubernatorem, supremi Confessus apud Indos Belgas Senatorem Extraordinarium, nunc vero Equestris Ordinis nomine Illustribus ac Praepotentibus Provinciae Ultrajectinae Proceribus adscriptum, Notis adauxit, & Commentariis illustravit Joannes Commelinus In ordinem redegit, & latinitate donavit Abrahamus à Poot M. D.* Amstelaedami. Viduae Joannis van Someren, Haeredum Joannis van Dyck, Henrici & Viduae Theodori Boom. 111 + [1] (of them, 59 plates).

RHEDE. Redi, 1664.

Roberg. diss. de lue vener. Linder, 1705.

Rondel. insect.; Rond. pisc.; Rondeletius. Rondelet, 1554.

Rondelet, G. 1554. *Libri de piscibus marinis, in quibus verae piscium effigies expressae sunt. Quae in tota piscium historia contineantur, indicat elenchus pagina nona & decima*. Lugduni. Matthiam Bonhomme. [15] + 1-583 + [23].

Rondel. pisc. Rondelet, 1555.

Rondelet, G. 1555. *Universae aquatilium historiae pars altera, cum veris ipsorum imaginibus*. Lugduni. Matthiam Bonhomme. [11] + 1-242 + [9].

Rotheram, J. 1775. *Medicamenta Purgantia*. Upsaliae. Johan. Edman, Reg. Acad. Typogr. 26 pp.

Roy. lugdb. Royen, 1740.

Royen, A. van. 1740. *Florae Leydensis prodromus, Exhibens Plantas Quae in Horto Academico Lugduno-Batavo Aluntur.* Lugduni Batavorum. Samuelem Luchtmans. [70] + 1-538 + [30] + 3 tab.

Rumpf, G. E. 1741. *Herbarium Amboinense, plurimas complectens arbores, frutices, herbas, plantas terrestres & aquaticas, quae in Amboina, et adjacentibus reperiuntur insulis, adcuratissime descriptas juxta earum formas, cum diversis denominationibus, cultura, usu, ac virtutibus. Quod & insuper exhibet varia insectorum animaliumque genera, plurima cum naturalibus eorum figuris depicta. (...). Pars secunda*. Amstelaedami. Franciscum Changuion, Johannem Catuffe, Hermannum Uytwerf. / Hagae Comitis. Petrum Grosse, Johannem Neaulme, Adrianum Moetjens, Antonium van Dole. / Ultrajecti. Stephanum Neaulme. Pars secunda. 1-270.

Salomon, E. D. 1775. *Specimen Inaugurale de Scorbuto*. Upsaliae. Apud Joh. Edman, Regiae Acad. Typogr. 24 pp.

Scanam. Linnaeus, 1751.

Scheuchzer, J. J. 1723. *ΟΥΡΕΣΙΦΟΙΤΗΣ Helveticus, sive Itinera per Helvetiae alpinas regiones, facta annis MDCCII. MDCCIII. MDCCIV. MDCCVI. MDCCVII. MDCCXI. MDCCX. MDCCXI. In quatuor tomos distincta*. Lugduni Batavorum. Petri Vander. Four volumes.

Schott. Phys. Schott, 1662.

Schott, G. 1662. *Physicae Curiose Pars II. Complectens sex posteriores libros, videlicet VII. Mirabilia Animalium in genere. VIII. Mirabilia Animalium Terrestrium. IX. Mirabilia Animalium Volatilium. X. Mirabilia Animalium Aquatilium. XI. Mirabilia Meteororum. XII. Mirabilia Miscellanea*. Herbipoli. Johannis Andree Endteri & Wolff Gangi Jun Haeredum. 771-1583 + [23] + XXIII-LVII plates.

Schreber, J.C.D. (Ed.). 1785. *Caroli a Linné Equit. aur. de Stella polari Archiatri Regii Med. & Botan. Profess. Upsal. Acad. Parisin. Petrop. &c. Soc. Amoenitates Academicae seu dissertationes variae physicae, medicae botanicae antehac seorsim editae nunc collectae et auctae cum tabulis aeneis. Edidit Jo. Christianus Daniel Schreberus Ser. Marggr. Brandenb. Onold. & Culmb. Consil. Aul. Med. Bot. Hist. nat. & Oec. P. P. O. in Acad. Erlangensi.* Volumen Octavum, [3] + 2-332 + [1] + 8 pl. Erlangae. Jo. Jacobi Palm.

Schwenck. rept. [il]. Schwenkfeld, 1603.

Schwenckfeld, C. 1603. *Theriotropheum Silesiae, in quo animalium, hoc est, quadrupedum, reptilium, avium piscium, insectorum natura, vis & usus sex libris perstringuntur*. Lignicii. Davidis Alberti. [22] + 1-563 + [4].

Scopoli, J. A. 1763. *Entomologia Carniolica exhibens insecta Carnioliae indigena et distributa in ordines, genera, species, varietates. Methodo Linnaeana*. Vindobonae. Ioannis Thomae Trattner. [1-35] + 1-420 + [1].

Seb. thes. 1.; Seb. thes. [aur]. t. [om]. I.; Seba. Seba, A. 1734

Seba, A. 1734. *Locupletissimi rerum naturalium thesauri accurata descriptio, et iconibus artificiosissimis expressio, per universam physices historiam. Opus, cui, in hoc rerum genere, nullum par exstitit / Ex toto terrarum orbe collegit, digesit, descripsit, et depingendum curavit Albertus Seba, Etzela Oosterisius, Academiae Caesareae Leopoldino Carolinae Naturae Curiosorum collega Xenocrates dictus; Societatis Regiae Anglicanae, et Instituti Bononiensis, sodalis. Tomus I.* Amstelaedami. Janssonio-Waesbergios, & J. Wetstenium, & Gul. Smith. [33] + 1-178 + 111 plates.

Seb. thes. 2. Seb. thes. [aur]. t. [om]. II. Seba, A. 1735.

Seba, A. 1735. *Locupletissimi rerum naturalium thesauri accurata descriptio, et iconibus artificiosissimis expressio, per universam physices historiam: opus, cui, in hoc rerum genere, nullum par exstitit / Ex toto terrarum orbe collegit, digesit, descripsit, et depingendum curavit Albertus Seba, etzela oosterisius, Academiae Caesareae Leopoldino Carolinae Naturae Curiosorum collega Xenocrates dictus ; Societatis Regiae Anglicanae, et Instituti Bononiensis, sodalis. Tomus II.* Amstelodami. Wetstenium, & Gul. Smith, & Janssonio-Waesbergios. [34] + 1-154 + 114 plates.

Sibb. mus. Sibbald, 1697.

Sibbald, R. 1697. *Auctarium Musaei Balfouriani, e Musaeo Sibbaldiano; sive Enumeratio & Descriptio Rerum Rariorum, tàm Naturalium quàm Artificialium; tàm Domesticarum quàm Exoticarum; quas Robertus Sibbaldus M.D. Eques Auratus, Academiae Edinburgenae Donavit.*

Quae quasi Manuductio brevis est, Ad Historiam Naturalem. Edinburgi. Academiae Typographum, Sumptibus Academiae. [15] + 1-216.

Sidrén, J. 1750. *Materia medica in regno animali.* Upsaliae. S. n. [2] + 3-20.

Sidrén, J. 1751. Materia medica in regno animali. *Amoenitates Academicae,* 2: 307-331. (Holmiae edition).

Sloan. hist. 2. Sloan. hist. II. Sloan. Jam. 2. Sloane, H. 1725.

Sloan. jam. Sloane, 1696.

Sloane, H. 1696. *Catalogus plantarum quae in Insula Jamaica sponte proveniunt, vel vulgò coluntur, cum earundem synonymis & locis natalibus; adjectis aliis quibusdam quae in insulis Maderae, Barbados, Nieves, & Sancti Christophori nascuntur. Seu prodromi Historiae Naturalis Jamaicae. Pars prima.* Londini. D. Brown. [5] + 1-232 + [43].

Sloane, H. 1725. *A voyage to the islands Madera, Barbadoes, Nieves, St Christophers, and Jamaica; with the natural history of the herbs and trees, four-footed beasts, fishes, birds, insects, reptiles, &c. of the last of those islands. To which is prefix'd, an introduction, wherein is an account of the inhabitants, air, waters, diseases, trade, &c. of that place; with some relations concerning the neighbouring continent, and islands of America. Illustrated with the figures of the things described, which have not been heretofore engraved. In large copper-plates as big as the life. In two volumes. Vol. II.* London. Author. [1] + 1-499 + plates 157-274.

Smit, P. 1979. The zoological dissertations of Linnaeus. *Svenska Linnésällskapets Årsskrift,* 1978: 118–136.

Smith, J. E. 1821. *A selection of the correspondence of Linnaeus and other naturalists, from the original manuscripts. In two volumes. Vol. I.* London. Longman, Hurst, Rees, Orme, and Brown. I-XIV + 1 - 605.

Söderberg, O. 1748. *D. F. G. Specimen academicum De Curiositate Naturali..* Holmiae. Laurentii Salvii. [1] + 1-25 + [1].

Söderberg, O. 1749. Curiositas naturalis. *Amoenitates Academicae,* 1: 540-563. (Holmiae edition).

Söderberg, D. H. 1771. *Pandora et Flora Rybyensis.* Upsaliae. Joh. Edman. [4] + 6-23 + [1].

Sparrman, A. 1768. *Sistens iter in Chinam.* Upsaliae. Joh. Edman. [3] + 5-16.

Sparrman, A. 1769. Iter in Chinam. *Amoenitates Academicae,* 7: 487-506. (Holmiae edition)

Stafleu, F. A. 1971. *Linnaeus and the Linnaeans: the Spreading of their Ideas in Systematic Botany, 1735–1789.* Utrecht. International Association for Plant Taxonomy. i-xvi + 1-386.

Stearn, W. T. 1966. The use of bibliography in Natural History. In: Buchman TR, editor. *Bibliography & Natural History. Essays presented at a Conference in June 1964.* Lawrence (USA): University of Kansas Libraries. p. 1–26.

Stejneger, L. 1933. The ophidian generic names *Ahaetulla* and *Dendrophis. Copeia,* 1933 (4): 199-203.

Sundius, P. 1748. *Surinamensia Grilliana.* Holmiae. Laurentii Salvii. [5] + 1-34 + 1 pl.

Sundius, P. 1749. Surinamensia Grilliana. *Amoenitates Academicae,* 1: 403-508 + pl. XVII. (Holmiae & Lipsiae edition).

Surinamens. Grill. Sundius, 1748.

Swammerdam, J. 1693. *Historia Insectorum Generalis, in qua verissimae mutationem, seu lentae in membra epigenescos rationes, duce experientia, redduntur, recepta vulgo insectorum metamorphosis solide refutatur: plurimi errores auctorum circa hoc argumentum deteguntur ac corriguntur: & tandem omnes insectorum mutationes ad quatuor ordines distinctos summa industria descriptos revocantur, subjecto singulis ordinibus indice insectorum eo spectantium: adjicitur dilucidatio, qua specialia cujusvis ordinis exempla figuris accuratissime, tam naturali magnitudine, quam ope microscopii aucta illustrantur; summa in singulis harmonia, quae & sanguineis animantibus, nec non plantis conspirare, itidem figurarum ope, specialibus exemplis, docetur. Obiter in toto opera hinc inde varia insecta, auctoribus physicis aut neglecta, aut male descripta, additis figuris, aeri tam naturali, quam aucta microscopia magnitude, nitidissime incisis, accuratius describuntur. Cum indic. necessariis. Ex Belgica Latinam fecit Henricus Christianus Henninius. Editio secunda.* Ultrajecti. Otthonis de Vries. [17] + 1-212 + [17] + 13 plates.

Syst. Nat. 6. - ***System. N. 6.*** Linnaeus, 1748a.

Systema Nat. 216, N:o 173. Linnaeus, 1758.

Tennent. Tennent, 1738.

Tennent, J. 1738. *An epistle to Dr. Richard Mead, concerning the epidemical diseases of Virginia, particularly a pleurisy and peripneumony: Wherein is shewn the surprising efficacy of the Seneca Rattle-Snake Root, in diseases owing to a viscidity and coagulation of the blood; such as pleurisies and peripneumonies, these being epidemick, and very mortal in Virginia, and other colonies on the continent of America, and also the Lee-Ward Islands. To which is prefixt, a cut of the most valuable plant: and an appendix annexed, demonstrating the highest probability, that this root will be of more extensive use than any medicine in the whole Materia Medica, and of curing the Gout, Rheumatism, Dropsy, and many nervous diseases.* Edinburgh. P. Matthie. 3-102.

Thunberg, C.P. 1767. *D. D. Dissertationem Physiologicam De Venis Resorbentibus.* Upsaliae. Johan Edman, Reg. Acad. Typogr. 10 pp.

Thyson. act. Tyson, 1683.

Tillaeus, P.C. 1771. *Dissertatio Medica De Varia Febrium Intermittentium Curatione.* Upsaliae. Typis Edmannianis. 58 pp.

Tournef. inst. Tournefort, 1700.

Tournefort, J.P. de. 1700. *Institutiones rei herbariae. Editio Altera, Gallica Longe Auctior. Quingentis circiter Tabulis aeneis adornata. Tomus Primus. + Corollarium Institutionum Rei Herbariae, in quo Plantae 1356 munificentiâ Ludovici Magni in Orientalibus Regionibus observatae recensentur, & ad genera sua revocantur.* Parisiis. Typographia Regia. [17] + 2-697 + [6] + 1-54 + [5].

Tozzi, L. 1693. *In Hippocratis aphorismos commentaria: ubi universae medicinae, cum theoretice, tum practicae celebriores quaestiones perpenduntur, atque nedum recentiorum inventis, sed & genuine ejusdem Hippocratis menti congruentes, quam dilucide explicantur*. Neap. Ant. Parrino & Michaelis Aloysii Mutti / Haeredum Cosimi Fioravanti. 2 vols. [18] + 1-325 + [47] and [4] + 1-377 + [41].

Tozzi. Tozzi, 1747.

Tozzi, L. 1747. *In Hippocratis aphorismos commentaria. Pars prima. Ubi universae medicinae, cum theoretice, tum practicae celebriores quaestiones perpenduntur, atque nedum recentiorum inventis, sed & genuine ejusdem Hippocratis menti congruentes, quam dilucide explicantur. Tomus secundus*. Venetiis. Nicolaum Pezzana. 1-156.

Tyson. Tyson, 1683.

Tyson, E. 1683. Vipera caudi-sona Americana, or the anatomy of a rattle-snake, dissected at the repository of the Royal Society in January, 1682/3. *Philosophical Transactions*, 13 (144): 24-46 + 12 figs.

Uddman, I. 1765. *Lepra*. Upsaliae. S. n. 1-14.

Uddman, I. 1769. Lepra. *Amoenitates Academicae*, 7: 94-108. (Holmiae edition).

Urban, S. 1749. Historical Chronicle, September 1749. *The Gentleman's Magazine and Historical Chronicle. Volume XIX for the year M DCC XLIX: 428* (unnumbered).

Vaill. paris. Vaillant. 1727.

Vaillant. S. 1727. *Botanicon Parisiense ou denombrement par ordre alphabetique des plantes qui se trouvent aux environs de Paris compris dans la carte de la prévôté & de l'election de la dite ville le sieur Danet Gendre année MDCCXXII. Avec plusieurs descriptions des plantes, leurs synonymes, les tems de fleurir & de grainer et une critique des auteurs de botanique (...). Enrichi de plus de trois cents figures, dessinées par le sieur Claude Aubriet, peintre du Cabinet du Roy*. Leide & Amsterdam. Jean & Herman Verbeek et Balthazar Lakeman. [31] + i-vii + 1-205 + [34] + 33 pl.

***Vincent catal.* Vincent, 1726a.**

Vincent, L. 1726a. *Catalogus et descriptio animalium volatilium, reptilium, & aquatilium. Uti & Animalium quadrupedum, tam oviparum quam viviparorum, aliarumque creaturarum rarissimarum, ex quatuor mundi plagis, quae in liquoribus ad vivum conservant. Levinus Vincent, Regiae Societatis Anglicae Membrum. catalog et description des animaux volatils, aquatils, & des reptiles. Comme aussi des Animaux à quatre pieds, tant ceux qui porte les oeufs, que ceux qui nourrisent leur foetus de leur substance, & autres Creatures tres rares des quatre parties du monde, toutes conservés dans leur naturel dans les Liqueurs, par Levin Vincent, membre de la Société Royale de Londres*. La Haye: Pierre de Hondt. [8] + 1-72.

Vincent, L. 1726b. *Descriptio pipae, / Seu / Bufonis Aquatici Surinamensis, / Foetus enitentis in dorso, unde videntur prodire; / Addita Succincta / Descriptione Ranarum, / Earumque successiva mutatione; cui accedit Descriptio omnium / diversorum generum Bufonum et Ranarum, tam Orientalium / quam Occidentalium, conservatorum in Gazophylacio / Levini Vincent, Membri Societatis Regalis. Description du Pipa, / Ou / Crapéau Aquatil de Surinam, / Qui engendre ses petits par le dos, & qui semble les en faire éclore; / Avec une courte / Description des Grenouilles, / Touchant leurs changemens par degrés suivie d'une Description de toutes les / differentes espèces de Crapeaux et de Grenouilles des Indes, contenus / dans le Cabinet de Levinus Vincent, Membre de la Société Royale*. Haarlem. Autheur-Pierre de Hondt. [26] + 6-63 + [1] + 1-16 + 2 plates.

Virid. cliff. Linnaeus, 1737b.

Wahlbon, J. G. 1746. *Sponsalia plantarum*. Stockholmiae. Laurentii Salvi. [12] + 8-60 + 1 plate + [2].

Wahlbon, J. G. 1749. Sponsalia plantarum. *Amoenitates Academicae*, 1: 327-380, plate XV. (Holmiae & Lipsiae edition).

Wahlgren, R. 2011. Carl Linnaeus and the Amphibia. *Bibliotheca Herpetologica*, 9 (1-2): 5-37.

Wählin, A. 1756. *D. D. Dissertatio Medica De Pulsu Intermittente*. Upsaliae. L. M. Hojer, Reg. Acad. Typogr. 18 pp.

Wallin, L. 1992. Linnaean specimens in the Zoological Museum of Uppsala University. *Archives of Natural History*, 19 (2): 219-230.

Wallin, L. 2001. *Catalogue of type specimens. 4. Linnaean specimens*. Uppsala University, Museum of Evolution, Zoology section. 128 pp.

Wänman, C.H. 1768. *De Morbis Nautorum Indiae*. Upsaliae. Joh. Edman. [2] + 3-4.

Wännman, C. H. 1759. *Flora capensis*. Upsaliae. S. n. [3] + 1-19 + [1].

Wännman, C. H. 1760. Flora capensis. *Amoenitates Academicae*, 5: 353-370. (Holmiae edition).

Wedenberg, A. F. 1767. *De Varietate ciborum*. Upsaliae. Johan. Edman, Reg. Acad. Tipogr. [2] + 4-20.

Wedenberg, A. F. 1769. Varietas ciborum. *Amoenitates Academicae*, 7: 197-213. (Holmiae edition).

Wilcke, H. C. D. 1760. *Politia naturae*. Upsaliae. S. n. [1] + 1-22.

Wilcke, H. C. D. 1763. Politia naturae. *Amoenitates Academicae*, 6: 17-39. (Holmiae edition).

Williams, S. 1738. Extract of a letter from Stephen Williams, M. B. F. R. S. concerning the viper-catchers, (mention'd in no 443. of these Transactions) and the efficacy of oil of olives in curing the bite of vipers. *Philosophical Transactions*, 40 (1737-1738): 26-27.

Willugb.; Willugb. icht. Willughbeii, 1686.

Willughbeii, F. 1686. *De Historia Piscium Libri Quatuor, Jussu & Sumptibus Societatis Regiae Londinensis editi. In quibus non tantum De Piscibus in genere agitur, Sed & species omnes, tum ab aliis traditae, tum novae & nondum editae bene multae, naturae ductum servante Methodo dispositae, accurate describuntur. Earumque effigies, quotquot haberi potuere, vel ad vivum delineatae, vel ad optima exemplaria impressae; Artifici manu elegantissime in Aes incisae, ad descriptiones illustrandas exhibentur. Cum Appendice Historias & Observationes in supplementum Operis collatas complectente. Totum*

Opus Recognovit, Coaptavit, Supplevit, Librum etiam primum & secundum integros adjecit Johannes Raius e Societate Regia. Oxonii e Theatro Sheldoniano. [7] + 2-343 + 1-30 + [113] + Plates numbered by sections: A2 B9 C5 D5 E2 F8 G10 H7 I27 L7 M5 N13 O5 P8 Q10 R4 S22 T1 V12 X15; Appendix 10.

Worm. mus. Worm, O. 1655.

Worm, O. 1655. *Museum Wormianum. Seu historia rerum rariorum, tam naturalium, quam artificialium, tam domesticarum, quam exoticarum, quae Hafniae Danorum in aedibus authoris servantur. Variis & accuratis iconibus illustrata.* Amstelodami. Ludovicum & Danielem Elzevirios. [10] + 1-389 + [4].

Zetzell, P. 1754. D. D. *Consectaria Electrico-Medica.* Upsaliae. L. M. Hojer, Reg. Acad. Typ. 8 pp.

Zwölfern, J. 1692. *Oder Dispensatorium. Das ist: Neu-bereicherter ganz-angefüllter Schatz-Kasten Der außerlesensten Artzneyen, so wol inner- als äusserlichen, einfachen, als zusammen-gesetzten, so wol Chymischen, als Galenischen, so wol ausländisch- als innländischen, so wol alten, als neuen, so wol für die Menschen, als für das Vieh, in allerhand Zufällen und Kranckheiten, auch wunden, Schäden, Brüchen, Verrenckungen, Podagrischen- und andern Schmertzhafften Gliedern, so wol zur Verhütung, als Cur und Linderung derselben, mit sonderbarem Fleiß und Mühe zusammen getragen, und in richtige Ordnung verfasset, auch mit vielen schönen, nützlichen und practischen Anmerckungen, wie auch allerhand raren Geheimnissen, so wol aus Chymischen, als Galenischen Büchern, und andern bewehrten Autoribus, auch eigenen langwierigen Experientz und Practic zusammen gezogen, und dem gemeinen Nutzen zum besten in Druck gegeben.* Nuremberg. Johann Hoffmann. [6] + 1-780 + [26].

TANTUM
[That's all]

www.ingramcontent.com/pod-product-compliance
Lightning Source LLC
LaVergne TN
LVHW071343210926
848486LV00003B/223